The Botanical Society of Britain & Ireland (BSBI) is for everyone who cares about our wild plants. We help train botanists, support recording and research projects across Britain and Ireland, and build engagement to help drive a passion for wild plants.

Botanical Society of Britain & Ireland

As one of the world's largest contributors of biological records, we supply the scientific data used by policy-makers, conservation bodies and academia. We also publish newsletters, handbooks, local Floras, national distribution atlases and a scientific journal to advance botanical knowledge.

BSBI promotes the study, understanding and enjoyment of British and Irish botany - find out more on our website: www.bsbi.org

T0190742

BRITISH & IRISH WILD FLOWERS AND PLANTS

A POCKET GUIDE

RACHEL HAMILTON,
CHRIS GIBSON & ROBERT STILL

PRINCETON
press.princeton.edu

Published by Princeton University Press,
41 William Street, Princeton, New Jersey 08540
99 Banbury Road, Oxford OX2 6JX
press.princeton.edu

British Library Cataloging-in-Publication Data is available

Library of Congress Control Number 2023940414
ISBN 978-0-691-24540-9
Ebook ISBN 978-0-691-24541-6

Editorial: Chris Gibson, Rob Still and Megan Mendonça
Cover Design: Rob Still
Production: Ruthie Rosenstock
Publicity: William Pagdatoon and Caitlyn Robson
Copyeditor: Jude Gibson

Printed in Italy

Cover images (left to right): Wood Horsetail, Hybrid Bluebell,
Autumn Hawkbit, Wall Barley. Photos by Rob Still

10 9 8 7 6 5 4 3 2 1

Contents

Introducing the book 4

Why and how are plants classified and named? 6

The structure of the book 7

How to use this book 8

Looking at plants: some practicalities 9

Getting to the right group of plants 11

The broad groups used in this book 12

Plants: knowing the parts 14
Pteridophytes 14
Gymnosperms 15
Grasses, sedges, rushes 15
Flowering plants 16
    Leaf shapes and features 18
    Fruit names, shapes and types 18
    Vegetative and growth form terms 20

How to identify plants 21
    The broad groups roadmap key 26
Plants not green 27
Aquatic plants 28
    AQUATIC PLANT GALLERY 29
Woody plants 31
    TREES AND SHRUBS GALLERIES 34
        By flowers 34
        By leaves 38
        By fruit 40
        By winter twigs 42
Herbaceous flowering plants 46
    Conspicuous flower identification 48
    FLOWERING PLANT GALLERIES 50
    Characters of flowering plants in the book 64

THE SPECIES ACCOUNTS 65

Index 309

Acknowledgements 320

Introducing the book

Flowers and other plants are all around us. They are diverse and fascinating – and sometimes a little confusing. Our aim is to introduce you to the craft, technicalities and delights of finding and identifying, and so getting to know, common British and Irish plants – to show you what they look like, and what to look for.

This book is not comprehensive: it is small enough to go in a pocket and be carried around, handy when an unfamiliar plant is found. Inevitably, you will find plants that are not included as we have covered only those that are regarded as widespread (see *p. 6*) along with brief details of similar species, plus a few plants that are very common in the uncommon habitats where they are found (*e.g.* saltmarshes).

Introducing plants

A few vascular plants do not have chlorophyll and look superficially like fungi. Equally, there are some **mosses and liverworts** (bryophytes) and **green algae** (stoneworts in particular) that resemble vascular plants. **Fungi** and **algae** (see *opposite*) are distinct from **plants** and are each classified in their own kingdom of life (Fungi, Protista and Plantae respectively) by most authorities.

The types of plant

The diagram below sets out plant types in simplified evolutionary order with the most ancient first. The boxes have notes on their distinctive features, although experience will soon enable you to distinguish generally the main groups, such as horsetails, ferns, conifers and flowering plants.

	NON-VASCULAR PLANTS (not included in this book)		
	Bryophytes	**Mosses, liverworts and hornworts**: SEXUAL REPRODUCTION via single-celled spores; VASCULAR + ROOT SYSTEMS rudimentary	
	VASCULAR PLANTS – non-flowering		
	Pteridophytes	**Ferns, horsetails, clubmosses, spike-mosses and quillworts**: SEXUAL REPRODUCTION via single-celled spores; VASCULAR + ROOT SYSTEMS moderately developed	
	Gymnosperms	**Conifers**: SEXUAL REPRODUCTION via seeds not enclosed in an ovary, and that are borne in woody or fleshy cones; wind-pollinated; VASCULAR + ROOT SYSTEMS well-developed	
Plants	**VASCULAR PLANTS – flowering**		
	Angiosperms	**Flowering plants**: SEXUAL REPRODUCTION via flowers, producing seeds within an ovary that develops into a fruit; VASCULAR + ROOT SYSTEMS well-developed; complex	
	Pre-dicots [7 spp. B&I] A group that evolved prior to the divergence of the **monocots** and **dicots**. They have 'primitive' features such as numerous petals in spirals that transition to stamens.	**Dicots** [± 1,750 spp. B&I] SEEDS contain two embryonic leaves; STEM bundles of vascular tubes in a cylindrical pattern; FLOWER PARTS in whorls of 4 or 5; LEAF-VEINS typically form a network.	**Monocots** [± 570 spp. B&I] SEEDS contain one embryonic leaf; STEM bundles of vascular tubes randomly scattered; FLOWER PARTS in whorls of 3; LEAF-VEINS typically parallel.

Not-plants and plants compared

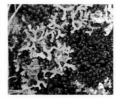

Fungi (Fungi) lack chlorophyll. They use spreading root-like threads to acquire nutrients from their surroundings. Fungal cell walls contain chitin, absent in plants but found in insects and some other invertebrates.

Lichens are actually a close association between at least two organisms from different kingdoms: a green alga or a blue-green alga (**Monera**) that inhabits the tissues of one or more species of **fungus**.

Algae (Protista) are predominantly aquatic, with perhaps the most well-known being the hair-like blanket weed of garden ponds and the conspicuous seaweeds.

There are a few plants that also lack chlorophyll which look superficially similar to some fungi – see *p. 27*.

Lichens are often treated as part of the kingdom Fungi as the fungus is the dominant partner.

Stoneworts (*above*) are multicellular green algae that resemble small-leaved stalked aquatic plants – see *p. 30*.

MAT-FORMING LIVERWORT | LEAFY LIVERWORT | FEATHER-MOSS | THREAD-MOSS

Non-vascular plants (bryophytes) have rudimentary root-like structures which are used for attachment and, at most, a very limited nutrient uptake. Bryophytes reproduce both vegetatively and by means of spores. **Liverworts and hornworts** are either mat-forming or leafy, while **mosses** have green shoots with simple leaves arranged spirally on the stem.

A few leafy liverworts and mosses can resemble vascular plants, particularly spike-mosses, clubmosses, a few rare ferns and the immature stages of common ones.

PTERIDOPHYTE (FERN) | GYMNOSPERM (PINE) | ANGIOSPERM (DICOT) | ANGIOSPERM (MONOCOT)

Vascular plant stems contain two forms of vascular tissue – **xylem** (which transports water and mineral nutrients up from the roots) and **phloem** (which distributes sugars around the plant). These, and the rigid cellulose cell walls, have allowed the evolution of a huge diversity of forms: from tiny ground-huggers to climbers and tall woody trees; from robust insect-pollinated to fragile wind-pollinated flowers; and from juicy fruits to tiny dust-like seeds.

Why and how are plants classified and named?

This book is about the identification of plants. If we identify something, it is good to give it a name that others can recognize. Plants, as with all organisms, are known by a two-word **scientific ('Latin') name** e.g. **Daisy** has the scientific name *Bellis perennis*. This standardized two-part binomial comprises the **genus name** (with initial capital) followed by the **species name**. The binomial is not just a name: it can also give information useful in identification if the words are descriptive. For example, *Bellis perennis* tells you it a perennial, and **Great Willowherb** *Epilobium hirsutum* is hairy. A knowledge of rudimentary Latin and Greek helps but is not essential: you will soon learn the meanings of regularly used specific names.

Daisy *Bellis perennis* (*p. 214*) is a pretty perennial

Plants also, of cours,e have **common names** relevant to the culture in which the plant is important. In this book, we use **English names** alongside the scientific. These are not governed by strict rules of application, and some plants have many different ones: we have followed almost exclusively those used by the Botanical Society of Britain and Ireland (BSBI). In the accounts we give species both names, the common name with initial capitals followed by the scientific name in italics, with the genus capitalized e.g. Silver Birch *Betula pendula*.

Great Willowherb *Epilobium hirsutum* (*p. 127*) is hairy if looked at closely

Related species (abbreviated to sp. and spp. (plural)) are grouped into a **genus** (plural genera), related genera into a **family**: *Bellis perennis* is thus in the daisy family Asteraceae – yes, families have scientific and common names too! For family English names, we have normally used the names of one or two prominent members as convenient 'flags'. The species accounts that follow are subdivided into families. With practice, knowledge of the family can be a first step to identifying a species: "*that flower has five showy yellow petals with lots of bits in the middle, so let's try the buttercup family*". The Galleries that are a key part of the identification process in this book, show representatives of each family sorted according to their features.

Silver Birch *Betula pendula* (*p. 110*) has somewhat pendulous (drooping) branches, and helps distinguish it from the more pubescent (softly hairy) **Downy Birch** *Betula pubescens*

The above is only a small part of the hierarchy of **biological naming and classification (taxonomy)** as it relates to this book and its purpose: to help you identify plants. Don't be put off by the unfamiliar words, and their pronunciation is irrelevant to making a correct identification!

Plants included in this book

The majority of plants in this book are geographically widespread: all species that have been recorded from one-third or more of the approximately 3,000 hectads (10 km × 10 km squares) in Britain and Ireland. Some less widely occurring species that are, for example, conspicuous, or are reasonably common and close in appearance to widespread ones are also included. In addition, there are a few plants that are common and likely to be encountered in scarcer habitats e.g. saltmarshes, even though their habitat preferences mean they are not widespread across the countries.

The structure of the book

This book has been designed with those near the start of their botanical journey in mind. It starts with an introduction to plants in general and to vascular plants in particular. Technical jargon is kept to a minimum and clear, illustrated explanations are given of any essential botanical terms. There are keys, but they are 'different' to those found in most other guides. The many illustrations in the keys play a vital role, helping to ensure the identification process is easy to follow. There is also a selection of 'Galleries', suites of images that group together families with similar features and show the necessary detail to make confident matches, with subsquent species accounts that provide confirmation.

The diagram below lays out the structure of this book and presents a visual guide to the identification process. ❶ denotes a **group introduction**; Ⓖ denotes a **gallery**:

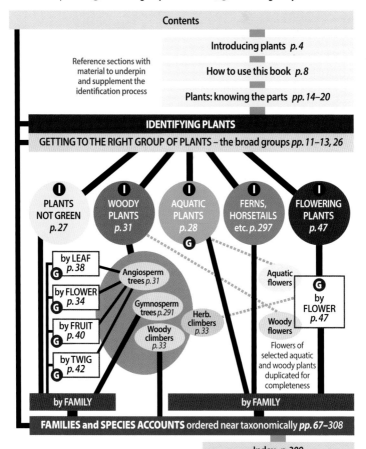

Contents

Introducing plants *p. 4*

How to use this book *p. 8*

Plants: knowing the parts *pp. 14–20*

Reference sections with material to underpin and supplement the identification process

IDENTIFYING PLANTS

GETTING TO THE RIGHT GROUP OF PLANTS – the broad groups *pp. 11–13, 26*

❶ PLANTS NOT GREEN *p. 27*

❶ WOODY PLANTS *p. 31* Ⓖ

❶ AQUATIC PLANTS *p. 28*

❶ FERNS, HORSETAILS etc. *p. 297*

❶ FLOWERING PLANTS *p. 47*

by LEAF *p. 38* Ⓖ

by FLOWER *p. 34* Ⓖ

by FRUIT *p. 40* Ⓖ

by TWIG *p. 42* Ⓖ

Angiosperm trees *p. 31*

Gymnosperm trees *p. 291*

Woody climbers *p. 33*

Herb. climbers *p. 33*

Aquatic flowers Ⓖ

Woody flowers

by FLOWER *p. 47*

Flowers of selected aquatic and woody plants duplicated for completeness

by FAMILY

by FAMILY

FAMILIES and SPECIES ACCOUNTS ordered near taxonomically *pp. 67–308*

Index *p. 309*

How to use this book

This book has been designed with identification (rather than description) in mind. 21 **Galleries** (*pages 29–63*) group together species with similar features, for ease of comparison, while the **Species Accounts** (*pages 67–308*) are grouped by family. It is structured so a minimal number of steps are needed to put a name to a plant. Where possible the identification features given will enable identification on a single visit, although in a few cases a further visit (*e.g.* when in fruit) to examine other key details may be needed. The pathway employs a pragmatic approach to descriptions and uses specialist botanical words only where necessary.

However, remember that these terms are essential to the identification of some groups or species and they do form the basis of good field botany. It is worth learning them (even if a little daunting at first!), as this will only enhance your time 'in the wild' but also help greatly when faced with a wholly unfamiliar plant.

The identification process

1 When you first see an unfamiliar plant, go to **Getting to the right group of plants** (*p. 11*) and the **overview spread** (*pp. 12–13*). This may take you directly to a relevant **broad group introduction** (see **2** below) or to the **Broad groups roadmap key** (*p. 26*), a step-wise process asking such simple questions as "is it woody?". This stage has more significance than may appear at first glance, because even though broad group differentiation uses straightforward appearance criteria, this may be based on features, such as shape and form, which are not necessarily aligned to taxonomy, or botanical differences which do have taxonomic importance.

2 The cross-reference from your chosen 'best-match' image will take you to the relevant **broad group introduction** (*pages 27, 28, 31, 47, 291, 297*). For larger groupings this introduction may refer you to a sub-group with images and, if needed, further subdivisions with more detailed identification information before referring you to a **family account**.

3 By now you should be in the part of this book that has a reasonably close match for your plant. Details in the **species accounts** associated with that family should enable you to put a name to the species you are looking at. In a few cases the process will arrive at a species group that requires such detailed differentiation (*e.g.* at the microscopic level) as to be beyond the scope of this book.

Limitations of the process

The main limitations to successful identification might be hasty picture matching or unconventional interpretation of the images and descriptions. Commonly occurring pitfalls are illustrated and/or annotated throughout, cross-referenced to any relevant pages.

If something is not right, double-check the features and revisit the key as well as information on commonly confused groups or species – or it may be that you have found a scarcer species not covered by this book.

Example identification

A 'pavement plant' is not woody and has flowers. This suggests checking the **herbaceous plant** sub-groups first (Gallery pages *50–63*).

flowerhead (capitulum)

Looking in the galleries the flowers may be found via **Gallery 20** (*p. 62*), in the section covering the Daisy family (Asteraceae), or more likely in **Gallery 10** (*p. 52*) as a member of the Daisy-type group (with tiny rays). This leads, via a page reference, to the **Daisy family introduction** (*p. 194*) where it

single fully open flower; others just open/in bud

is found that the plant is one of several closely related species in group **3** that have tiny white to reddish rays. An examination of the critical comparative features shown and described in the species accounts allows the conclusion to be drawn that the plant is a **Bilbao Fleabane** (*p. 216*).

Looking at plants: some practicalities

Magnifying and measuring key features

The best way to observe some important identification features is to magnify them. Digital cameras with a viewfinder may offer a good alternative to the botanist's trusty hand lens, sometimes sold as a jeweller's loupe.

To use a lens, it should be as close to the eye as possible, with the plant as close to the other side of the lens as needed for it to be in focus. This may result in poor light conditions which make it difficult to see critical features, such as hairs. Some lenses avoid this by having integral LED lighting – although remembering to turn the light off is an issue! Plastic lenses are easily scratched and glass ones are preferable, with a 'triplet' lens providing the highest quality. In the absence of a lens it is perfectly possible to get a magnified image by using binoculars the 'wrong way round' – looking through the larger lens and moving close to the plant until in focus.

Measurements, while inherently variable, are often useful identification features. A millimetre scale is printed on the book's cover flaps. Submillimetre measurements are rarely used in this book.

A good 10× or 20× hand lens (*top*) is the preference for most botanists in the field. Alternatively most cameras' viewfinders allow a user to zoom in to fine detail (*bottom*).

Poisonous plants

Most plants are harmless but you should NEVER eat any part of a plant unless you are absolutely sure of its identity and that it is safe. Furthermore, some plants are dangerous because they can cause localized skin 'burning' and sensitivity to sunlight through skin contact. **The most poisonous plants in the species accounts are marked ✕, but NOTE absence of the ✕ does NOT imply that any plant is safe to eat.**

Supporting identification features

There are some details that can be used to help identify plants but which are at best only suggestive as to identity. These include **date**, **habitat** and **location**, which can be helpful as back-up criteria but, as distributions and flowering periods are changing in response to *e.g.* climate change, it is better not to rely on these parameters alone.

Date

Some similar plants flower at completely different times or have only a narrow overlap *e.g.* **Cow Parsley**, **Rough Chervil** and **Upright Hedge-parsley** (see *p. 222*), and this can be useful supporting evidence to an identification. Additionally, some plant groups, *e.g.* many trees, have a short flowering period and are more often identified using other features such as leaves, fruit, twigs or bark.

Lilac (*p. 178*) and Butterfly-bush (*p. 179*) have rather similar flower spikes; apart from structural botanical differences Lilac (May–June) will have finished flowering before **Butterfly-bush** starts (July–September).

Location

You are much more likely to find a plant previously known in a locality than one new to the area. Most counties have a detailed Flora which presents local distribution, and the whole of Britain and Ireland is covered by the BSBI's, comprehensive *Plant Atlas 2020* and accompanying website *plantatlas2020.org*. However plants can be overlooked and distributions change, so your leap into the world of botany could well lead to new discoveries.

Habitat

Some species are found in very particular conditions, related to *e.g.* geology, soil type, altitude or moisture; others grow almost anywhere. Identifications based on parameters such as extreme soil pH or localized geology carry more weight than others such as altitude.

Wild plants can be found in a wide range of situations, from mountain-top to coastal saltmarsh, modern city centre to ancient woodland. The **species accounts** include basic habitat information. Although these are simplified descriptions of broad categories there is enough detail to use them as supporting evidence for an identification, bearing in mind that plants can occur in locations outside of their normal habitat and geographical range.

A patch of chalk grassland contains plenty of flowering plant species to keep a botanist busy.

Habitats, as for plants themselves, come with many technical terms. We have mostly used familiar, albeit imprecise, terms, though a few are worthy of explanation:

calcareous – *e.g.* of grassland, influenced by chalk or limestone rocks (otherwise limy, basic, base-rich, alkaline or high pH)

acid – the ecological opposite of calcareous: low pH, influenced by underlying granite, sandstone, peat or sand

brackish – transitional waters, intermediate between fresh and salty.

There are books to help in the interpretation of habitats

The problem with hybrids

Plants can hybridize, some with particular abandon. Hybrids typically have a more vigorous growth form, and show features intermediate between those of the parents and this possibility should be borne in mind if an unusual plant is encountered. Not all hybrids can be identified and sometimes it is a case of just walking on by!

This plant, with the larger flowers of **Primrose** on the end of the longer stalk of a **Cowslip**, is the hybrid of the two species, known as '**False Oxlip**' on account of its similarity to the rare **Oxlip** *Primula elatior*. The hybrid has its own scientific name *Primula* × *polyantha* – note the '×' that indicates its hybrid origin.

Getting to the right group of plants

The broad groups used in this book spread (*pp. 12–13*) compares 'types', or categories, of plants, based on a broad assessment of size, shape, growth form, habitat and the presence or absence of flowers, and groups them together for ease of comparison.

Alternatively, the **broad groups roadmap key** (*p. 26*), offers a question-based step-wise approach to the same end – moving the identification process on to the next stage.

When faced with an unfamiliar plant, rather than initially looking for fine details, first assess its overall size and form. If in doubt, check any additional information in the **broad groups roadmap key** (*p. 26*). From there go to the relevant section and read the introduction. This should help to confirm that you are in the right area and, by following a pathway that involves a closer examination of features, you should ultimately be able to put a name to the plant.

In some cases a plant may fit into more than one group. For example, **Heather** (ABOVE, growing with Bracken, a **pteridophyte**) could fall into either the **shrub**, or **flowering plant** categories, depending on the age and size of the specimen. Similarly a **water-crowfoot** (BELOW) could be regarded as either an **aquatic** or **flowering plant** depending on where it is growing. In cases such as these, alternatives are given to provide a steer in the right direction, though it is impossible to cover all situations and potential pitfalls and so it is recommended to not always apply a rigid application of the grouping.

The first steps in plant differentiation
Some of the broad groups used in this book – **trees and shrubs, flowering plants** (including grasses) and **aquatic plants** can be readily identified here in this New Forest scene. Those with more experience may also spot **Water Horsetail** (a **pteridophyte**) and rushes (**flowering plants**).

The broad groups used in this book

The initial distinction, between **non-flowering** and **flowering** plants, is based on biology. **Non-flowering plants** comprise the **Pteridophytes** and **Gymnosperms**; **flowering plants** contains just the **Angiosperms**. Within this book, to help in identification, there are three further groups – based on non-taxonomic criteria: **plants not green** (a few angiosperms and pteridophytes (horsetails)); aquatic plants (some angiosperms; one pteridophyte) and woody plants comprising *trees and shrubs* found within both gymnosperms (all) and angiosperms (some) and *climbers* (all angiosperms).

PTERIDOPHYTES
▶ *p. 297*

NON-FLOWERING PLANTS

FLOWERING PLANTS

A 'step-by-step' approach to finding the broad groups can be found on *p. 26*.

ANGIOSPERMS
▶ *p. 47*

Plants Not Green
▶ *p. 27*

Within the angiosperms (flowering plants) there are three broad groups that are defined by genetic and seed details (see *p. 4*).
These are colour-coded throughout this book: pre-dicots; dicots; and monocots.

Aquatic Plants
▶ *p. 28*

comprising the ferns and near-relatives (13 [20 B&I] families | 29 [87 B&I] species) – one truly aquatic species

GYMNOSPERMS
▶ *p. 291*

comprising pines, spruces, larches, firs, cypresses, Juniper and Yew (3 [5 B&I] families | 14 [40 B&I] species)

NOTE: the number of taxa covered in this book is given, together with [in square brackets] the number found in Great Britain and Ireland

TREES and SHRUBS

comprising the vast majority of plant species in Great Britain and Ireland (94 [143 B&I] families | 723 [2,280 B&I] species)

Grasses, Rushes, Sedges *p. 245*

Shown above is just a small sample of the highly diverse range of flowering plants, from those with showy, petalled flowers; those with small flowers in flowerheads; to those that lack petals entirely.

Woody Plants
▶ *p. 31*

WOODY CLIMBERS

▶ Use the flowering plants key and galleries if further checks are required *pp. 47–63*

Plants: knowing the parts

How terms are used in this book

Our aim in this book is to keep terminology and jargon to a minimum. A raft of unfamiliar words which need frequent cross-reference to a glossary is not particularly useful when learning to identify plants. The following seven pages introduce the terms we are using throughout this book in **bold type** (usually by means of an illustration or photo). In some cases, these terms are accompanied, in *italic type*, with the botanical terms that will be encountered in other literature. In a few cases possibly unfamiliar botanical terms are actually the best way to describe some features and these are fully explained where relevant.

It is recognized that this approach may be seen by some as oversimplistic, and lacking a degree of botanical precision, but the purpose of this book is as an introduction to plant identification and it is expected that many readers will learn the detailed terms and outgrow the content of this book.

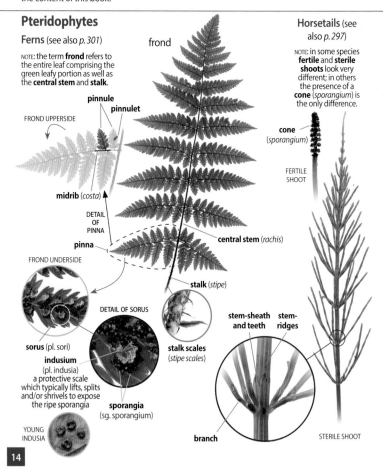

Pteridophytes

Ferns (see also *p. 301*)

NOTE: the term **frond** refers to the entire leaf comprising the green leafy portion as well as the **central stem** and **stalk**.

frond

FROND UPPERSIDE

pinnule

pinnulet

midrib (*costa*)

DETAIL OF PINNA

pinna

FROND UNDERSIDE

central stem (*rachis*)

stalk (*stipe*)

DETAIL OF SORUS

sorus (pl. sori)

indusium (pl. indusia) a protective scale which typically lifts, splits and/or shrivels to expose the ripe sporangia

stalk scales (*stipe scales*)

sporangia (sg. sporangium)

YOUNG INDUSIA

Horsetails (see also *p. 297*)

NOTE: in some species **fertile** and **sterile shoots** look very different; in others the presence of a **cone** (*sporangium*) is the only difference.

cone (*sporangium*)

FERTILE SHOOT

stem-sheath and teeth

stem-ridges

branch

STERILE SHOOT

Gymnosperms

(see also p. 291)

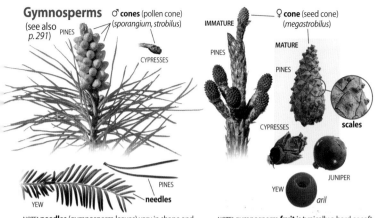

♂ **cones** (pollen cone)
(*sporangium, strobilus*)

PINES

CYPRESSES

♀ **cone** (seed cone)
(*megastrobilus*)

IMMATURE

PINES

MATURE

PINES

CYPRESSES

scales

JUNIPER

YEW

aril

PINES

YEW

needles

NOTE: **needles** (gymnosperm leaves) vary in shape and arrangement but are typically much longer than wide, *i.e.* needle-shaped!

NOTE: gymnosperm **fruit** is typically a hard or soft cone. **Yew** is an exception, being a single seed almost surrounded by soft flesh known as an *aril*.

Grasses, sedges, rushes (see also p. 245)

Grasses (see also p. 266)

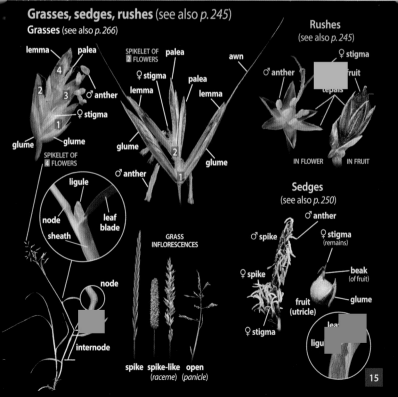

lemma

palea

4

2

3

♂ **anther**

♀ **stigma**

1

glume

glume

SPIKELET OF **4** FLOWERS

SPIKELET OF **2** FLOWERS

palea

♀ **stigma**

lemma

glume

♂ **anther**

palea

lemma

awn

2

1

glume

ligule

node

sheath

leaf blade

node

internode

GRASS INFLORESCENCES

spike   spike-like   open
        (*raceme*)   (*panicle*)

Rushes

(see also p. 245)

♀ **stigma**

♂ **anther**

fruit

tepals

IN FLOWER   IN FRUIT

Sedges

(see also p. 250)

♂ **anther**

♂ **spike**

♀ **stigma**
(remains)

♀ **spike**

♀ **stigma**

beak
(of fruit)

fruit
(utricle)

glume

leaf

ligule

15

Flowering plants

Flower parts | NOTE: not all of the features shown below are present in all species, or in plants with unisexual flowers.

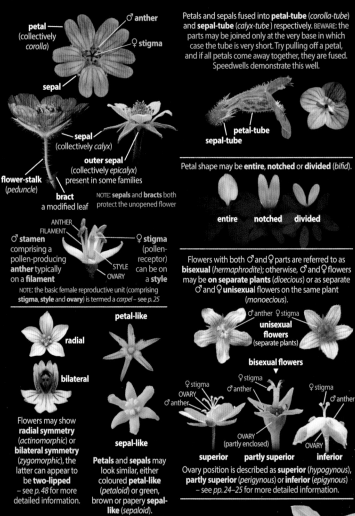

petal
(collectively *corolla*)

♂ **anther**

♀ **stigma**

sepal

sepal
(collectively *calyx*)

outer sepal
(collectively *epicalyx*)
present in some families

flower-stalk
(*peduncle*)

bract
a modified leaf

NOTE: **sepals** and **bracts** both protect the unopened flower

Petals and sepals fused into **petal-tube** (*corolla-tube*) and **sepal-tube** (*calyx-tube*) respectively. BEWARE: the parts may be joined only at the very base in which case the tube is very short. Try pulling off a petal, and if all petals come away together, they are fused. Speedwells demonstrate this well.

petal-tube

sepal-tube

Petal shape may be **entire**, **notched** or **divided** (*bifid*).

entire   **notched**   **divided**

ANTHER
FILAMENT

♂ **stamen**
comprising a pollen-producing **anther** typically on a **filament**

♀ **stigma**
(pollen-receptor)
can be on a **style**

STYLE

OVARY

NOTE: the basic female reproductive unit (comprising **stigma**, **style** and **ovary**) is termed a *carpel* – see p. 25

Flowers with both ♂ and ♀ parts are referred to as **bisexual** (*hermaphrodite*); otherwise, ♂ and ♀ flowers may be **on separate plants** (*dioecious*) or as separate ♂ and ♀ **unisexual** flowers on the same plant (*monoecious*).

♂ anther ♀ stigma

unisexual flowers
(separate plants)

bisexual flowers ▼

radial

bilateral

Flowers may show **radial symmetry** (*actinomorphic*) or **bilateral symmetry** (*zygomorphic*), the latter can appear to be **two-lipped** – see *p. 48* for more detailed information.

petal-like

sepal-like

Petals and **sepals** may look similar, either coloured **petal-like** (*petaloid*) or green, brown or papery **sepal-like** (*sepaloid*).

♀ stigma
OVARY
♂ anther

♀ stigma
♂ anther

♀ stigma
♂ anther

OVARY
(partly enclosed)

OVARY

superior   **partly superior**   **inferior**

Ovary position is described as **superior** (*hypogynous*), **partly superior** (*perigynous*) or **inferior** (*epigynous*) – see *pp. 24–25* for more detailed information.

There are also a number of specialized terms restricted to certain families or genera (*e.g.* **corona** in Daffodil (*p. 241*)). Such terms are used only when necessary and are defined or illustrated in the relevant section.

corona

Flowering plants – inflorescence types

Flowers can either be **solitary** (*right*), or arranged in groups with a recognizable architecture. Although there are many classifications and variants the vast majority of inflorescences, especially within the scope of this book, fall into the four broad types (**solitary**, **flowerhead**, **spike or spike-like**, and **clustered**) described here. Inflorescence shape can be useful in identification, both to family *e.g.* the carrot (see *p. 222*) and daisy families (see *p. 194*) and for differentiating members within a family (*e.g.* Tufted Vetch in the pea family (*p. 80*)).

SOLITARY

FLOWERHEAD	SPIKE or SPIKE-LIKE
A group of unstalked flowers densely packed together on a flat-topped or domed receptacle (see *p. 52*) – almost exclusive to the daisy family.	Flowers in an elongated, **unbranched** group. Structures within this category include spikes and racemes. Catkins are a form of spike.

RECEPTACLE

SPIKE

RACEME

Flowerheads themselves can be in spike-like or cluster inflorescences. See below and Gallery 10 *p. 53*

CLUSTER	Flowers in branched or unbranched groups but all appearing to be clusters to some extent. Structures within this category include *corymbs*, *cymes* and *panicles*.

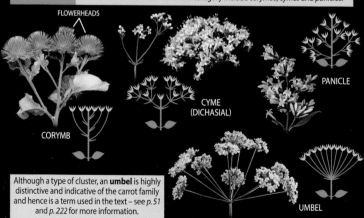

FLOWERHEADS

PANICLE

CYME
(DICHASIAL)

CORYMB

Although a type of cluster, an **umbel** is highly distinctive and indicative of the carrot family and hence is a term used in the text – see *p. 51* and *p. 222* for more information.

UMBEL

For differentiation within groups, descriptors covering the character of the inflorescence are used. For example **shape** (*e.g.* pyramidal; flat-topped), **density** (*e.g.* loose; spreading; dense) and **appearance** (*e.g.* drooping; whorled; 1-sided) are also used.

Flowering plants – leaf shapes and features

There are a huge number of very specific terms for leaf shape *e.g. crenate, cuneate, hastate*. However, such is the variability in leaf shape as a result of genetics, environment, herbivory amongst other factors only the following are used, as shown here:

Leaf arrangement

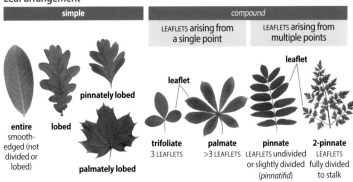

simple	compound	
	LEAFLETS arising from a single point	LEAFLETS arising from multiple points

leaflet

pinnately lobed

leaflet

leaflet

entire
smooth-edged (not divided or lobed)

lobed

palmately lobed

trifoliate
3 LEAFLETS

palmate
>3 LEAFLETS

pinnate
LEAFLETS undivided or slightly divided (*pinnatifid*)

2-pinnate
LEAFLETS fully divided to stalk

Flowering plants – fruit names and shapes

The large numbers of botanical terms used for fruits and seeds have been simplified by grouping them into the following broad categories. Many plant families contain multiple fruit types – the Rose family (*p. 89*), in particular, being exceptionally diverse. Inevitably some of the category boundaries are somewhat blurred (as between **seed** and **nut**, and **pod** and **capsule**) but these distinctions are rarely important for identification.

Familiar genus- or species-specific terms | In this book species-specific or genus-specific terms in common usage are used instead of botanical terms. Below are some examples together with a more detailed botanical description of their structure.

Fruit shapes |
3D shape can be a help in identification, as in the examples shown here:

WAYFARING-TREE

egg-shaped
ovoid

acorn (OAKS)
a **nut** (a dry 1-seeded fruit that has a woody wall), typically in a 'cup' made up of fused vegetative scales

hip (ROSES)
a false fruit – a fleshy part of the flower-tube that contains a number of dry 1-seeded fruits that have a papery wall

haw (HAWTHORNS)
a **berry-like** structure (*drupe*) consisting of small fruitstones surrounded by flesh with a tough outer skin

conker
(HORSE-CHESTNUT)
a dry **capsule** (spiky in some) containing large **seeds** (conkers)

GUELDER-ROSE

globular
globose

Leaf shape

triangle, triangular	**heart-shaped** *cordate*	**round, rounded** *orbicular*	**broadly oval** *ovate*	**narrowly oval** *ovate, lanceolate*	**linear, grass-like**

Leaf details | The margins and tip of a leaf in particular may have diagnostic features:

LEAFLETS

STIPULES (can be leaf-like, as here)

entire, untoothed smooth-edged (not divided or lobed)	**toothed** *serrate*	**wavy** *undulate*	**pointed** *acute*	**with drawn-out point** *acuminate*	**leaf-stalk** *petiole*	**stipule** – structures at base of **leaf-stalk** (in some families)

Fruit types

nut
a hard-shelled (often large) dry fruit

SEED

SEEDS

PAPPUS
SEED

pod (*follicle, siliqua*)
a structure with a long and slender outer coat, within which are the seeds

nutlet

FRUITING HEAD

WING

capsule (*silicle*)
a structure with a shorter and wider outer coat than a **pod**, within which are the seeds

seed (*e.g. achene, schizocarp, mericarp, nutlet*)
a hard, dry single fertilised unit, often wrapped within an outer coat. *Nutlet* is used traditionally for seeds in *e.g.* the Dock + Knotweed (*p. 140*) and Borage (*p. 169*) families. Seeds that are wind-dispersed often have terms associated with this: *e.g.* **wings** are flaps of thin tissue that increase the surface area (markedly winged seeds of *e.g.* Sycamore are termed *samara*), and many members of the Daisy family have a feathery extension known as a **pappus**

berry or **fleshy** are both used for fruits that are significantly fleshy, spongy or succulent. This grouping includes many other terms, which largely depend on which part of the flower is enlarged and fleshy in fruit – such as *drupe, drupelet, pome,* or *aril* – see *p. 295*):

TOMATO

PLUM
drupelets
WALNUT

APPLE

GOOSEBERRY

BRAMBLE

ROWAN

berry – succulent outer layer enclosing (typically >1) seeds that lack a stony coat

drupe – dry or succulent outer layer enclosing a typically single stony (nut-like) seed. A single *drupe* in an aggregation is termed a *drupelet*

pome – thick, fleshy outer layer surrounding a central capsule that encloses the (typically 5) seeds

NOTE: some of the technical terms that are included in these simplified groupings are shown in *italic text*.

Vegetative and growth form terms

Form | The **overall form** of a plant can be described in many ways. A fundamental feature is whether they are **woody** or **non-woody** (also referred to as **herbaceous**) – see also *pp. 31–33*.

Non-woody or herbaceous plants, are described according to their life-span and reproductive strategies as follows:	
Perennial	lives for > two years, and reproduces more than once; may have a **woody** lower stem and so merges with dwarf shrubs (*box left*)
Biennial *monocarpic*	lives for ≥ two years; reproducing only once
Annual	lives for less than a year, dies after setting seed; **winter annuals**, germinate in autumn, flower, set seed and die the following year

NOTE: the term *herb* is often used synonymously with herbaceous; in this book *herb* is only used in the culinary sense.

Woody plants in diminishing order of stature (although rather blurred boundaries) are:	
Tree	typically >5 m tall, usually with a single trunk
Shrub	typically <5 m tall, usually with multiple stems
Dwarf shrub	typically <100 cm tall, without a trunk

Growth habit | Terms used (in increasing order of uprightness) are **prostrate**, **trailing**, **sprawling**, or **erect**.

prostrate	trailing	sprawling (*sub-erect*)	erect

Vegetative spread | Some species spread by means of above-ground horizontal stems that root at least at the tip (*stolons*) or by underground or surface horizontal stems which usually root (*rhizomes*). As these look somewhat similar (and to avoid potentially damaging excavation of soil), in this book both these are termed as **runners**; the habit as **creeping**; and the result as **patch-forming**. Woody plants may produce new shoots from their roots. These shoots, which can be some distance from the main trunk or shoots are termed **suckers** (*right, top*).

suckers

Positional | Terms used are usually familiar and easily understood *e.g.* **terminal** (*apical*), **lateral**, **basal.** One less-well-known, but frequently used, term is **axil** (adj. *axillary*), which means something located within an angle. The most common use is **leaf-axil** – the angle between where the leaf, or leaf-stalk meets the stem (*right, bottom*).

Texture and colour | The terms covering the hairiness, texture, colour and characteristics of vegetative parts are relatively standard. Terms used in this book which have been simplified are as follows: COLOURS **blue-green** used as an equivalent to *glaucous*; TEXTURES **papery** is used as an equivalent for *membranous* and *scarious*; HAIRINESS (and other leaf and stem characters) use terms as shown below with further terms describing the shape and orientation of hairs:

a **flower-stalk** arising from a **leaf-axil**

hairless *glabrous*	downy *pubescent*	appressed held close to stem	spreading held away from stem	glandular 'blob'-ended	bristly *hispid, scabrid*	spiny (straight and narrow)	prickly (often curved, with expanded base)
			hairy *pilose*				

How to identify plants

Introduction

There are many features, used singly or in combination, that can be used in identification; all have potential pitfalls, hence the need to consider as many features as possible.

Equally, some methods of identification can appear daunting due to reliance on detailed botanical examination. Although these methods are not needed for the identification of many common plants, they are more consistent when compared with less 'botanically robust' approaches and can be used with greater confidence when faced with a completely unknown, maybe rare, species. As a result, learning and practising these methods will eventually lead to greater insight and the ability to recognize more species more easily.

The following pages look at these groups of identification features, and highlight some of their pitfalls, as well as some more detailed botanical features that support confident identification.

Size and form

The differences between a towering Scots Pine and a diminutive duckweed are evident. However the size and form between Scots and Black Pines, or the duckweed species are not so clear. As well as *e.g.* the age of a plant, size can be linked to growing conditions. Plants with optimal light, water and nutrients will be more typical of their species than those that have grown in more stressful conditions: such specimens can present identification challenges.

Form can also vary within a species, such as forms adapted for particular conditions (*e.g.* fleshy leaves in salty habitats).

Size and form are useful for broad differentiation but may be less reliable at a more detailed level.

Growth habit

The way a plant grows can be a clue as to its identity, although some care is needed as plants may show different growth forms in different places (*e.g.* those affected by grazing, cutting or trampling).

Growth habit is useful for identification but beware of individuals atypical of the species.

SIZE AND FORM: **Lesser Stitchwort** (*left* – to 80 cm) often grows taller than **Greater Stitchwort** (*right* – to 60 cm). Fortunately their petal lengths do not overlap and their stems differ – see *p. 148*.

GROWTH HABIT: **Knotgrass** (*p. 142*) can be prostrate (*left*), particularly in trampled areas; or can be found scrambling or even upright (*right*) – very different looks for the same species.

Annual or perennial?

Simplistically, annuals are typically less robust than perennials and have shallower, more delicate roots. However, remember it is illegal to uproot most wild plants without the landowners' permission!

Life-cycle is most useful in those families that contain similar-looking annual and perennial species.

Flowers or not?

Ferns, horsetails and conifers lack flowers, but are for the most part easily differentiated to group by sight. Although most species in this book are plants with conspicuous flowers, within this group there are many that appear to lack flowers, at least at certain times of year.

Recognition of whether a plant is not flowering, in bud, in fruit or simply has inconspicuous flowers is a useful skill to develop.

Flower form and symmetry

The shape of a flower, whether it is alone or in a group (inflorescence), and whether that group is loose, tight or has a definable form all helps identification.

Even in grouped flowers it is good practice to note features of an individual flower as these, together with inflorescence form, provide clues about identity.

For any flower, its symmetry is important. A 'circular' flower will more than likely have radial symmetry; an 'uneven' flower will probably have bilateral symmetry (see p. 16).

Although flower form and structure vary, often there will be enough flowers such that atypical examples can be disregarded. Also, symmetry in flowers is rarely mathematically perfect so a little imagination may be required!

The table on *p. 66* show all the families included in this book and the symmetry that can be found in that family.

Petals – free or fused?

Whether petals are separated (free) or fused together is helpful in identification. In simple terms free petals will fall from a plant as individual units, whereas fused petals will fall from the plant as a complete unit.

The table on *p. 66* show all the families included in this book and the petal arrangement that can be found in that family.

Petals – quantity and colour

It is easy to think that, for identifying conspicuous flowers, petal count, shape and colour are the only pieces of information needed. However, colour can vary (*e.g.* Devil's-bit Scabious (*p. 220*) from pink to blue) and any plant with petals can have some missing or some that are not 'perfect'. In addition, some species, such as stitchworts (*p. 148*), can

FLOWERS OR NOT?: Some unrelated plants, such as this **arrowgrass** (*p. 233*) and **plantain** (*p. 174*) seemingly lack flowers. Here, a knowledge of botanical features helps recognition.

FLOWER FORM: An umbel inflorescence typical of the carrot family (**Apiaceae** *p. 222*) and distinctive enough to identify the family by.

FLOWER SYMMETRY: From the front a typical **pea** (*top (p. 80)*) shows bilateral symmetry, while that of a mouse-ear (*bottom (p. 149)*) is radial.

have petals divided to the base, thus confounding a true count. Length or diameter are also variable quantities, and so should be used only with caution.

Petal count, colour and size are straightforward in most cases, but bear in mind that they are variable and could mislead.

Vegetative parts

Many features (*e.g.* leaf shape, the presence or absence of bracts, sepal length, stem shape and hairiness, stipule and glume form and colour) are central to accurate identification. The salient features vary depending on the group involved, so it is worth having knowledge of all parts – see *pp. 14–20*.

In the keys that follow, plants (or groups) most likely to be identified by their non-flowering features are shown with the green leaf icon 🌿.

Botanical details – the right features for a group

The parameters outlined on the previous pages are all good for identifying plants in general. Frequently, though, assessment of detailed features is needed to be fully confident of an identification. Such features vary from group to group: differences may be easy to see in some, though not so easy in others!

Knowledge of which features are required for each group (clearly indicated in this book) helps considerably when botanizing.

Ovary position

Pages 24–25 provide a visual explanation of one of the important details of a flower (important taxonomically and practically in the keys), the position of the ovary in relation to the attachment point of other parts.

PEA (BIRD'S-FOOT-TREFOIL)

COMMON COW-WHEAT

PETALS FREE OR FUSED: Although superficially similar, **pea** (*top* (*p. 80*)) flowers have separate petals and those of the **Common Cow-wheat** (*bottom* (*p. 190*)) are fused.

PETAL COUNT: One could be forgiven for thinking the **Common Chickweed** here has ten separate petals. In fact it has five; each divided almost to the base.

MISTLETOE

MARSH PENNYWORT

VEGETATIVE PARTS: **Mistletoe** (*p. 131*) and **Marsh Pennywort** (*p. 221*) are examples of species most easily recognized by non-flowering features, *e.g.* growth form or leaf shape.

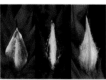

BOTANICAL DETAILS: Those important for identification vary between group. For example, **pinnule shape** (*top*) is frequently used for ferns; glume shape and colour (*bottom*) in sedges.

BOTANICAL DETAILS: Although not needed to separate lime trees (*left*) from mallows (*right*), the stamens fused around the female parts (*top right*) are diagnostic and could help identify an unfamiliar plant as from the mallow family.

23

Ovary position

Although relatively technical, the positioning of a plant's ovary forms a solid basis for plant identification. In the field many plants, particularly with experience, can be confidently identified without referencing ovary position. However, when faced with an unknown plant this information, in conjunction with other features of the flower, can be very helpful at the outset of the identification process.

Superior ovary

– located above the attachment point of any other parts of the flower

Some species, such as this fumitory, have a well-hidden ovary when in flower, which can make assessing its position difficult. However, on fruiting, the flower detaches and in this image the fruiting ovary can be seen to be located above the flower's point of attachment to the stem.

FUMITORY – PAPAVERACEAE

OVARY

POINT OF ATTACHMENT

STIGMA

OVARY

STYLE

STAMEN

TEPALS (CLOSED IN FRUIT)

STAMEN

BUTTERCUP – RANUNCULACEAE

OVARIES

OVARY

Petals and sepals have fallen off.

TEPALS (OPEN IN FLOWER)

RUSH – JUNCACEAE

The ovary can be singular, or in groups, such as in *Ranunculus* buttercups – most easily seen when in fruit.

Inferior ovary

– located below the attachment point of any other parts of the flower

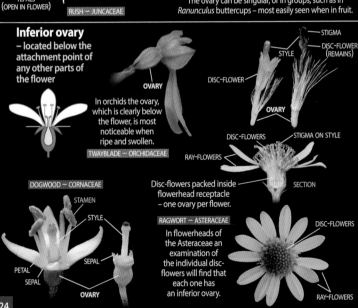

In orchids the ovary, which is clearly below the flower, is most noticeable when ripe and swollen.

OVARY

TWAYBLADE – ORCHIDACEAE

STIGMA

DISC-FLOWER (REMAINS)

STYLE

DISC-FLOWER

OVARY

DISC-FLOWERS

STIGMA ON STYLE

RAY-FLOWERS

SECTION

Disc-flowers packed inside flowerhead receptacle – one ovary per flower.

DOGWOOD – CORNACEAE

STAMEN

STYLE

SEPAL

PETAL

SEPAL

OVARY

RAGWORT – ASTERACEAE

In flowerheads of the Asteraceae an examination of the individual disc-flowers will find that each one has an inferior ovary.

DISC-FLOWERS

RAY-FLOWERS

Partly superior ovary

– located within a receptacle that typically has the petals attached to its rim

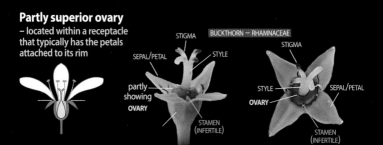

BUCKTHORN ~ RHAMNACEAE

STIGMA
SEPAL/PETAL
STYLE
partly showing
OVARY
STAMEN (INFERTILE)

STIGMA
STYLE
OVARY
SEPAL/PETAL
STAMEN (INFERTILE)

Partly superior ovary | Rosaceae

Rose family species exhibit a range of partly superior ovary positions; from almost fully superior (avens – *top*) to almost fully inferior (roses – *bottom*)

In avens, multiple ovaries are fused within a shallow receptacle; most of the carpels are above the highest point of petal attachment, but the lowest are below.

WOOD AVENS

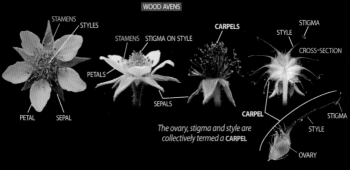

STAMENS
STYLES
STAMENS STIGMA ON STYLE
CARPELS
STYLE
STIGMA
CROSS-SECTION
PETALS
SEPALS
CARPEL
STIGMA
STYLE
PETAL SEPAL
OVARY

The ovary, stigma and style are collectively termed a CARPEL

In roses, a deep protective receptacle encloses multiple ovaries; their individual styles are fused into one that protrudes above the petals, as can be seen in the cross-section (*below*). The style can persist after the petals and sepals have fallen.

ROSE

STYLE STAMENS
STAMENS (REMAINS OF)
STYLE
SEPAL
petals have fallen
RECEPTACLE
STYLE
OVARY
SEED
Ripe seeds inside a rosehip
SECTION

The receptacle enclosing the ovaries swells as they ripen becoming the fruit known as a rosehip.

The broad groups roadmap key

If there is any uncertainty about where to start, the following flowchart is designed to guide a user to a starting point for a confident identification.

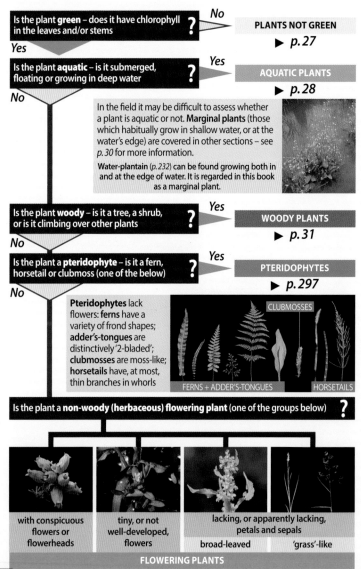

Is the plant green – does it have chlorophyll in the leaves and/or stems **?**

No ▶ **PLANTS NOT GREEN** ▶ *p. 27*

Yes

Is the plant aquatic – is it submerged, floating or growing in deep water **?**

Yes ▶ **AQUATIC PLANTS** ▶ *p. 28*

No

In the field it may be difficult to assess whether a plant is aquatic or not. **Marginal plants** (those which habitually grow in shallow water, or at the water's edge) are covered in other sections – see *p. 30* for more information.

Water-plantain *(p. 232)* can be found growing both in and at the edge of water. It is regarded in this book as a marginal plant.

Is the plant woody – is it a tree, a shrub, or is it climbing over other plants **?**

Yes ▶ **WOODY PLANTS** ▶ *p. 31*

No

Is the plant a pteridophyte – is it a fern, horsetail or clubmoss (one of the below) **?**

Yes ▶ **PTERIDOPHYTES** ▶ *p. 297*

No

Pteridophytes lack flowers: **ferns** have a variety of frond shapes; **adder's-tongues** are distinctively '2-bladed'; **clubmosses** are moss-like; **horsetails** have, at most, thin branches in whorls

CLUBMOSSES

FERNS + ADDER'S-TONGUES   HORSETAILS

Is the plant a non-woody (herbaceous) flowering plant (one of the groups below) **?**

with conspicuous flowers or flowerheads

tiny, or not well-developed, flowers

lacking, or apparently lacking, petals and sepals

broad-leaved   'grass'-like

FLOWERING PLANTS

▶ *p. 47*

PLANTS NOT GREEN

Plants that are regarded as **not green** include:

1 Flowering plants without chlorophyll *i.e.* **saprophytes** that feed on dead material or plants fully **parasitic** on other plants
2 Flowering plants apparently without chlorophyll – with **flowers only** and no obvious leaves
3 Fertile spring (March–April) stages of **horsetails** that lack, or have very little, chlorophyll
4 **Winter trees or shrubs**, without leaves.

SAPROPHYTIC

Yellow Bird's-nest
Hypopitys monotropa (Ericaceae) is one of three saprophytes found in Britain & Ireland

NB **Fungi** (*left*) and **lichens** (*right*) also lack chlorophyll but are outside the scope of this book – see *p. 5*.

PARASITIC

BROOMRAPES
64 Orobanchaceae
▶ *p. 192*

Dodder
54 Convolvulaceae
▶ *p. 163*

Horsetails (Mar–Apr)	Woody trees/shrubs without leaves (winter)	Leafless non-woody plants with flowers only

Identify by stem and sheath characters ▶ *p. 298*	See winter twig gallery ▶ *p. 42*	Identify using flower characters in the flowering plant galleries ▶ *pp. 50–63*

A AQUATIC PLANTS

The categorization and identification of plants on the basis of habitat selection is fraught with difficulty, not least because some species have broad preferences and many occupy the transition zones between contrasting habitats.

One case in point is aquatic *versus* terrestrial plants, where there are all sorts of intergrades. In this book we are treating as aquatics all those species that live predominantly in or on a waterbody, whether rooted or not. If any structures stick out of the water it is only by a few centimetres at most. Aquatics, signified by AQ in the species accounts, are generally not tied to the shallower margins of a waterbody or course; marginal plants are discussed further on *p. 30*.

Aquatics can be grouped into three forms:

Free-floating | on the water surface (*e.g.* **duckweeds** (*p. 231*)) or in the water column | (*e.g.* **hornworts** (*p. 67*)). If these have roots at all, they simply hang in the water; the flowers are usually tiny and inconspicuous.

Rooting, leaves wholly submerged | as above, generally without showy flowers (*e.g.* **water-milfoils** (*p. 80*) and **waterweeds** (*p. 232*)) although there are exceptions (*e.g.* **bladderworts** (*p. 193*)).

Rooting, leaves wholly or partly floating on the water surface | many of these have inflorescences raised above the water surface on short stalks (*e.g.* some **pondweeds** (*p. 233*)); some have showy flowers (*e.g.* **water-lilies** (*p. 67*)).

Aquatic plant sampling

As the leaves and shoots of aquatics are supported by the water in which they live, their leaves are typically rather flaccid and formless. Viewing them in the water is potentially dangerous, often difficult because of limited water clarity or by the presence of overlapping swathes of vegetation; in addition, staring at water even through polarizing lenses (*e.g.* the more expensive sunglasses) is likely to be hampered by surface glare.

One way to sample aquatic plants, particularly submerged ones, is to use a pond-net or a grapnel to drag small amounts of vegetation ashore for closer examination. To see the structural form in a true light, any vegetation removed in this way should be placed into a container of clear water.

Please always return any material to the water as soon as you have finished so as not to kill any pond-life therein, and remember not to get your hands in water if you have any open wounds because of the risk of Weil's disease (leptospirosis).

TOP A grapnel being used to extract aquatic plant samples safely; BOTTOM aquatic haul for identification.

Gallery 1

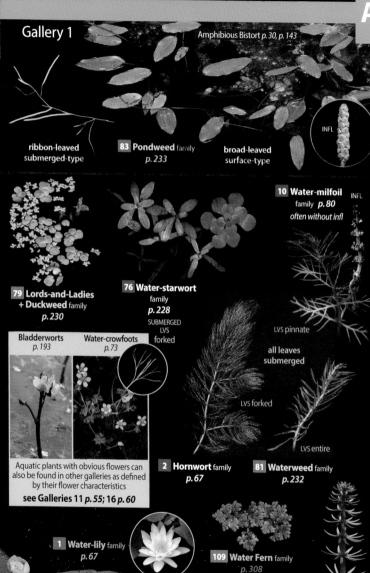

Amphibious Bistort *p. 30, p. 143*

83 Pondweed family
p. 233

ribbon-leaved
submerged-type

broad-leaved
surface-type

INFL

10 Water-milfoil
family ***p. 80***
often without infl

INFL

**79 Lords-and-Ladies
+ Duckweed** family
p. 230

76 Water-starwort
family
p. 228
SUBMERGED
LVS
forked

LVS pinnate

all leaves
submerged

LVS forked

LVS entire

Bladderworts
p. 193

Water-crowfoots
p. 73

Aquatic plants with obvious flowers can
also be found in other galleries as defined
by their flower characteristics
see Galleries 11 *p. 55*; 16 *p. 60*

2 Hornwort family
p. 67

81 Waterweed family
p. 232

1 Water-lily family
p. 67

109 Water Fern family
p. 308

60 Mare's-tail *p. 180*

Horsetails *p. 298*

A AQUATIC PLANTS

A note about stoneworts

Stoneworts (order Charales) are multicellular green algae. There are 33 species found in Britain and Ireland, in two families. They bear a superficial resemblance to some higher plants. Stoneworts are typically found in fresh water, submerged and attached by simple root-like structures to muddy sediments. Structurally, stoneworts are up to 100 cm long with regular whorls of approximately 8 cylindrical branchlets that may be unbranched (horsetail-like) or divided (hornwort-like). Fruiting bodies are distinctive, egg-like, and unlike anything found in higher plants. They are important in the conservation monitoring of aquatic habitats and as such are often grouped with higher plants (*e.g.* in the *BSBI Plant Atlas 2020*).

Delicate Stonewort *Chara virgata* and the egg-like fruiting bodies (*inset*) that distinguish the stoneworts.

Marginal plants

There are a great number of species that for preference, exclusively for some, grow rooted in or around shallow water. These are regarded as marginals in this book with their habitat descriptions indicating clearly their dependence on water or wetlands.

Most marginal species, when growing in a waterbody, produce emergent shoots and leaves, which can rise several metres above the water surface – *e.g.* **Common Reed** (*p. 287*), which also famously forms extensive beds. Although many marginals are within the grass, sedge and rush families there are others with showy flowers *e.g.* **Water-plantain** (*p. 232*) and **Yellow Iris** (*p. 241*). Marginal plants are not in the aquatic plant gallery (*p. 29*), but are found in the relevant places (determined by their floral characteristics) in the main herbaceous flowering plant galleries.

There are even a few species that cross between aquatic and terrestrial realms very extensively: one is **Amphibious Bistort** (*p. 143*), which is equally happy rooted in deep water with floating leaves and emergent flowers, and another is **New Zealand Pigmyweed** (*p. 78*), mainly a plant of exposed muddy margins, but frequently forming either a dense mat on open water or a creeping blanket on dry ground. We treat the former as an aquatic, but the latter as a marginal.

The invasive mat-forming marginal **New Zealand Pigmyweed**

Given that these growth forms and preferences are a subjective continuum, such a pragmatic distinction between aquatics and marginals might seem simplistic. However, the habitat descriptions allocated to these in the book will help in successful identification.

An extensive area of **Amphibious Bistort** (detail *inset*) growing with **Water Horsetail** (➡ *p. 298*) – another marginal plant with an aquatic disposition

Woody plants are those with stiff, hard stems that are rich in lignin – an organic polymer which does not rot easily, and provides rigidity to the cell walls of woody plants.

In Great Britain and Ireland this group of plants includes trees, shrubs and some climbing plants. There are other plants which climb or scramble over other plants and items, but are not woody – see *Woody Climbers* below for more details.

From an identification perspective, something that woody plants have in common is that, whether evergreen or deciduous, they have buds that overwinter above ground (unlike herbaceous plants which die back after flowering). Because of this, woody plants can be readily identified all year round.

During winter, the oppositely arranged buds of a seemingly dead trailing stem of **Honeysuckle** (*p. 218*) are distinctive.

TREES and SHRUBS ▶ *p. 32*

Tree, shrub or woody-based perennial?
Although there is no hard and fast botanical definition of either a 'tree' or a 'shrub', the general distinction is that trees are woody plants with a single trunk that reaches a height of 5 m or more, whereas shrubs have multiple stems typically shorter than 5 m.

At the other end of the scale, shrubs also overlap in stature with woody-based perennials. We have taken **heathers** (*p. 159*) as our notional lower shrub limit, with *e.g.* Sea-purslane (*p. 154*), **restharrows**, (*p.82*) and **thymes** (*p. 189*) falling below the limit.

Hazel (*p. 109*) is a prime example of the tree/shrub conundrum. Most people regard it as a tree but, as it has several stems, it is a shrub. Exceptional examples over 10 m tall have been recorded.

For identification purposes, these distinctions rarely matter. Trees and shrubs are both in the woody section, and lower shrubs and woody perennials feature in the relevant flowering plant sections based on flower details.

Trees transcend reproduction strategies
Trees are found in the spore-producing **Gymnosperms** (Conifers) and the flowering **Angiosperms**. Other than the 'berry'-bearing Yew and Juniper the distinction between the two groups is straightforward. In many trees that bloom, the flowers are on show for a short time and so other factors, such as leaf shape and fruit characters, are relevant to identification; this is reflected in the keys that follow.

For some trees and for all low-growing shrubs woodiness is unlikely to be the primary identifier and these species can also be keyed-out using flower, leaf and fruit characteristics.

WOODY CLIMBERS ▶ *p. 33*

There are only a few woody climbers found wild in Britain and Ireland. Although their scrambling growth form is distinctive, and especially apparent in the winter, they are just as likely to be primarily differentiated using flower and leaf characteristics. They are included with trees and shrubs for completeness, but also feature in the main key.

Non-woody climbers (*p. 33*) | Other species, such as vetches and bindweeds, which also have climbing and scrambling growth forms are briefly covered in this section for comparison; but as they are most likely to be primarily noticed by their flowers they are covered fully within the flowering plants section (*pp. 50–63*).

Woody climbers, such as the highly invasive and fast-growing **Russian-vine** (*p. 144*) can form extensive dense patches, covering everything it climbs over such that it appears to be a shrub in its own right.

Gymnosperms (Conifers) | 3 [5 B&I] families; 14 [40 B&I] spp. ▶ *p. 291*

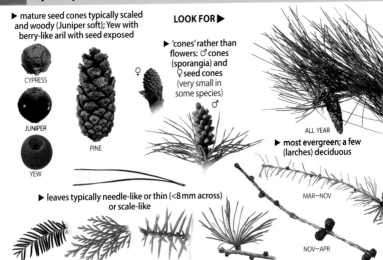

▶ mature seed cones typically scaled and woody (Juniper soft); Yew with berry-like aril with seed exposed

CYPRESS

JUNIPER

YEW

PINE

LOOK FOR ▶

▶ 'cones' rather than flowers: ♂ cones (sporangia) and ♀ seed cones (very small in some species)

♀

♂

ALL YEAR

▶ most evergreen; a few (larches) deciduous

MAR–NOV

NOV–APR

▶ leaves typically needle-like or thin (<8 mm across) or scale-like

Angiosperms (Flowering plants) | 93 [143 B&I] families; 723 [2,280 B&I] spp.

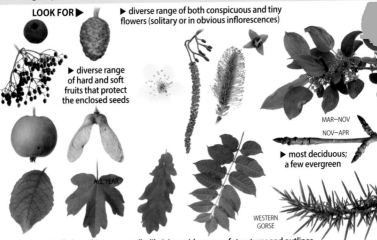

LOOK FOR ▶

▶ diverse range of both conspicuous and tiny flowers (solitary or in obvious inflorescences)

▶ diverse range of hard and soft fruits that protect the enclosed seeds

ALL YEAR

MAR–NOV

NOV–APR

▶ most deciduous; a few evergreen

WESTERN GORSE

▶ leaves typically broad (gorses needle-like), in a wide range of structures and outlines

▶ Tree leaves, flowers and fruits occur in diverse forms. One, some, or all of these features can be used as the basis of identification as set out on the following pages.

IDENTIFY FLOWERING TREES AND SHRUBS TO FAMILY BY

32   ▶ flower *pp. 34–37* | ▶ leaf shape *p. 38* | fruit *p. 40* | twig + bud *p. 42*

WOODY CLIMBERS

Woody

Stems rooted in the ground;
woody, at least at the base.
Young stems long and flexible,
twisting around branches.
Old stems less flexible. Ivy has
specialized hairs that stick to
the supporting surface. Woody
plants *e.g.* **Bittersweet** (*p.177*)
and some **roses** (*p.94*) scramble
over other plants, but do not
twine or attach to them. **Hop** is
rather intermediate between
woody and non-woody.

Traveller's-joy
5 Ranunculaceae (*p.73*)

IVIES
73 Araliaceae (*p.221*)

Hop
16 Cannabaceae (*p.103*)

Honeysuckle
70 Caprifoliaceae (*p.218*)

Russian-vine
41 Polygonaceae (*p.144*)

Non-woody

Herbaceous climbing plants,
often with adaptations
such as tendrils, or twining
spiralling stems. Other weak-
stemmed species (*e.g.* some
bedstraws (*p.160*)) use other
vegetation for support: sticky
hooks on their stems are
among their adaptations for
climbing.

White Bryony (*p.114*)

Black Bryony (*p.236*)

Climbing Corydalis (*p.70*)

Dodder (*p.163*)

PEAS (*p.80*)

Black-bindweed (*p.144*)

BINDWEEDS (*p.162*)

▶ **Use the flowering plants key and galleries if further checks are required** *pp.47–63*

W TREES and SHRUBS

Angiosperm tree and shrub flower identification
Those included in this book are grouped here as follows:

▶ flowers conspicuous; many together in a tight to slightly open inflorescence (*below*)

▶ flowers conspicuous; single flowers or grouped in a loose inflorescence (*p. 36*)

▶ FLOWERS small, PETALS + SEPALS absent or not well-developed; in catkins, catkin-like structures or clusters (*p. 37*)

Gallery 2   Flowers conspicuous; many; in a tight to slightly open inflorescence

PETALS FREE

SYMMETRY radial

OVARY superior   OVARY partially superior

35 Mallow + Lime   *p. 129*

LIMES

conspicuous bracteole

INFL ± clustered in lf-axils

34 Maple + Horse-chestnut family *p. 128*

4 Barberry family *p. 71*

15 Buckthorn family *p. 103*

Sycamore, MAPLES

PETALS FUSED

SYMMETRY bilateral   SYMMETRY radial

OVARY superior

59 Figwort + Mullein family *p. 179*

Lilac

58 Ash family *p. 178*

PRIVETS

Butterfly-bush

Flowering trees and shrubs flowers 1/2

- Conspicuous flowers are arranged using details of their petal structure, flower symmetry and ovary position so as to be comparable with the main flowering plant key – see *p. 24* for full details.

- How tightly flowers are clustered can be variable, or may be open to interpretation and it may be necessary to also check *p. 36*

| SYMMETRY radial | SYMMETRY bilateral | OVARY superior | OVARY part superior | OVARY inferior |

- For woody plants that are typically low-growing (*e.g.* Aubretia) and are more herbaceous-like plants in their habit check the flowering plant galleries *pp. 50–63*)

- Families for which leaf shape is a primary or significant identification feature are marked 🍃.

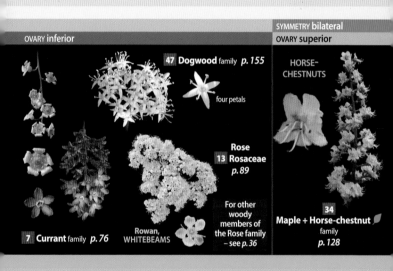

OVARY inferior

47 Dogwood family *p. 155*

four petals

Rose
13 Rosaceae
p. 89

For other woody members of the Rose family – see *p. 36*

7 Currant family *p. 76*

Rowan, **WHITEBEAMS**

SYMMETRY bilateral
OVARY superior

HORSE-CHESTNUTS

34
Maple + Horse-chestnut 🍃
family
p. 128

OVARY inferior

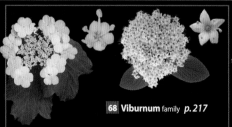

68 Viburnum family *p. 217*

69 Elder family *p. 217* 🍃

W TREES and SHRUBS

Gallery 3 Flowers conspicuous; single flowers or grouped in a loose inflorescence

PETALS FREE

SYMMETRY radial

OVARY inferior

33 Willowherb family *p. 125*

Fuchsia

FRUIT

♂ ♀

38 Mistletoe family *p. 131*

COTONEASTERS (some)

erect petals give 'closed' flower look

OVARY partially superior

'FRUIT' TREES and SHRUBS (Hawthorns, plums, apples cherries, brambles etc.)

13 Rose family **Rosaceae** *p. 89*

ROSES

Although the rose family contains a diverse range of inflorescence structures, individual flowers are typically 5-petalled with multiple stamens see Gallery 12 (*p. 56*)

OVARY superior

26 Spindle family *p. 115*

petals ± fused at the extreme base, and appear to be free

♀

♂

54 Holly family *p. 169*

Tutsan

30 St John's-wort family *p. 120*

SYMMETRY bilateral

OVARY superior

11 Pea family *p. 80*

Broom, GORSES

Restharrows *p. 61*

PETALS FUSED

SYMMETRY radial

OVARY inferior

70 Honeysuckle family *p. 218*

Snowberry

OVARY superior

57 Nightshade family *p. 175*

SYMMETRY bilateral

RHODODENDRONS

'BILBERRIES' and HEATHERS

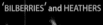

49 Heather family *p. 158*

Crowberry has tiny flowers with 3 free petals – see *p. 63*

36

Flowering trees and shrubs flowers 2/2 **W**

Gallery 4   **Petals + sepals small/not well-developed, and/or in catkins or clusters**

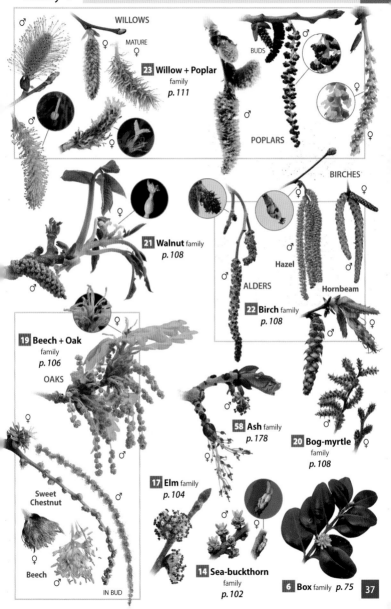

WILLOWS

MATURE ♀

23 Willow + Poplar family *p. 111*

BUDS

♂

♀

POPLARS

BIRCHES

21 Walnut family *p. 108*

Hazel

ALDERS

Hornbeam

22 Birch family *p. 108*

19 Beech + Oak family *p. 106*

OAKS

58 Ash family *p. 178*

20 Bog-myrtle family *p. 108*

Sweet Chestnut

17 Elm family *p. 104*

Beech

IN BUD

14 Sea-buckthorn family *p. 102*

6 Box family *p. 75*

37

Gallery 5

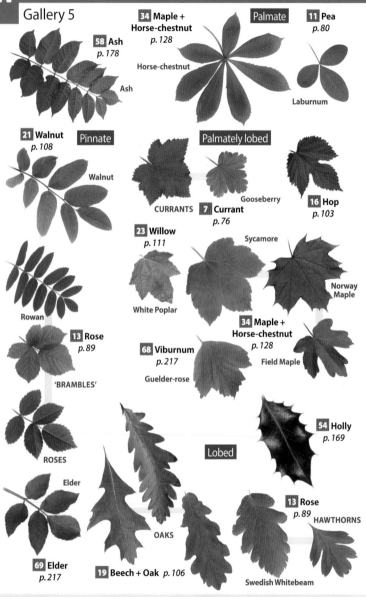

58 Ash p. 178

Ash

34 Maple + Horse-chestnut p. 128

Palmate

Horse-chestnut

11 Pea p. 80

Laburnum

21 Walnut p. 108

Pinnate

Walnut

Palmately lobed

CURRANTS

7 Currant p. 76

Gooseberry

16 Hop p. 103

23 Willow p. 111

Sycamore

Norway Maple

Rowan

White Poplar

34 Maple + Horse-chestnut p. 128

68 Viburnum p. 217

Field Maple

Guelder-rose

13 Rose p. 89

'BRAMBLES'

ROSES

Elder

Lobed

54 Holly p. 169

13 Rose p. 89

HAWTHORNS

OAKS

69 Elder p. 217

19 Beech + Oak p. 106

Swedish Whitebeam

► Leaf arrangement, shape and other details **pp. 18–19**

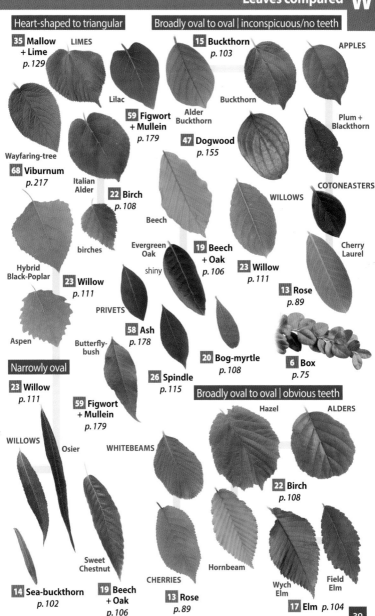

Heart-shaped to triangular

35 Mallow + Lime *p. 129*

LIMES

Lilac

Wayfaring-tree

68 Viburnum *p. 217*

Italian Alder

22 Birch *p. 108*

birches

Hybrid Black-Poplar

23 Willow *p. 111*

Aspen

Narrowly oval

23 Willow *p. 111*

WILLOWS

Osier

PRIVETS

Butterfly-bush

58 Ash *p. 178*

26 Spindle *p. 115*

59 Figwort + Mullein *p. 179*

Sweet Chestnut

14 Sea-buckthorn *p. 102*

19 Beech + Oak *p. 106*

Broadly oval to oval | inconspicuous/no teeth

15 Buckthorn *p. 103*

APPLES

Buckthorn

Alder Buckthorn

59 Figwort + Mullein *p. 179*

47 Dogwood *p. 155*

Plum + Blackthorn

Beech

Evergreen Oak

shiny

19 Beech + Oak *p. 106*

WILLOWS

23 Willow *p. 111*

COTONEASTERS

Cherry Laurel

13 Rose *p. 89*

20 Bog-myrtle *p. 108*

6 Box *p. 75*

Broadly oval to oval | obvious teeth

Hazel

ALDERS

WHITEBEAMS

22 Birch *p. 108*

Hornbeam

CHERRIES

13 Rose *p. 89*

Wych Elm

Field Elm

17 Elm *p. 104*

Gallery 6

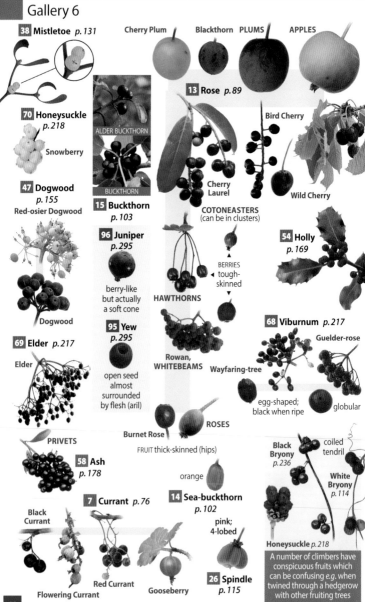

38 Mistletoe p. 131

Cherry Plum Blackthorn PLUMS APPLES

70 Honeysuckle p. 218

Snowberry

ALDER BUCKTHORN

BUCKTHORN

13 Rose p. 89

Bird Cherry

Cherry Laurel

Wild Cherry

47 Dogwood p. 155

Red-osier Dogwood

15 Buckthorn p. 103

COTONEASTERS
(can be in clusters)

54 Holly p. 169

96 Juniper p. 295

berry-like but actually a soft cone

BERRIES tough-skinned

Dogwood

HAWTHORNS

95 Yew p. 295

open seed almost surrounded by flesh (aril)

68 Viburnum p. 217

Guelder-rose

69 Elder p. 217

Elder

Rowan, WHITEBEAMS

Wayfaring-tree

egg-shaped; black when ripe

globular

ROSES

Burnet Rose

FRUIT thick-skinned (hips)

Black Bryony p. 236

coiled tendril

White Bryony p. 114

PRIVETS

58 Ash p. 178

orange

7 Currant p. 76

14 Sea-buckthorn p. 102

Black Currant

pink; 4-lobed

Honeysuckle p. 218

Red Currant

Gooseberry

26 Spindle p. 115

Flowering Currant

A number of climbers have conspicuous fruits which can be confusing e.g. when twined through a hedgerow with other fruiting trees

11 Pea
p. 80

Laburnum

58 Ash
p. 178

Ash

SEED
1-winged

typical
pea pod

22 Birch
p. 108

SEED CLUSTER

Hornbeam

17 Elm
p. 104

SEED with
3-lobed
bracts

BIRCHES
SEED winged

Downy Birch

Silver Birch

Wych Elm
SEED **nearer
base**

'Field Elm'
SEED nearer
tip

34 Maple + Horse-Chestnut
p. 128

FIELD
MAPLE

Norway
Maple

Sycamore

ALDERS
SEED in cone-like structures

YOUNG
FRUIT

OLD
FRUIT

Pine

Larch

Spruce

Cypress

Horse-
Chestnuts

FRUIT (capsule)
stiff spines

NUT (conker)

Horse-chestnut capsule
has much fewer, stiffer,
spines than the unrelated
Sweet Chestnut

FRUIT (chestnut)
enclosed in scaly cup
with soft spines

Hazel
NUT surrounded
by deeply
lobed bracts

With some similarity in
appearance to alder fruit;
the cones of pines, spruces,
larches and cypresses
(gymnosperms) are
relatively straightforward to
identify to species
▶ *p. 291*

OAKS
FRUIT (acorn)
in scaly cup

Turkey
Oak

FRUIT
enclosed in
a scaly cup

FRUIT
points up

**35 Mallow
+ Lime**
p. 129

Lime

Small-leaved Lime

FRUIT
hangs down

FRUIT
CAPSULE

SEED

19 Beech + Oak
p. 106

Sweet Chestnut

NUT (chestnut)

Beech

NUT (mast)

open scaly
cup exposing
FRUIT (nut)

21 Walnut
p. 108

41

W TREES and SHRUBS

Gallery 7 1/2
Identify trees and shrubs without leaves by twig and bud features as follows: are the **lateral buds** arranged oppositely or alternately?; is there a **scar ridge** (indistinct in some) that connects the **leaf-scars**?; then by the shape and angle of the **terminal** and **lateral bud** and details of the **bud scales** (if present).

Lateral buds ± opposite

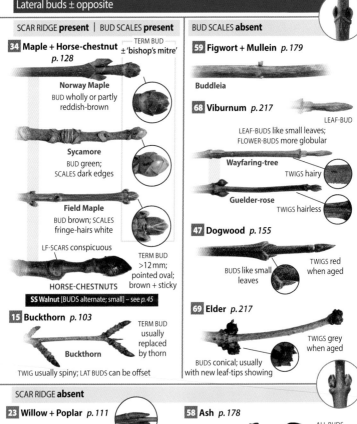

SCAR RIDGE **present** | BUD SCALES **present**

34 Maple + Horse-chestnut p. 128
TERM BUD ± 'bishop's mitre'

Norway Maple
BUD wholly or partly reddish-brown

Sycamore
BUD green; SCALES dark edges

Field Maple
BUD brown; SCALES fringe-hairs white

LF-SCARS conspicuous

HORSE-CHESTNUTS
TERM BUD >12 mm; pointed oval; brown + sticky

SS Walnut [BUDS alternate; small] – see p. 45

15 Buckthorn p. 103
TERM BUD usually replaced by thorn

Buckthorn

TWIG usually spiny; LAT BUDS can be offset

BUD SCALES **absent**

59 Figwort + Mullein p. 179

Buddleia

68 Viburnum p. 217
LEAF-BUD

LEAF-BUDS like small leaves; FLOWER-BUDS more globular

Wayfaring-tree
TWIGS hairy

Guelder-rose
TWIGS hairless

47 Dogwood p. 155
BUDS like small leaves
TWIGS red when aged

69 Elder p. 217
TWIGS grey when aged
BUDS conical; usually with new leaf-tips showing

SCAR RIDGE **absent**

23 Willow + Poplar p. 111

Purple Willow
only willow typically with opposite lateral buds

LAT BUDS close to twig; BUD SCALES 1

26 Spindle p. 115

TWIG green; 4-ridged when older

LAT BUDS close to twig; BUD SCALES 2–6 pairs

58 Ash p. 178

Ash
ALL BUDS large + black

'TERM' BUD 2 large; green + red; LAT BUDS green + red

Lilac

TWIG pale greenish-brown

'TERM' BUD small; green- to purplish-brown

PRIVETS

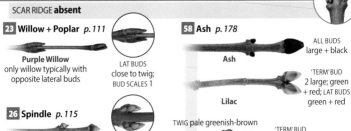

42

Winter twig terms

LATERAL BUD (LAT BUD)   TERMINAL BUD (TERM BUD)

BUD SCALES

SCAR RIDGE   LEAF-SCAR

Lateral buds alternate

1/2

| BUD SCALES 0 | BUD SCALES 1 | BUD SCALES 2 or 3 | | BUD SCALES ≥3 | |
		'boxing glove'	± same size	long + pointed	pointed oval to rounded

BUD SCALES 0

15 Buckthorn *p. 103*   **Alder Buckthorn**

LEAF-BUDS covered in dense hairs

BUD SCALES 1 | 23 Willow + Poplar *p. 112* [STRIAE are thin, lengthwise ridges under the bark]

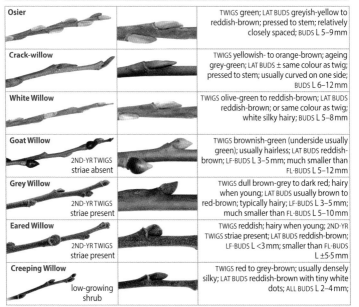

Osier		TWIGS green; LAT BUDS greyish-yellow to reddish-brown; pressed to stem; relatively closely spaced; BUDS L 5–9 mm
Crack-willow		TWIGS yellowish- to orange-brown; ageing grey-green; LAT BUDS ± same colour as twig; pressed to stem; usually curved on one side; BUDS L 6–12 mm
White Willow		TWIGS olive-green to reddish-brown; LAT BUDS reddish-brown; or same colour as twig; white silky hairy; BUDS L 5–8 mm
Goat Willow 2ND-YR TWIGS striae absent		TWIGS brownish-green (underside usually green); usually hairless; LAT BUDS reddish-brown; LF-BUDS L 3–5 mm; much smaller than FL-BUDS L 5–12 mm
Grey Willow 2ND-YR TWIGS striae present		TWIGS dull brown-grey to dark red; hairy when young; LAT BUDS usually brown to red-brown; LF-BUDS L 3–5 mm; much smaller than FL-BUDS L 5–10 mm
Eared Willow 2ND-YR TWIGS striae present		TWIGS reddish; hairy when young; 2ND-YR TWIGS striae present; LAT BUDS reddish-brown; LF-BUDS L <3 mm; smaller than FL-BUDS L ±5·5 mm
Creeping Willow low-growing shrub		TWIGS red to grey-brown; usually densely silky; LAT BUDS reddish-brown with tiny white dots; ALL BUDS L 2–4 mm;

BUDS with 2–3 scales | 'boxing glove'

35 Mallow + Lime *p. 129*   **Lime**   BUDS those of **Small-leaved Lime** are smaller, rounder and spaced more closely on the twig than those of **Lime**

BUDS L 5–10 mm   **Small-leaved Lime**   BUDS L 3–6 mm

SS Hazel [SCALES 6–8; green – see *p. 45*]

W TREES and SHRUBS

Gallery 7 2/2 — Lateral buds alternate

BUDS with 2–3 scales | ± same size `SS Sweet Chestnut (opposite) buds can have 2 scales`

22 Birch p. 108

Alder
TWIGS hairless

Grey Alder
TWIGS hairy

Italian Alder
BUDS greener, smaller and on longer stalk than those of other alders

	Grey Alder	Alder	Italian Alder
	TWIGS **hairy**; BUDS L 6–10mm; on **hairy** stalk (L 2–8mm)	TWIGS hairless; BUDS L 6–10mm; on **hairless** stalk (L 2–4mm)	TWIGS hairless; BUDS L 4–8mm; on hairless stalk (L 3–10mm)

14 Sea-buckthorn p. 102

TWIG **thorny**

BUDS ♀ with 2–4 scales; ♂ with 6–8

BUDS with ≥3 scales | long + pointed

23 Willow + Poplar p. 111

Aspen
TERM BUD L 6–12mm; SCALES dark reddish-brown with pale edges

LAT BUD SCALES Aspen 2 or 3; Hybrid Black-poplar 3–5

Hybrid Black-poplar
TERM BUD L 12–20mm; SCALES yellowish- to reddish-brown

19 Beech + Oak p. 106

Beech
LAT BUDS held away from stem; SCALES ± 20

Hornbeam
LAT BUDS pressed against stem; SCALES 8–16

BUCKTHORN
Buckthorn buds could be regarded as alternately arranged – see p. 42

BUDS with ≥3 scales | long + pointed to pointed-oval

13 Rose p. 89

Bird Cherry
TWIG dark brown; shiny

BUD SCALES base dark brown; tip pale

TWIG greyish- to reddish-brown; dull, hairless

Rowan
TERM BUD purplish-brown with **dense white hairs**

23 Willow + Poplar p. 111

White Poplar
TERM BUD L 3–8mm; SCALES 3–5

TWIG white-felted; LF-SCAR dark brown

Grey Poplar [TERM BUD L 5–10mm (SCALES 6–12); LF-SCAR pale, brown] – see p. 111

22 Birch p. 108

TWIG usually ± shiny and hairless; BUDS generally more pointed and greener than in **Downy**

Silver Birch

TWIG usually dull and hairy; BUDS generally more rounded and browner than in **Silver**

Downy Birch

BUDS **usually with ≥3 scales** | **pointed-oval to rounded**

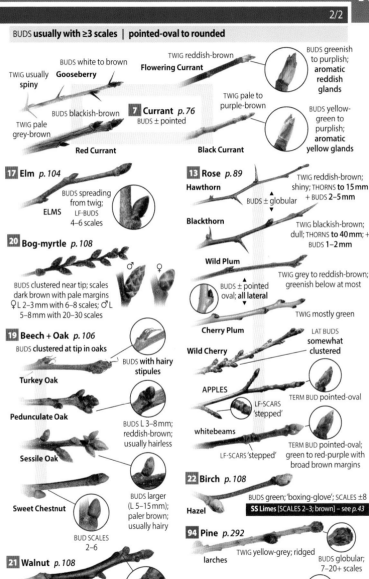

BUDS white to brown **Gooseberry**

TWIG usually **spiny**

TWIG reddish-brown **Flowering Currant**

BUDS greenish to purplish; **aromatic reddish glands**

BUDS blackish-brown

7 Currant *p. 76*
BUDS ± pointed

TWIG pale grey-brown

Red Currant

TWIG pale to purple-brown

BUDS yellow-green to purplish; **aromatic yellow glands**

Black Currant

17 Elm *p. 104*

ELMS

BUDS spreading from twig; LF-BUDS 4–6 scales

20 Bog-myrtle *p. 108*

BUDS clustered near tip; scales dark brown with pale margins ♀ L 2–3 mm with 6–8 scales; ♂ L 5–8 mm with 20–30 scales

19 Beech + Oak *p. 106*

BUDS **clustered at tip in oaks**

BUDS with hairy stipules

Turkey Oak

Pedunculate Oak

BUDS L 3–8 mm; reddish-brown; usually hairless

Sessile Oak

BUDS larger (L 5–15 mm); paler brown; usually hairy

Sweet Chestnut

BUD SCALES 2–6

21 Walnut *p. 108*

LF-SCARS **large**

SS Horse-chestnut [BUDS large (*p. 42*)]

13 Rose *p. 89*
Hawthorn

TWIG reddish-brown; shiny; THORNS **to 15 mm** + BUDS **2–5 mm**

BUDS ± globular

Blackthorn

TWIG blackish-brown; dull; THORNS **to 40 mm**; + BUDS **1–2 mm**

Wild Plum

TWIG grey to reddish-brown; greenish below at most

BUDS ± pointed oval; **all lateral**

TWIG mostly green

Cherry Plum

Wild Cherry

LAT BUDS somewhat clustered

APPLES

LF-SCARS 'stepped'

TERM BUD pointed-oval

whitebeams

LF-SCARS 'stepped'

TERM BUD pointed-oval; green to red-purple with broad brown margins

22 Birch *p. 108*
Hazel

BUDS green; 'boxing-glove'; SCALES ±8

SS Limes [SCALES 2–3; brown] – see *p. 43*

94 Pine *p. 292*

larches

TWIG yellow-grey; ridged

BUDS globular; 7–20+ scales

11 Pea *p. 80*

Laburnum   TWIG green

BUDS silky hairy

H

HERBACEOUS FLOWERING PLANTS

Flowering plants consist of the **Pre-dicots** and the **Angiosperms**. The pre-dicots are a small group of ancient families that include magnolias and Bay and are represented naturally in B&I only by the aquatic water-lilies (*p. 67*).

The angiosperms comprise the vast majority of plants in Great Britain and Ireland and include just over 140 families with almost 2,300 native or naturalized species.

This book covers the 93 families (723 species) most likely to be encountered.

Rather than simply being petals, a flower is the reproductive structure which, in the simplest form, consists of male anthers and female stigma(s) and ovary. All flowering plants have these basic features, with either male and female parts within the same flower (**bisexual**) or on the same plant (**unisexual**, *monoecious*) or on different plants (*dioecious*).

Generally speaking those plants that rely on the attraction of insects for pollination have showy petals (and scents); those with petals much reduced or absent typically use movement (such as the wind) for pollination.

In addition to basic reproductive parts, many have additional structures. For example, the **bracts** and **sepals** of showy flowering plants, and the **lemma** and **palea** in grasses, all serve to protect the reproductive parts of the flower.

From an identification perspective, improved confidence in identification goes hand-in-hand with a greater understanding of a flower structure and the diverse range of additional parts. A significant number of species' identifications rely on being able to locate and assess these features – ovary position, lemma shape and number of stigmas are examples.

Although perhaps not the most apparent difference, Common Chickweed (*left*) has 3–8 stamens and Common Mouse-ear (*right*) has 10.

Although these may seem daunting at first, it is worth persevering – full details can be found on *pp. 14–20* and are referenced at appropriate points in the key that follows.

Flowering plant identification

Herbaceous flowering plants can be split into three broad groups based on how conspicuous (or not) the flowers are. Species that are representative of each of the families included in these groups are depicted in a range of feature-based galleries.

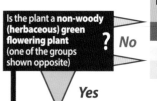

Is the plant a **non-woody (herbaceous) green flowering plant** (one of the groups shown opposite) **?** **No**

If the plant has flowers but is woody (*e.g.* a tree or a shrub); a better identification pathway is:

WOODY PLANTS ▶ *p. 31*

If the plant is not green (lacks chlorophyll):

PLANTS NOT GREEN ▶ *p. 27*

Yes

Assign the plant to one of the groups defined opposite and then scan through the relevant gallery indicated in order to find a close match for your plant

This key is based upon the characteristics of the individual flower; for compact, tightly clustered or composite flowerheads close examination of an individual flower may be needed.

Does the plant have **conspicuous** flowers ?

Do the 'flowers' appear large but are actually a tight collection of small flowers in a head?

Are individual flowers small and in a flat or clustered inflorescence?

Are the flowers large and single (or a few in a group)?*

▶ *Gallery 9 p. 51*   ▶ *Galls 11–19 p. 54–61*

▶ *Gallery 10 p. 52*

*pp. 48–49 contain essential underpinning information and descriptions of the differentiating features needed in the identification of plants with conspicuous flowers

Does the plant have **inconspicuous** flowers ?

Are the flowers fully formed but tiny?...

...or are the flowers not well-developed?

▶ *Gallery 20 p. 62*   ▶ *Gallery 21 p. 62*

Rushes could key out here as structurally they are very similar, but a close examination will see that they have flowers with petals and sepals that look the same (tepals)

see Galleries 11 (*p. 54*) and 21 (*p. 63*)

Does the plant lack, or seem to be **lacking**, petals + sepals ?

Does the plant have broad leaves?

Does the plant have long, narrow leaves (grass-like)?

Gallery 21 (left page)
▶ *p. 62*

Gallery 21 (right page)
▶ *p. 63*

47

Conspicuous flower identification

In the plants with conspicuous flowers there are a number of characters which, once assessed, form the basis for a confident identification to family level at least. In some cases only one or two characters from the set will be enough; in others more may be needed.

The pathway below orders the features to check, with the most apparent first, together with annotated images of these features. Feature sets and their galleries are given here and in a look-up table, covering all set combinations, on *p. 64*.

Conspicuous flower identification – the questions to ask

▶ are the petals **FREE** (separated); or **FUSED** (either significantly or just at the base) **?**

▶ what is the **SYMMETRY** of the flower – **RADIAL;** or **BILATERAL** **?**

▶ where is the **OVARY LOCATED*** in relation to the other flower parts **?**

PETALS FREE vs. FUSED

free (petals separated) · CINQUEFOIL

fused, but only at the base · LOOSESTRIFE

significantly fused · BUTTERFLY-BUSH

Galleries 11–13; 18–19 *pp. 54–57, 61*

Galleries 14–17 *pp. 58–60*

▶ In plants with **free** petals, each petal is separate and will fall individually from the plant.

▶ In plants with **fused** petals, the petals are conjoined to a greater or lesser extent; and will fall from the plant as a whole unit.

▶ Some plants have petals **fused, but only at the base**, and, on occasion, these can be mistaken for free petals unless examined closely.

FLOWER SYMMETRY

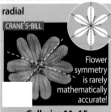

radial · CRANE'S-BILL

Flower symmetry is rarely mathematically accurate!

Galleries 11–15 *pp. 54–58*

bilateral · EYEBRIGHT

Galleries 16–19 *pp. 60–p. 61*

Diagrams of the botanical characters shown on these pages, together with more detailed information can be found on *pp. 14–20*

▶ In plants with **radial symmetry** any petals and sepals are arranged in a regular pattern around the centre, such that the flower can be **divided into similar halves along more than one plane** passing through the centre.

▶ In plants with **bilateral symmetry** any petals and sepals are arranged on opposite sides of a central axis, such that the flower can only be **divided into similar halves along that single plane.**

OVARY POSITION – see *p. 24* for more information

superior	partly superior	inferior

PEARLWORT

OVARY

Galleries 11, 14, 16, 18
pp. 54, 58, 60, 61

BUCKTHORN

PARTLY
SHOWING

PARTLY
ENCLOSED

Gallery 12
p. 56

BEDSTRAW

OVARY

Galleries 13, 15, 17, 19
pp. 57, 58, 60, 61

* not applicable to male dioecious flowers.

▶ In plants with a **superior ovary** it is located **above the attachment point of any other part of the flower.**

▶ In plants with a **partly superior ovary** it is located **within a receptacle that typically has the petals attached to its rim.**

▶ In plants with an **inferior ovary** it is located **below the attachment point of any other part of the flower.**

Conspicuous flower identification – next steps

Once a suite of characters has been assessed a search of the relevant gallery should prove successful in finding the family that the plant in question belongs to. The galleries can be found using the page numbers shown here, or by using the look-up table, which groups families by each combination of features. Note that throughout the galleries the key character groups are colour-coded and accompanied by diagrams (as on these pages) as an additional navigation aid.

▶ **Look-up table by features** *p. 64*   ▶ **The Galleries** *pp. 50–63*

Most flowers have petals clearly different from sepals in colour and/or size and shape. In those that have similar petals and sepals, they may be termed tepals. But in a further complication, petals may be sepal-like (sepaloid) *e.g.* **Black Bryony** or sepals petal-like (petaloid) *e.g.* **Wood Anemone** (see also *p. 16*). To avoid such complications, we do not use the similarity/difference between sepals and petals as a differentiator.

WOOD-RUSH

P1
S3
P3
S1
S2
P2

similar in shape, size and colour– Two sets: three petals (inner [P1–P3]) and sepals (outer [S1–S3])

SEA-SPURREY

S5   P1
        S1
P5
        P2
S4
        S2
P4
    P3
    S3

obviously different in shape, size and/or colour Two sets: five petals (inner [P1–P5]) and sepals (outer [S1–S5])

WOOD ANEMONE

6 SEPALS
0 PETALS

0 SEPALS; 6 PETALS

BLACK BRYONY

petal-like sepals (*left*) and **sepal-like petals** (*right*)

H Conspicuous flowerhead or inflorescence of similar flowers

Gallery 8 There are two groups of plants of which examples are shown in this gallery: 1) **highly distinctive flowers** – in some cases very different from the 'normal' flowers associated with that family and 2) **'bell'-shaped flowers**. For both groups the defining flower features can be difficult to establish without a close examination.

Fuchsia

Enchanter's-nightshade

88 Daffodil + Onion family *p. 241*

Wild Onion

33 Willowherb family *p. 125*

EVENING-PRIMROSES

Wild Daffodil

Montbretia

87 Iris family *p. 241*

70 Honeysuckle family *p. 218*

Snowdrop

Columbine

Traveller's-joy

5 Buttercup family **Gallery 11** *p. 55* *p. 71*

Water Avens

77 Bellflower family *p. 229*

59 Figwort + Mullein family Gallery 16 *p. 60* *p. 179*

FIGWORTS

13 Rose family Gallery 12 *p. 56* *p. 89*

BLUEBELLS

Navelwort

GRAPE-HYACINTHS

9 Stonecrop family Gallery 14 *p. 59* *p. 78*

89 Bluebell [Asparagus] family *p. 243*

49 Heather family *p. 158*

50 ▶ Plants are not to scale and may not be the same colour as the examples shown here – flower shape and structure are the important features to note when looking for a match.

Gallery 9 Although in this book flower identification is based on the characters of an individual flower, the families, groups or species here and some of those opposite are primarily recognized by their conspicuous inflorescences.

FLOWERS IN AN UMBEL – multiple flower-stalks (spokes) that arise from a single point at the top of the stem and forming a flat- or round-topped inflorescence.

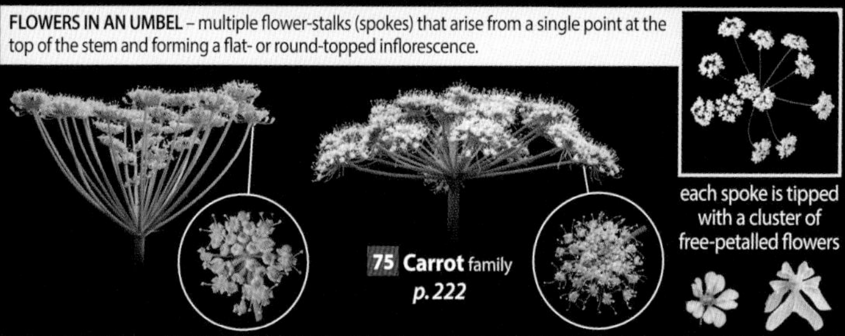

each spoke is tipped with a cluster of free-petalled flowers

75 Carrot family *p. 222*

Carrot family umbels Individual flowers have free petals that differ from the sepals. Flower symmetry in most species is radial, although in some (especially the outer flowers) are bilateral. The ovary is inferior.

FLOWERS IN OTHER ARRANGEMENTS – see *p. 17* for more details.

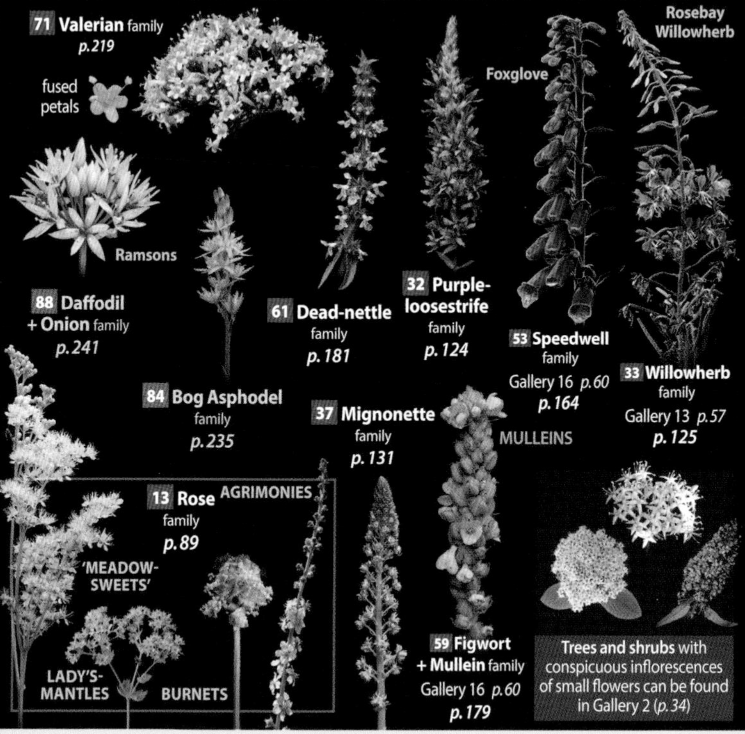

71 Valerian family *p. 219*

fused petals

Rosebay Willowherb

Foxglove

Ramsons

88 Daffodil + Onion family *p. 241*

61 Dead-nettle family *p. 181*

32 Purple-loosestrife family *p. 124*

53 Speedwell family Gallery 16 *p. 60* *p. 164*

33 Willowherb family Gallery 13 *p. 57* *p. 125*

84 Bog Asphodel family *p. 235*

37 Mignonette family *p. 131*

MULLEINS

13 Rose AGRIMONIES family *p. 89*

'MEADOW-SWEETS'

59 Figwort + Mullein family Gallery 16 *p. 60* *p. 179*

Trees and shrubs with conspicuous inflorescences of small flowers can be found in Gallery 2 (*p. 34*)

LADY'S-MANTLES BURNETS

► Some members of the daisy family have conspicuous inflorescences – see Gallery 10 *p. 53*.
► Individual flowers of the plants shown above are also shown in their relevant gallery.

51

H Conspicuous heads of small flowers | heads single or clustered

Gallery 10

Plants in this group exhibit a high diversity of individual flower shape as well as a wide variety of how individual flowers (even if small) are grouped to form a conspicuous flowerhead (or inflorescence) – **see p. 17 for more information.**

Daisy family individual flowers |

Observing individual flowers requires a close examination of the flowerhead (*capitulum*). Flower symmetry varies depending on the flower: there are two flower types – strap-like **ray-flowers** and tubular **disc-flowers**. Flowerheads can consist of disc-flowers only, ray-flowers only or both disc- and ray-flowers. The unstalked flowers are packed tightly on a specialized flat-topped or domed receptacle that is formed from the swollen top of the inflorescence stalk, surrounded by whorls of bracts (the *involucre*) that protects the developing flowers. Symmetry in the individual tiny flowers may be either radial or bilateral. Disc-flowers are usually radially symmetrical though they are occasionally slightly irregular. Ray-flowers are always bilaterally symmetrical. Understanding these detailed characters will help you to make sense of this very large family – **see p. 24 for further details.**

67 Daisy + Thistle family *p. 194*

All species shown here have fused petals and an inferior ovary

DISC-FL
RAY-FL
DISC-FLOWERS   RAY-FLOWERS
RECEPTACLE SURROUNDED BY BRACTS

Common Ragwort has a loose cluster of flowerheads

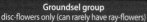

| **Thistle group** | **Dandelion-type** |
| disc-flowers only | ray-flowers only |

DISC-FL

RAY-FL

| **Groundsel group** | **Daisy-type** |
| disc-flowers only (can rarely have ray-flowers) | disc- and ray- flowers (rays can be tiny) |

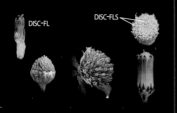

DISC-FL
DISC-FLS

DISC-FL
RAY-FL (tiny)
DISC-FL
RAY-FL

▶ **Plants are not to scale and may not be the same colour as the examples shown here – flower shape and structure are the important features to note when looking for a match.**

Clustered flowerhead group
Inflorescence consisting of clusters of flowerheads;
themselves each containing multiple individual flowers

Flowerhead with both disc- and ray-flowers	Flowerhead with disc-flowers only

'YARROWS'

'WORMWOODS'

CUDWEEDS

'HELIOTROPES'

Tansy

GOLDENRODS

Hemp-agrimony

Mountain-everlasting   ♂ ♀

67 **Daisy + Thistle** family
p. 194

---

72 **Teasel + Scabious** family
p. 220

SCABIOUSES

Thrift

Sheep's-bit

Teasel

40 **Thrift + Sea-lavender** family
Gallery 14 *p. 58*
p. 140

77 **Bellflower** family
Gallery 15 *p. 58*
p. 229

SS two look-alike clustered flowerheads

▶ **Flowers with radial symmetry can be found in Galleries 11–15 (*pp. 54–59*).**

H Conspicuous flowers

Gallery 11

Plants in this group exhibit a high diversity of individual flower/petal shapes as well as a wide variety of inflorescence types (see *p. 17*).

Species with tiny flowers which fit into this category and are more likely to be noticed by other features are Hop (Cannabaceae) *p. 103*; Nettle (Urticaceae) *p. 105*; Bog-myrtle (Myricaceae) *p. 108*; and Mercuries (Euphorbiaceae) *p. 117* and are covered in Galleries 4 (*p. 37*), 20 and 21 (*pp. 62–63*).

Greater Celandine

four petals

3 Poppy + Fumitory family *p. 68*

POPPIES

four petals

39 Cabbage family *p. 132*

31 Crane's-bill family *p. 122*

CRANE'S-BILLS

STORK'S-BILLS

27 Flax family *p. 115*

46 Blinks family *p. 154*

25 Wood-sorrel family *p. 114*

three petals

43 Sundew family *p. 145*

80 Water-plantain family *p. 232*

MALLOWS

35 Mallow + Lime family *p. 129*

91 Rush family *p. 245*

A number of woody plant families have flowers in this category – **Spindle, Holly, Elm, Lime** (Malvaceae) and **Barberry** – see Galleries 2 and 3 (*pp. 34–36*)

LIMES

26 Spindle family *p. 115*

4 Barberry family *p. 71*

54 Holly family ♀ *p. 169*

88 Daffodil + Onion family *p. 241*

► Plants are not to scale and may not be the same colour as the examples shown here – flower shape and structure are the important features to note when looking for a match.

PETALS **free**   SYMMETRY **radial**   OVARY **superior**

44 Campion [Pink]
family
p. 145

84 Bog Asphodel family
p. 235

30 St John's-wort family
p. 120

9 Stonecrop family
p. 78

SS some **roses** (Rosaceae)
[OVARY partly superior]
Gallery 12 (*p. 56*)

**SS pimpernels and
loosestrifes** (Primulaceae)
[PETALS fused at the base]
Gallery 14 (*p. 58*)

36 Rock-rose
family
p. 130

5 Buttercup
family
p. 71

▶ **Other flowers with radial symmetry can be found in Galleries 12–15 (*pp. 56–59*).**

H Conspicuous flowers

Gallery 12

PETALS **free**   SYMMETRY **radial**   OVARY **pt superior**

SAXIFRAGES

8 Saxifrage family
p.77

32 Purple-loosestrife
family
p.124

SS **rock-roses** (Cistaceae); **buttercups** (Ranunculaceae); and
pimpernels (Primulaceae) [all with OVARY superior];
Gallery 11 (*p.54*) and 14 (*p.58*)

**34 Maple +
Horse-chestnut**
family
p.128

15 Buckthorn
family
p.103

13 Rose family
p.89

A number of woody trees, shrubs and climbers have
flowers in this category – **Maples** (Sapindaceae),
Buckthorns (Rhamnaceae) and members of the **Rose**
family (Rosaceae) – **see Galleries 2 and 3** (*pp.34–36*).

Some trees with catkins or catkin-like inflorescences
(**Beech** (Fagaceae), **Walnut** (Juglandaceae), **Sea-
buckthorn** (Elaeagnaceae) **and birches** (Betulaceae))
have free-petalled male flowers with radial symmetry
– **see Gallery 4** (*p.37*).

▶ Plants are not to scale and may not be the same colour as the examples shown here –
flower shape and structure are the important features to note when looking for a match.

Gallery 13   PETALS **free**   SYMMETRY **radial**   OVARY **inferior**

33 Willowherb family
p. 125

Enchanter's-nightshade

Fuchsia

EVENING-PRIMROSES

8 Saxifrage
family
p. 77

GOLDEN-SAXIFRAGES

74 Pennywort
family
p. 221

38 Mistletoe
family
p. 131

♀   ♂

73 Ivy family
p. 221

SNOWDROPS

**88 Daffodil
+ Onion** family
p. 241

7 Currant family
p. 76

symmetry bilateral in some flowers

75 Carrot family
p. 222

SS Yarrow and
Sneezewort
(Asteraceae)
Gallery 10 (*p. 53*)

13 Rose family
p. 89

47 Dogwood family
p. 155

Mistletoe and a number of woody trees,
shrubs and climbers have flowers in this
category – **Currants** (Grossulariaceae),
Rowan and **whitebeams** (Rosaceae), and
Dogwoods (Cornaceae) – **see Galleries 2
and 3** (*pp. 34–36*)

▶ Other flowers with radial symmetry can be found in Galleries 11,14,15 (*pp. 54, 58–59*).

H Conspicuous flowers

Gallery 14

48 Primrose family *p. 155* petal-like sepals

52 Periwinkle family *p. 163*

GRAPE-HYACINTHS

BLUEBELLS

89 Bluebell [Asparagus] family *p. 243*

62 Gentian family *p. 189*

78 Bogbean family *p. 230*

Thrift

SEA-LAVENDERS

55 Borage family *p. 169*

40 Thrift + Sea-lavender family *p. 140*

Gallery 15

sepal-like petals

24 White Bryony [Gourd] family *p. 114*

85 Black Bryony family *p. 236*

50 Bedstraw family *p. 160*

77 Bellflower family *p. 229*

► Plants are not to scale and may not be the same colour as the examples shown here – flower shape and structure are the important features to note when looking for a match.

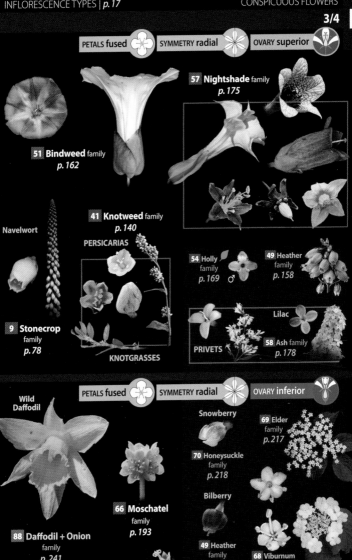

PETALS **fused** SYMMETRY **radial** OVARY **superior**

51 Bindweed family
p. 162

57 Nightshade family
p. 175

Navelwort

41 Knotweed family
p. 140

PERSICARIAS

9 Stonecrop
family
p. 78

54 Holly
family
p. 169 ♂

49 Heather
family
p. 158

Lilac

PRIVETS

58 Ash family
p. 178

KNOTGRASSES

PETALS **fused** SYMMETRY **radial** OVARY **inferior**

Wild Daffodil

Snowberry

69 Elder
family
p. 217

70 Honeysuckle
family
p. 218

Bilberry

66 Moschatel
family
p. 193

88 Daffodil + Onion
family
p. 241

The disc-flowers of some
members of the **Daisy family**
(Asteraceae) fall into this
category; see Gallery 10 (*p. 52*)

49 Heather
family
p. 158

68 Viburnum
family
p. 217

A number of woody trees, shrubs and climbers
have fused petals and radial symmetry
– see Galleries 2 and 3 (*pp. 34–36*)

▶ Other flowers with radial symmetry can be found in Galleries 11–13 (*pp. 54–57*).

H Conspicuous flowers

Gallery 16 | PETALS **fused** | SYMMETRY **bilateral** | OVARY **superior**

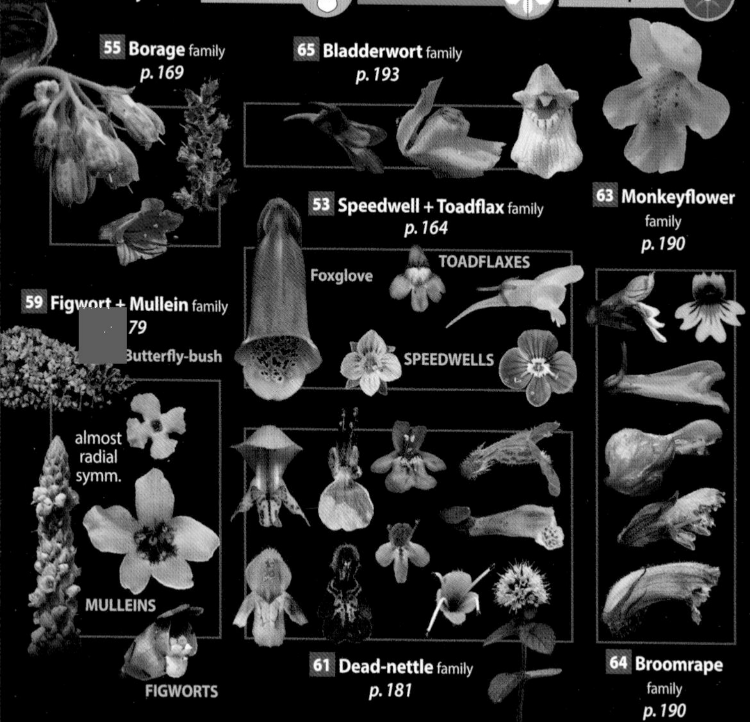

55 Borage family
p. 169

65 Bladderwort family
p. 193

63 Monkeyflower
family
p. 190

53 Speedwell + Toadflax family
p. 164

Foxglove

TOADFLAXES

59 Figwort + Mullein family
p. 179

Butterfly-bush

SPEEDWELLS

almost
radial
symm.

MULLEINS

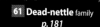

FIGWORTS

61 Dead-nettle family
p. 181

64 Broomrape
family
p. 190

Gallery 17 | PETALS **fused** | SYMMETRY **bilateral** | OVARY **inferior**

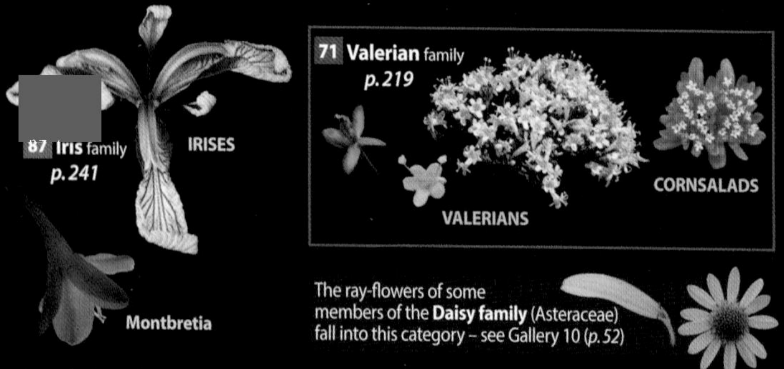

71 Valerian family
p. 219

87 Iris family
p. 241

IRISES

CORNSALADS

VALERIANS

Montbretia

The ray-flowers of some
members of the **Daisy family** (Asteraceae)
fall into this category – see Gallery 10 (*p. 52*)

► Plants are not to scale and may not be the same colour as the examples shown here –
flower shape and structure are the important features to note when looking for a match.

Gallery 18 · PETALS **free** · SYMMETRY **bilateral** · OVARY **superior**

**3 Poppy +
Fumitory** family
p. 68

29 Violet family
p. 118

42 Balsam family
p. 144

37 Mignonette family
p. 131

12 Milkwort family
p. 88

side petals
fused in
pairs

Horse-chestnut
(p. 129)

RESTHARROWS
woody-based
perennials

GORSES
shrubs

11 Pea
family
p. 80

Rhododendron (p. 159)

Gorses, Horse-chestnut (Sapindaceae) and
Rhododendron (Ericaceae) have flowers in this
category – see Galleries 2 and 3 (*pp. 34–36*).

Gallery 19 · PETALS **free** · SYMMETRY **bilateral** · OVARY **inferior**

86 Orchid family *p. 236*

33 Willowherb
family
p. 125

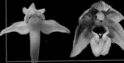

Rosebay Willowherb

75 Carrot family *p. 222*

Carrots
symmetry radial in
some flowers

SS Yarrow
and
Sneezewort
(Asteraceae)
Gallery 10
(*p. 53*)

▶ Flowers with radial symmetry can be found in Galleries 11–15 (*pp. 54–59*).

H Flowers with tiny/inconspicuous petals and/or sepals

Gallery 20

Plants in this group do have flowers with petals/sepals but can also appear to be lacking flowers – check both galleries on these pages. Scale bars indicate the actual max/min diameter of flowers.

New Zealand Pigmyweed

9 Stonecrop family *p. 78*

46 Blinks family *p. 154*

74 Pennywort family *p. 221*

38 Mistletoe family *p. 131*

Mossy Stonecrop

10 Water-milfoil family *p. 80*

PEARLWORTS

44 Campion [Pink] family *p. 145*

Water-purslane

32 Purple-loosestrife family *p. 124*

Individual flowers, particularly disc-flowers can be tiny (D <1 mm) in some members of the **Daisy family** – see Gallery 10 (*p. 52*)

H Flowers lacking or appearing to be lacking petals and/or sepals

Gallery 21

Plants in this group may have tiny flowers with petals/sepals but which often appear to be absent – check both galleries on these pages.

PLANTS WITH BROADER LEAVES

28 Spurge family *p. 116*

SPURGES

13 Rose family *p. 89*

PARSLEY-PIERTS

16 Hop family *p. 103*

— FLS

18 Nettle family *p. 105*

41 Knotweed family *p. 140*

13 Rose family *p. 89*

BURNETS

SS Pondweeds in fruit (*p. 233*)

in flower   *in fruit*

GOOSEFOOT

ORACHE

79 Lords-and-Ladies family *p. 230*

SS Adder's-tongue (*p. 300*)

56 Plantain family *p. 174*

45 Goosefoot family *p. 151*

► **Plants are not to scale and may not be the same colour as the examples shown here – flower shape and structure are the important features to note when looking for a match.**

A number of aquatic plants have flowers in these categories – **see Gallery 1 (*p. 29*)**

13 Rose family *p. 89*

LADY'S-MANTLES

♀ ♂

MERCURIES

28 Spurge family *p. 116*

37 Mignonette family *p. 131*

51 Bindweed + Dodder family *p. 162*

Dodder

FL-HEAD

16 Hop family *p. 103*

♂

41 Knotweed family *p. 140*

flowers often closed

SWINE-CRESSES

39 Cabbage family *p. 132*

A number of trees and shrubs have flowers in these categories – **see Gallery 4 (*p. 37*)**

H

PLANTS WITH NARROWER LEAVES

♀ ♂

Crowberry

88 Daffodil + Onion family *p. 241*

can have small flowers

Wild Onion

91 Rush family *p. 245*

TEPALS

in fruit

Rushes look similar but have tepals

49 Heather family *p. 158*

fleshy

82 Arrowgrass family *p. 233*

♂

45 Goosefoot family *p. 151*

GLASSWORTS

92 Sedge family *p. 250*

FRUIT

GLUME

REEDMACES

Shoreweed

56 Plantain family *p. 174*

90 Reedmace family *p. 244*

93 Grass family *p. 266*

BUR-REEDS

▶ **See *p. 245* for more detailed differentiation between grasses, sedges and rushes**

Characters of flowering plants in this book

This table shows the groups, subdivisions and numbers of families and species in each identification category in this book. If the features of a plant have been noted, the table will give a good idea of the number of identification options with the page numbers of the relevant galleries.

Group			Dicots		Monocots		
WOODY PLANTS			Families	Species	Families	Species	Page
Woody climbers			5	6	none in this book		33
Woody plants with tiny flowers or catkins			9	32	none in this book		37
Flowering trees and shrubs							
PETALS	SYMMETRY	OVARY					
free	radial	superior	5	6	none in this book		34, 36
		part. superior	3	20			34, 36
		inferior	5	17			34, 36
	bilateral or irregular	superior	1	4			36
		inferior	1	1			34
fused	radial	superior	2	4			34, 36
		inferior	4	13			34, 36
	bilateral or irregular	superior	1	1			34
		inferior	1	1			36
AQUATIC PLANTS			8	14	3	12	28
PLANTS LACKING CHLOROPHYLL			2	4	none in this book		27
HERBACEOUS FLOWERING PLANTS							
Lacking or seemingly lacking petals and sepals			9	38	6	113	62
Tiny, inconspicuous flowers			12	29	none in this book		62
Dense flowerheads of small flowers			2	80	none in this book		52
Conspicuous flowers							
PETALS	SYMMETRY	OVARY					
free	radial	superior	13	100	3	17	54
		part. superior	3	20	none in this book		56
		inferior	5	33	none in this book		57
	bilateral or irregular	superior	5	45	none in this book		61
		inferior	2	5	1	15	61
fused	radial	superior	10	51	1	4	58
		inferior	4	15	2	2	58
	bilateral or irregular	superior	7	66	none in this book		60
		inferior	1	5	1	3	60

The species accounts

These follow a consistent format, with text, labels and annotations boxes as follows:

ⓟ English name *Scientific name* **✕** WT

H to 90 cm. **Form** branched; spreading. **Fls** D 15–25 mm in spreading, stalked, loose clusters of up to 6. **Fr** bottle-shaped. **Hab** hedgerows. **SS** Yellow-juiced Poppy *P. lecoqii* (N/i) [STEM SAP yellow].

STEM contains abundant orange sap

LFLTS rounded

D J F M A M J J A S O N D

NB not related to, or even similar to, Lesser Celandine (buttercup family – *p. 72*).

FAMILY AND SPECIES NUMBERS

n The order in which families are presented is not taxonomic and each family is assigned a number to help in locating them.

n **spp.** | *n* spp. B&I The number of species with accounts in this book and the number of species included in the printed *BSBI Atlas*.

ACCOUNT TEXTS

English name and *Scientific name* – English names are those in popular usage in the context of Britain and Ireland and follow those used by the Botanical Society of the British Isles (BSBI) with a few variations; scientific names and taxonomy are those used by BSBI.

EASE OF IDENTIFICATION and PLANT LIFESTYLE – a coloured lozenge preceding the name indicates **identification difficulty**: ● = easy; ● = care needed; ● = difficult. Within each lozenge is a code for **plant lifestyle**: **A** = annual; **B** = biennial; **P** = perennial.

SIZE – is given for plant height (**H**), and for flowers (**Fls**) where D = diameter and L = length. Measurement units are broadly as follows: 1–10 cm in mm; 10–200 cm in cm; >200 cm in m. For some species where flower measurements are important for identification scale bars are given showing the minimum (▬) and maximum (⟻) dimensions.

IDENTIFICATION DETAILS (species account) summarizes those features necessary for confident identification. Subheadings within this section cover the information relevant to each species as follows:

Form general details of the whole plant, including structure, growth habitat and vegetative spread.

Relevant details, such as colour, shape and size is given for the following:

Fl, **Fls** [FLOWERS]

Infl [INFLORESCENCE]; **Spklt** [SPIKELET]

Lf, **Lvs** [LEAVES]; **Lflts** [LEAFLETS]

Fr [FRUIT]; **Utr** [UTRICLE]

Hab [HABITAT] briefly describes in which habitat(s) the plant can be found.

SS [SIMILAR SPECIES] – lists species that may present identification confusion together with a concise list of differences. Species not illustrated are designated (N/i).

Other relevant subheadings are used for particular groups (*e.g.* **Stem**, **Spathe**). These are explained where required. Within these paragraphs any other terms used (*e.g.* FL-STALK, ANTHERS) are either written in full or follow the abbreviations as used in the species accounts.

IMAGE ANNOTATIONS

These highlight key identification features using the same abbreviations as found in the species account; caveats and information relating to other species are in purple text.

ICONS

are used for information as follows:

✕ poisonous plant (see *p. 9*); AQ aquatic plant; C herbaceous climber; WC woody climber; WS woody shrub; WT woody tree; ♂ male; ♀ female

COMPARISON SPREADS and TABLES

Comparison image spreads and tables giving key identification features for similar species are provided where useful. Look at these in conjunction with the individual species accounts, as these often add important information that cannot be covered entirely on the species pages.

DISTRIBUTION MAPS and PHENOLOGY

The maps and phenology are based on data from the BSBI *Plant Atlas 2020*.

Phenology charts show at which time of year a plant can typically be found:

■ in leaf; ▨ in flower; ▨ in fruit.

PHOTOGRAPHS

Key features of a plant are shown as cutouts or magnified insets. These are accompanied by images of the plant in the wild in most cases.

LIST OF FAMILIES

Flower characteristics of families in this book

KEY | ● = occurs; O = occurs rarely; –●– = no differentiation
PET (petals): Se = Separate (free); Fu = Fused
SYM (symmetry): R = Radial; Bi = Bilateral
OV (Ovary position): S = Superior (* = partially superior); I = Inferior

#	Family	PET Se	PET Fu	SYM R	SYM Bi	OV S	OV I
1	Water-lily Nymphaeaceae	●		●		●	
2	Hornwort Ceratophyllaceae	●		●		●	
3	Poppy + Fumitory Papaveraceae	●		●	●	●	
4	Barberry Berberidaceae	●		●		●	
5	Buttercup Ranunculaceae	●		●	●	●	
6	Box Buxaceae	●		●		●	
7	Currant Grossulariaceae	●		●			●
8	Saxifrage Saxifragaceae	●		●		●*	
9	Stonecrop Crassulaceae	●	O	●		●	
10	Water-milfoil Haloragaceae	●		●			●
11	Pea Fabaceae	●	●		●	●	
12	Milkwort Polygalaceae		●		●	●	
13	Rose Rosaceae	●		●		●*	O
14	Sea-buckthorn Elaeagnaceae	sepals	●	●		●*	
15	Buckthorn Rhamnaceae	●		●		●	
16	Hop Cannabaceae	–●–		●			
17	Elm Ulmaceae	–●–		●		●	
18	Nettle Urticaceae	–●–		●		●	
19	Beech + Oak Fagaceae	–●–		●			●
20	Bog-myrtle Myricaceae	absent		●			
21	Walnut Juglandaceae	–●–		●			●
22	Birch Betulaceae	tiny / 0		●			●
23	Willow + Poplar Salicaceae	tiny		N/A	N/A	●	
24	White Bryony [Gourd] Cucurbitaceae		●	●			●
25	Wood-sorrel Oxalidaceae	●		●		●	
26	Spindle Celastraceae	●		●		●	
27	Flax Linaceae	●		●		●	
28	Spurge Euphorbiaceae	absent		N/A		●	
29	Violet Violaceae	●			●	●	
30	St John's-wort Hypericaceae	●		●		●	
31	Crane's-bill + Stork's-bill Geraniaceae	●		●	O	●	
32	Purple-loosestrife Lythraceae	●		●		●*	
33	Willowherb Onagraceae	●		●			●
34	Maple + Horse-chestnut Sapindaceae	●		●	O	●	
35	Mallow + Lime Malvaceae	●		●		●	
36	Rock-rose Cistaceae	●		●		●	
37	Mignonette Resedaceae	●			●	●	
38	Mistletoe Santalaceae	●		●			●
39	Cabbage Brassicaceae	●		●		●	
40	Thrift + Sea-lavender Plumbaginaceae		●	●		●	
41	Dock + Knotweed Polygonaceae	–●–		●		●	
42	Balsam Balsaminaceae	●		●		●	
43	Sundew Droseraceae	●		●		●	
44	Campion [Pink] Caryophyllaceae	●		●		●	
45	Goosefoot + Orache Amaranthaceae	–●–		●		●	
46	Blinks Montiaceae	●		●		●	
47	Dogwood Cornaceae	●		●			●
48	Primrose Primulaceae		●	●		●	
49	Heather Ericaceae	O	●	●	O	●	●
50	Bedstraw Rubiaceae		●	●			●
51	Bindweed + Dodder Convolvulaceae		●	●		●	
52	Periwinkle Apocynaceae		●	●		●	
53	Speedwell + Toadflax Veronicaceae		●		●	●	
54	Holly Aquifoliaceae	●	●	●		●	

#	Family	PET Se	PET Fu	SYM R	SYM Bi	OV S	OV I
55	Borage + Forget-me-not Boraginaceae		●	●	O	●	
56	Plantain Plantaginaceae		●	●		●	
57	Nightshade Solanaceae		●	●	O	●	
58	Ash Oleaceae [petals absent in some]	●	●	●		●	
59	Figwort + Mullein Scrophulariaceae		●	O	●	●	
60	Mare's-tail Hippuridaceae	tiny		N/A			●
61	Dead-nettle + Mint Lamiaceae		●		●	●	
62	Gentian + Centaury Gentianaceae		●	●		●	
63	Monkeyflower Phrymaceae		●		●	●	
64	Broomrape + Eyebright Orobanchaceae		●		●	●	
65	Bladderwort + Butterwort Lentibulariaceae		●		●	●	
66	Moschatel Adoxaceae		●	●			●
67	Daisy, Thistle + Cudweed Asteraceae		●	●			●
68	Viburnum Viburnaceae		●	●			●
69	Elder Sambucaceae		●	●			●
70	Honeysuckle Caprifoliaceae		●	●			●
71	Valerian Valerianaceae		●	●	O		●
72	Teasel + Scabious Dipsacaceae		●	●			●
73	Ivy Araliaceae	●		●			●
74	Pennywort Hydrocotylaceae	●		●			●
75	Carrot Apiaceae	●		●	●		●
76	Water-starwort Callitrichaceae	absent		N/A	N/A		●
77	Bellflower Campanulaceae		●	●			●
78	Bogbean Menyanthaceae		●	●		●	
79	Lords-and-Ladies + Duckweed Araceae	absent		●		●	
80	Water-plantain Alismataceae		●	●		●	
81	Frogbit Hydrocharitaceae	●		●			●
82	Arrowgrass Juncaginaceae	–●–		●		●	
83	Pondweed Potamogetonaceae	–●–		●		●	
84	Bog Asphodel Nartheciaceae	sepals	●	●		●	
85	Black Bryony Dioscoreaceae	–●–		●			●
86	Orchid Orchidaceae	●			●		●
87	Iris Iridaceae		●	●			●
88	Daffodil + Onion Amaryllidaceae	●	●	●			●
89	Bluebell [Asparagus] Asparagaceae	●	●	●		●	
90	Reedmace [Bulrush] Typhaceae	scales		N/A		●	
91	Rush + Wood-rush Juncaceae	–●–		●		●	
92	Sedge Cyperaceae	scales		N/A		●	
93	Grass Poaceae	scales		N/A		●	
94	Pine + Larch Pinaceae						
95	Yew Taxaceae						
96	Juniper + Cypress Cupressaceae						
97	Horsetail Equisetaceae						
98	Clubmoss Lycopodiaceae						
99	Spike-moss Selaginellaceae						
100	Adder's-tongue Ophioglossaceae						
101	Spleenwort Aspleniaceae						
102	Polypody Polypodiaceae						
103	Hard-fern Blechnaceae						
104	Royal Fern Osmundaceae						
105	Lady-fern Athyriaceae						
106	Marsh Fern Thelypteridaceae						
107	Male- + Buckler-fern Dryopteridaceae						
108	Bracken Dennstaedtiaceae						
109	Water Fern Salviniaceae						

Gymnosperms p. 291 (families 94–96)

Pteridophytes p. 297 (families 97–109)

PRE-DICOTS (see p.4)

1 Nymphaeaceae | **Water-lily** family   `2 spp. | 3 spp. B&I`

Form rooted aquatic. **Fls** many-petalled; at or above water surface. **Lvs** large, rounded, floating surface-leaves.

IDENTIFY BY: ► flower colour and general form
► leaf, stem and fruit details

ⓟ **White Water-lily** *Nymphaea alba* `AQ`

Fls D 10–20cm; on water surface; PET 20+, white. **Lvs** all floating; almost round; D to 30cm; veins **radiating from leaf/stalk join**, forming a net at the leaf-margin. **Fr** rounded. **Hab** still or very slow-moving fresh waters.

LF veins radiate from leaf/stalk join

J F M A M J J A S O N D

ⓟ **Yellow Water-lily** *Nuphar lutea* `AQ`

Fls D to 3–4cm; raised above water surface; SEP 5, **yellow**; PET many, smaller than sepals. **Lvs** L to 30cm; **veins in herringbone pattern**; SURFACE broadly oval, L>W; SUBMERGED more flaccid. **Fr** bottle-shaped. **Hab** still or flowing fresh waters. **SS Least Water-lily** *N. pumila* (N/I) [FL + LVS smaller; mainly upland]; **Fringed Water-lily** *Nymphoides peltata* (N/I) [PET fringed, LVS round].

LF herringbone vein-pattern

J F M A M J J A S O N D

DICOTS (see p.4)

RIGID HORNWORT

2 Ceratophyllaceae | **Hornwort** family   `2 spp. | 2 spp. B&I`

Form wholly submerged; not rooted, aquatic; STEMS L to 1m. **Lvs** forked; in whorls. **Fls** tiny; greenish; underwater; solitary leaf-axils; ♂ and ♀ at different nodes.

IDENTIFY BY: ► leaf forking ► leaf rigidity out of water

SS Water-milfoils (p.80); and **Water-violet** *Hottonia palustris* (N/I) [both have narrowly segmented, whorled leaves pinnately arranged and flowers above the water surface].

ⓟ **Rigid Hornwort** *Ceratophyllum demersum* `AQ`

Lvs dark green, **forked 1 or 2× only**; rather stiff – do not wholly collapse into a mush when out of water. **Hab** fresh to mildly brackish, still or very slow-moving waters.

J F M A M J J A S O N D

ⓟ **Soft Hornwort** *Ceratophyllum submersum* `AQ`

Lvs brighter green than Rigid Hornwort, **forked 3×**; more flimsy – collapsing into a mush when out of water. **Hab** predominantly coastal, more characteristic of brackish waters than Rigid Hornwort.

J F M A M J J A S O N D

LVS forked
1–2×

LVS forked
3×

67

3 Papaveraceae | **Poppy + fumitory** family

9 spp. | 28 spp. B&I

Divided into two main, very different-looking groups (subfamilies) that have often been treated as separate families.

- **Poppy subfamily** [13 spp. B&I]
Fls PETALS 4; SYMMETRY radial.

- **Fumitory subfamily** [15 spp. B&I]
Fls PETALS 4; SYMMETRY bilateral.

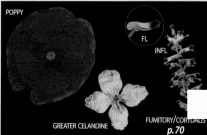

POPPY

FL

INFL

GREATER CELANDINE

FUMITORY/CORYDALIS
p.70

● **Poppies** and **Greater Celandine** | **Form** upright to bushy. **Fls** conspicuous; brightly coloured; obviously 4-petalled; SEPALS fall prior to flowering. **Fr** prominent capsule.

IDENTIFY BY: ► flower colour and general form ► leaf, stem and fruit details

Key to Poppies and Greater Celandine

Flower colour and form		Other features	Fruit *see opposite*	Species
Yellow	small; D 15–25 mm	SAP milky orange	long; pitted	**Greater Celandine**
	D 50–75 mm		narrowly oval	**Welsh Poppy**
Pale purple [white to red]	D 100–180 mm; much larger than other poppy flowers	STEM sparse bristles at most	rounded; hairless; much larger than that of Common	**Opium Poppy**
Red	bright scarlet, with or without a dark base to the petals; usually larger than other red-flowered poppies	STEM with spreading hairs	rounded; flat-topped; hairless	**Common Poppy**
	smaller and paler than Common Poppy; a few with a dark base to the petals	STEM with appressed hairs	long and narrow; hairless	**Long-headed Poppy**

P Greater Celandine *Chelidonium majus*

H to 90 cm. **Form** branched; spreading. **Fls** D 15–25 mm in spreading, stalked, loose clusters of up to 6. **Hab** hedgebanks, brownfield; often near buildings.

D J F M A M J J A S O N D

NOTE: not related to, or even similar to, Lesser Celandine (buttercup family – p. 72).

STEM contains abundant

LFLTS rounded

WELSH POPPY

P Welsh Poppy *Papaver cambricum*

H to 60 cm. **Form** erect, loosely tufted. **Fls** D 50–75 mm; solitary, on long stalks. **Hab** shady places: woods, rocky hillsides, streamsides, walls and roadsides.
SS Yellow Horned-poppy *Glaucium flavum* (N/i) [FRUIT very long (L 15–30 cm); coastal only].

J F M A M J J A S O N D

LFLTS pointed

FLS can be orange

D

Poppies are upright with a solitary flower on a long stalk

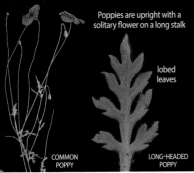

lobed leaves

COMMON POPPY

LONG-HEADED POPPY

LONG-HEADED | COMMON | OPIUM

J F M A M J J A S O N D | J F M A M J J A S O N D | J F M A M J J A S O N D

Lvs Common and Long-headed Poppies: greyish-green; lobed (lobes slightly broader in Common Poppy). Opium Poppy: less lobed; grey.

SS All red poppies have similarities and some care can be needed to differentiate.

Ⓐ Long-headed Poppy
Papaver dubium

H to 60 cm. **Form** erect; SAP white. **Fls** D 30–70 mm; ANTHERS violet. **Hab** waste ground, cultivated fields, especially on sandy soils. **SS** Yellow-juiced Poppy *P. lecoqii* (N/I) [SAP yellow].

STEM with appressed hairs

FL paler than those of Common Poppy

Ⓐ Common Poppy
Papaver rhoeas

H to 70 cm. **Form** erect; usually branched. **Fls** D 50–100 mm; ANTHERS bluish. **Hab** brownfield, cultivated fields. **SS** two rare red poppies *Roemeria* spp. (N/I) [CAPSULE bristly/prickly].

STEM with spreading hairs

FL bright scarlet to crimson; can have black base to petals (*right*)

Ⓐ Opium Poppy
Papaver somniferum

H to 100 cm. **Form** robust, erect. **Fls** D 100–180 mm. **Hab** brownfield. **SS** Oriental Poppy *P. setiferum* (N/I) [conspicuously hairy].

STEM hairless to sparsely bristly/hairy

FL typically pale purple, but ranges white to red (*right*)

Poppy fruit outlines compared

GREATER CELANDINE	WELSH POPPY	LONG-HEADED POPPY	COMMON POPPY	OPIUM POPPY
long, pitted cylindrical pod; L 30–50 mm	narrowly oval with persistent short style	oblong (L 2–3x W) or tapering from top to bottom	oval (L > W) with flattish top	rounded (L ± W), much larger than that of Common Poppy

(fruit reminiscent of some members of the cabbage family (*p. 132*))

69

● **Fumitories** and **Corydalis** | **Form** weakly erect herbaceous plants of disturbed ground, or scrambling on hedgebanks and walls. **Fls** distinctive; long, thin; 2-lipped. **Lvs** pinnate; green to bluish-green.

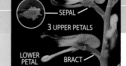

COMMON FUMITORY

SEPAL
3 UPPER PETALS
LOWER PETAL
BRACT

IDENTIFY BY: ► inflorescence details ► lower petal shape ► flower size/colour ► sepal shape/size ► leaf details ► fruit shape

FLOWERS **wholly yellow/yellowish | Corydalis**

Ⓟ Yellow Corydalis
Pseudofumaria lutea

H to 30 cm. **Form** erect to hanging. **Fls** L 12–18 mm; yellow; 5–10 flowers in a spike. **Hab** walls.

JFMAMJJASOND

LF 2-pinnate; no tendrils

Ⓐ Climbing Corydalis Ⓒ
Ceratocapnos claviculata

H to 75 cm. **Form** scrambling. **Fls** L 4–6 mm; cream; ± 6 flowers in a spike. **Hab** shady places, often on rocks, heathland.

JFMAMJJASOND

LF 2-pinnate; with tendrils

FLOWERS **pink/pinkish with darker tip | Fumitories**

The two widespread species are scrambling with flat, 1–2-pinnate leaves.

SS All fumitories, including seven rarer species, are very similar and identification needs care.

Ⓐ Common Fumitory
Fumaria officinalis

H to 30 cm. **Fls** L 7–9 mm; pink with dark tips; *key ID details below*. **Fr** egg-shaped; W > L; TIP truncate or slightly notched. **Hab** arable, brownfield, especially on sandy soils.

COMMON   C. RAMPING

JFMAMJJASOND   JFMAMJJASOND

Ⓐ Common Ramping-fumitory
Fumaria muralis

H to 70 cm. **Fls** L 8–12 mm; pink or white with dark tips; *key ID details below*. **Fr** smooth; globular to egg-shaped; L > W. **Hab** arable, brownfield, hedgebanks, walls.

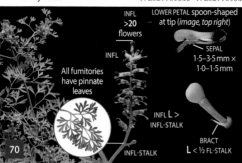

INFL >20 flowers

All fumitories have pinnate leaves

INFL

INFL-STALK

LOWER PETAL spoon-shaped at tip (*image, top right*)

SEPAL 1·5–3·5 mm × 1·0–1·5 mm

INFL L > INFL-STALK

BRACT L < ½ FL-STALK

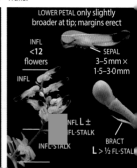

LOWER PETAL only slightly broader at tip; margins erect

INFL <12 flowers

INFL

SEPAL 3–5 mm × 1·5–3·0 mm

INFL L ± FL-STALK

INFL-STALK

BRACT L > ½ FL-STALK

4 Berberidaceae | **Barberry** family | 1 sp. | 7 spp. B&I

Form shrubs with yellow wood and spiny stems or leaves.
Fls flower parts in 4–5 whorls of 3 segments, usually yellow to orange; outer 2 whorls petal-like, spreading when mature.

outer 2 whorls of 3 petal-like segments

IDENTIFY BY: ▶ leaf, stem, twig and berry details

Ⓟ **Oregon-grape** *Mahonia aquifolium* WS

H to 120 cm. **Fls** D6–8 mm; yellow; almost globular; in dense spikes. **Lvs** evergreen; glossy; pinnate, with up to 8 spiny-margined leaflets.
Fr blue-black, grey-coated berry. **Hab** woods and other shaded habitats.

FR blue-black when ripe

LF pinnate; leaflets holly-like; bronzed in winter

SS Barberry *Berberis vulgaris* (N/I) [deciduous; LVS not spiny; FLS in hanging spikes; FR red berry]; leaves – **Holly** (p. 169) [LVS not pinnate; spines more robust].

taller, more erect *Mahonia* spp. can escape from gardens

5 Ranunculaceae | **Buttercup** family | 14 spp. | 50 spp. B&I

Form diverse range of different-looking annuals, biennials and woody climbers; all of which share a trait of 'primitive' flowers that have an indefinite and variable number of ♂ and ♀ parts.
Fls SEPALS petal-like in many, with petals reduced to nectar-producing organs (nectaries); PETALS (if present and showy) free and typically 5. **Fr** dry; 1- to many-seeded.

all family members have a variable, indefinite number of stamens and seeds

STAMENS surrounding SEEDS

IDENTIFY TO GROUP BY: ▶ number and colour of petals and/or sepals ▶ leaf shape

Buttercup groups and species identification

PETALS + SEPALS **both present**; PETALS **yellow or white**; SEPALS **greenish, not petal-like**

PETALS yellow	PETALS yellow	PETALS mostly white	

SPEARWORTS
SEPALS 5; PETALS 5;
LEAVES **unlobed**
p. 75

BUTTERCUPS
SEPALS 5; PETALS 5;
LEAVES lobed
p. 74

'CROWFOOTS'
SEPALS 5; PETALS 5;
LEAVES **lobed**
p. 73

Lesser Celandine
SEPALS **3**; PETALS 7–12,
yellow; narrow
p. 72

PETALS + SEPALS **violet** | PETALS **absent**; SEPALS **in a single whorl; petal-like**

STAMENS showy

Columbine
SEPALS 5 (flat) + PETALS 5
(tubular); both violet-blue
p. 72

flowers pink to white in garden varieties

Wood Anemone
SEPALS 6–12 petal-like,
white
p. 72

Marsh-marigold
SEPALS 5–8 petal-like,
yellow; broad
p. 72

Traveller's-joy
woody climber;
SEPALS 4, greenish
p. 73

● **Buttercups with non-typical 'buttercup' flowers** (not 5 yellow petals + 5 green sepals).

ⓟ Marsh-marigold *Caltha palustris*

H to 60 cm. **Fls** D 10–50 mm; PET **absent**; SEP **5–8**, golden-yellow. **Lvs** heart-shaped; toothed. **Fr** pods in star-like arrangement. **Hab** marshes, pond, ditch and stream edges. **SS** Globeflower *Trollius europeaus* (N/I) [FLS globular].

ⓟ Lesser Celandine *Ficaria verna*

H to 20 cm. **Form** patch-forming. **Fls** D 15–40 mm; PET **7–12**, shiny yellow; SEP **3**, green. **Lvs** heart-shaped, glossy, often marbled. **Fr** upright heads of seeds. **Hab** woods, hedgebanks, streamsides, damp grassland. **SS** Winter Aconite *Eranthis hyemalis* (N/I) [PETALS 0; SEPALS 6, yellow; ruff of leaves below flower].

ⓟ Wood Anemone *Anemone nemorosa*

H to 30 cm. **Form** patch-forming. **Fls** D 20–40 mm; PET **0**; SEP **6–12**, white to pale pink. **Lvs** trifoliate; LFLTS variously notched or divided. **Fr** drooping heads of seeds. **Hab** woodland, hedgebanks and grassland/heathland – especially in damper areas. **SS** leaves – Sanicle (*p. 228*) [LVS underside glossy]; Moschatel (*p. 193*) [ultimate lflts rounded].

ⓟ Columbine *Aquilegia vulgaris*

H to 100 cm. **Form** erect; basal rosette; stem usually branched. **Fls** D 30–50 mm, usually violet-blue; PET **5**, **tubular**, narrowing to long (to 25 mm), **hooked, backward-pointing spur**; SEP **5**, flat. **Lvs** trifoliate; LFLTS trifoliate; or 3-lobed; lobes broad. **Fr** 5 papery pods; each with a ± radiating point. **Hab** native plants in woodland, on calcareous grassland, screes. **SS** garden *Aquilegia* spp. and cultivars.

widespread as a garden escape in more urban areas

FL The family features are best seen viewed front on

℗ Traveller's-joy *Clematis vitalba* WC

L to 30 m. **Form** climbs over other plants and structures. **Fls** D 15–20 mm; clustered; PET absent; SEP 4, cream to green, hairy. **Hab** calcareous scrub, hedges, woodland edges. **SS** in winter – **Honeysuckle** (*p.218*) [stem twining].

J F M A M J J A S O N D

FR seeds with long feathery appendages

● Crowfoots and Water-crowfoots | White-flowered buttercups

associated with aquatic or damp habitats. In three sub-groups:
SHALLOW WATER/MUD – SURFACE LVS broad, rounded/lobed, SUBMERGED LVS absent or few (*e.g.* **Ivy-leaved Crowfoot**); **STILL WATER** – SURFACE LVS broad, rounded, SUBMERGED LVS finely divided (*e.g.* **Pond Water-crowfoot**); **FLOWING WATER** – SUBMERGED LVS finely divided, SURFACE LVS (rare) broad (*e.g.* **Stream Water-crowfoot**).

at base of petal

IDENTIFY BY: ▶ surface and/or submerged leaf detail ▶ sepal detail ▶ nectary shape

SS The 12 water-crowfoots can be difficult to separate as they are highly variable in response to environmental factors. All require an examination of the features above for identification.

ⓐℙ Ivy-leaved Crowfoot *Ranunculus hederaceus*

FL **small**, sepals flat; visible

H <20 mm. **Form** trailing. **Fls** D 3–6 mm; PET not overlapping. **Lvs** SURFACE 3–5 lobes. **Hab** shallow waters and muddy edges. **SS** Round-leaved Crowfoot *R. omiophyllus* (N/i) [FL D 8–12 mm; SEPALS downturned; LVS more rounded, lobes narrowest at base].

LF-LOBES broadest at base; divided < half depth

J F M A M J J A S O N D

ⓐℙ Pond Water-crowfoot *Ranunculus peltatus* AQ

NECTARY pear-shaped

H fruiting stalk L 30–150 mm. **Form** mostly submerged. **Fls** D 15–22 mm. **Lvs** SURFACE palmately 3–5-lobed; lobes divided > half lobe depth; SUBMERGED relatively short (L < internodes), segments often rigid. **Hab** shallow ponds, ditches, slow-moving waters. **SS** Common Water-crowfoot *R. aquatilis* (N/i) [NECTARY rounded].

SURF LVS divided > half depth

SUBM LVS short 'threads'

J F M A M J J A S O N D

℗ Stream Water-crowfoot AQ
Ranunculus penicillatus

NECTARY pear-shaped

H fruiting stalk L 20–30 mm. **Form** mostly submerged. **Fls** D 20–30 mm; NECTARY pear-shaped. **Lvs** SURFACE (rare) – few, palmately 3–5-lobed; SUBMERGED thread-like, divided >4×, L > internodes. **Hab** rivers and streams, often fast-flowing.

SURF LVS typically absent

SUBM LVS long 'threads'

J F M A M J J A S O N D

● **'Typical' buttercups** | **Fls** PETALS 5, yellow; SEPALS 5, greenish.
Lvs lobed. **Hab** varied, from dry meadows to wet pastures and shady woods.

IDENTIFY BY: ▶ flower, sepal and leaf details ▶ stem details

SS All typical buttercups are similar; possibly other yellow flowers – St John's
Worts *Hypericum*, **Rock-roses** *Helianthemum* and **Cinquefoils** *Potentilla* – see
Gallery 11 (p.55).

Typical 'meadow' buttercups: fruiting head a rounded cluster of seeds

Buttercup seed identification		Goldilocks	Creeping	Meadow	Bulbous
	Seeds	hairy	smooth		pitted
	Beak	curved		hooked	

BEAK

SEED

FRUITING HEAD

ⓟ Goldilocks Buttercup *Ranunculus auricomus*

J F M A M J J A S O N D

H to 40 cm. **Form** usually erect.
Fls D 15–25 mm; PET 0–5,
variably shaped; SEP erect.
Lvs variable; almost round to
deeply palmately lobed.
Hab deciduous woodland,
shady hedgebanks.

LVS **extremely variable**

FLS often look 'damaged'

ⓟ Creeping Buttercup *Ranunculus repens*

J F M A M J J A S O N D

H to 50 cm. **Form** creeping;
rooting at nodes. Fls
D 20–30 mm. **Lvs** outline
triangular; trifoliate; LFLTS
**trifoliate or lobed; central
lobe long-stalked. Hab** damp
meadows and pastures, open
woods, marshes.

SEP erect

LF **central lflt long-stalked**

FL-STALK grooved

STEM with **surface runners**

ⓟ Meadow Buttercup *Ranunculus acris*

J F M A M J J A S O N D

H to 100 cm. **Form** erect,
with branched stem. **Fls**
D 15–35 mm. **Lvs** outline
round; palmately 5–7-lobed;
LOBES **deeply cut; central lobe
unstalked. Hab** meadows and
pastures, often in calcareous
and/or damper environments.

SEP erect

LF **5–7-lobed;
central lobe
unstalked**

FL-STALK smooth

ⓟ Bulbous Buttercup *Ranunculus bulbosus*

J F M A M J J A S O N D

H to 40 cm. **Form** erect, with
swollen stem-base. **Fls** D 15–
30 mm. **Lvs** ± trifoliate; LFLTS/
LOBES 3-lobed, central lobe
unstalked or short-stalked.
Hab dry meadows and
pastures. **SS** Hairy Buttercup
R. sardous (N/I) [FLS lemon-
yellow; LVS yellow-green; no
swollen stem-base].

SEP **bent back**

LF **central lobe
unstalked or
short-stalked**

FL-STALK grooved

STEM-BASE **swollen**

FRUITING HEAD

Buttercup with elongated fruiting head

Ⓐ Celery-leaved Buttercup
Ranunculus sceleratus

H to 60 cm. **Form** erect; tufted. **Fls** D 5–10 mm; SEP **bent back** in flower. **Lvs** shiny; deeply palmately lobed. **Fr** seed-head elongates after flowering. **Hab** marshy fields and edges of waterbodies.

UPPER

LVS can vary from shown

LOWER

FR-HEAD **elongated**

J F M A M J J A S O N D

● **Spearworts** | **Fls** PETALS 5, yellow; SEPALS 5, greenish. **Lvs** unlobed, narrow. **Hab** wet, marshy habitats that are typically unshaded.

IDENTIFY BY: ► leaf details ► root form

Ⓟ Lesser Spearwort *Ranunculus flammula*

H to 50 cm. **Form** erect to prostrate; some rooting at nodes. **Fls** D 7–20 mm; STALKS furrowed. **Lvs** UPPER STEM **linear to narrowly oval**; LOWER STEM broader. **Fr** with short, straight beak. **SS Greater Spearwort** *R. lingua* (N/I) [taller; FL D 30–50 mm; STALK unfurrowed; FR with long hooked beak].

UPPER STEM-LF

LOWER STEM-LF

J F M A M J J A S O N D

---

⑥ Buxaceae | **Box**

`1 sp. B&I`

Form evergreen green-stemmed shrub or small tree. **Fls** petal-less; in clusters; ♂ with 4 stamens; ♀ with 3 styles, which persist in fruit as 3 'horns'.

LVS oval

IDENTIFY BY: ► evergreen leaves ► flowers

Ⓟ European Box *Buxus sempervirens* `WS`

H usually <5 m. **Fls** D 2 mm; yellow-green; clustered in leaf-axils; petal-less; single ♀ at apex, several ♂ below ♀. **Lvs** opposite; L to 25 mm; broadly oval; entire. **Hab** native in woodlands and calcareous scrub; often planted elsewhere. **SS** leaves of some shrubby honeysuckles *e.g.* **Wilson's Honeysuckle** (*p.218*) and **Box-leaved Honeysuckle** *Lonicera pileata* (N/I) are superficially similar but have hairy stem.

J F M A M J J A S O N D

♀ FL
3 styles

FR

♂ FL
4 stamens

Evergreen shrub

75

7 Grossulariaceae | Currant family 4 spp. | 7 spp. B&I

Form deciduous, erect berry-bearing shrubs. **Fls** usually greenish; in loose hanging clusters; PETALS + SEPALS arising from a rim above the ovary. **Lvs** palmately lobed.

IDENTIFY BY: ► fruit ► flower details ► leaf details

SS Two other species (both native in uplands) are similar to Red Currant: **Downy Currant** R. spicatum (N/I) [separated anther lobes, cup-shaped flowers]; **Mountain Currant** R. alpinum (N/I) [separate ♂ + ♀ plants, FL CLUSTERS more erect].

RED CURRANT

Currant bushes typically occur in patches; flowers can be unobtrusive amongst the foliage

P Gooseberry Ribes uva-crispa WS

H to 1·2m. **Form** spiny. **Fls** D6–12mm; 1–3, greenish-yellow (can be red-tinged). **Lvs** unscented. **Fr** L to 20mm; greenish-yellow (can be red-tinged). **Hab** garden escape into woodland, scrub, hedges.

J F M A M J J A S O N D

FR greenish-yellow

P Red Currant Ribes rubrum WS

H to 1·5m. **Fls** D4–6mm; greenish-yellow; saucer-shaped in hanging clusters. **Lvs** downy, unscented. **Fr** berry; D 6–10mm; **red** (can be whitish). **Hab** damp deciduous woodland; also widespread garden escape.

J F M A M J J A S O N D

FR red
FL 'open'
LF UND no glands

P Black Currant Ribes nigrum WS

H to 2m. **Fls** D6–10mm; greenish-yellow (can be purplish); saucer-shaped in hanging clusters. **Lvs** upperside hairless, **strongly scented** when crushed. **Fr** berry; D 10–15mm; black. **Hab** garden escape into damp deciduous woodland.

J F M A M J J A S O N D

FR black
FL 'closed'
LF UND glands

P Flowering Currant Ribes sanguineum WS

H to 2·5m. **Fls** D6–10mm; PET fused into a tube; **pink or red**; in hanging clusters. **Lvs** scented when crushed. **Fr** berry; D 6–10mm; **black with blue-grey coating**. **Hab** garden escape into woodland, scrub and hedges.

J F M A M J J A S O N D

FR blue-grey coating

8 Saxifragaceae | **Saxifrage** family

4 spp. | 24 spp. B&I

Fls PETALS 0 or 5; SEPALS 4 or 5 and STAMENS 3, 5, 8 or 10. **Fr** 2 pod-like; fused at least at the base – the only feature shared by all members of this family.

● **Saxifrages** | **Form** erect; many spp. stickily hairy.
Fls PETALS + SEPALS 5, joined at base; STAMENS 10. **Lvs** palmately lobed.

IDENTIFY BY: ▶ overall form ▶ flower details ▶ leaf details

Ⓐ **Rue-leaved Saxifrage** *Saxifraga tridactylites*

D

H to 12 cm. **Form** erect; stickily hairy. **Fls** D 3–6 mm; white; in few-flowered spike. **Lvs** 3 (–5) **deep lobes**; often turn reddish. **Hab** bare, dry sandy and stony ground; rocks and walls.

LF 3 (–5) deep lobes

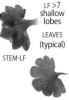

J F M A M J J A S O N D

Ⓟ **Meadow Saxifrage** *Saxifraga granulata*

H to 50 cm. **Form** basal rosette with brown bulblets. **Fls** D 10–20 mm; white; in branched spike. **Lvs** STEM outline round; >7 **shallow lobes**; BASAL long-stalked. **Hab** grassland. **SS** rarer, mostly upland, saxifrages and garden escapes (N/I) [LF details].

D

J F M A M J J A S O N D

LF >7 shallow lobes

LEAVES (typical)

STEM-LF

ROSETTE LF

● **Golden-saxifrages** | **Form** short; creeping; patch-forming.
Fls small, golden-yellow in branched heads; PETALS 0; SEPALS 4; STAMENS 8.

FL branched heads

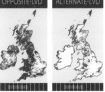

OPPOSITE-LVD GOLDEN-SAXIFRAGE

IDENTIFY BY: ▶ leaf arrangement and shape

Ⓟ **Opposite-leaved Golden-saxifrage**
D *Chrysosplenium oppositifolium*

H to 15 cm. **Fls** D 3–5 mm. **Lvs** NON-FL STEM **leafy**; FL-STEM LVS **opposite**, rounded, W to 20 mm. **Hab** wet areas and shady streamsides in woodland; mountain ledges.

Ⓟ **Alternate-leaved Golden-saxifrage**
D *Chrysosplenium alternifolium*

H to 20 cm. **Fls** D 5–6 mm. **Lvs** NON-FL STEM **leafless**; FL-STEM LVS **single**, rounded, W to 30 mm, lower leaves kidney-shaped. **Hab** wet areas and shady streamsides in woodland.

OPPOSITE-LVD   ALTERNATE-LVD

J F M A M J J A S O N D   J F M A M J J A S O N D

OPPOSITE-LEAVED

The bracts below the flowers are usually smaller on Opposite-leaved Golden-saxifrage

BRACTS

FLOWERING STEM

LVS
opposite

LVS
single

Often found growing together; Alternate-leaved is typically less abundant

FL 4 sepals; 8 stamens – [4 withered on this flower]

Both grow in patches in wetter areas

9 Crassulaceae | **Stonecrop** family

8 spp. | 20 spp. B&I ENGLISH STONECROP

Form mostly succulent species; many low-growing. **Fls** PETALS number varies between species, but typically 5; STAMENS 2× the number of petals in some. Beware, recent taxonomic research has created a number of unfamiliar generic names in this family.

typical flower with 5 petals and sepals

IDENTIFY BY: ► overall form ► flower details ► leaf details

Stonecrops with non-typical flowers (PETALS < 5 or > 8)

Ⓐ **Mossy Stonecrop** *Crassula tillaea*

Form low, appressed to ground; creeping. **Fls** tiny; PET 3; whitish. **Lvs** L to 2 mm, overlapping; opposite pairs fused at base; **turning strongly red. Hab** bare, especially compacted, sandy ground.

J F M A M J J A S O N D

Highly conspicuous once mature!

Ⓟ **New Zealand Pigmyweed** *Crassula helmsii*

Form mat-forming to 20 cm deep. **Fls** D 1–2 mm; PET 4; pink; SEP green (L < petals); on stalks arising from leaf-axils. **Lvs** L to 15 mm; succulent; opposite. **Hab** bare mud or in shallow water. **SS** water-starworts (*p. 228*) [leaves not fleshy].

J F M A M J J A S O N D

SEP < PET

FL 4 petals

Invasive, forming extensive mats

Ⓟ **Navelwort** *Umbilicus rupestris*

H to 40 cm. **Form** unbranched, succulent. **Fls** D 7–10 mm; PET 5; greenish (can be tinged brown); **tubular; in long spike. Lvs** L to 70 mm; mostly basal; round; shallowly toothed; stalk attached in middle of leaf, the upper surface having a corresponding 'navel'. **Hab** cliffs, rocks, walls, hedgebanks. **SS** leaves – **Marsh Pennywort** (*p. 221*) [wet habitats; flower spikes not showy].

J F M A M J J A S O N D

FL tubular

LF **round** with shallow teeth

Ⓟ **House-leek** *Sempervivum tectorum*

H to 50 cm. **Fls** D 15–30 mm; PET ±13; pink-purple. **Lvs** L to 40 mm; succulent; flattened; crowded into **basal rosette** and up stem; usually tinged red. **Hab** roofs, walls, quarries, sand dunes.

J F M A M J J A S O N D

FL large, showy

LVS large, basal rosette

Stonecrops with flowers having 5 or more free petals; STAMENS = 2× petal count

Key to 'typical' stonecrops

Flower colour	FLOWER FEATURES	Leaves	Species
White	**pink scales** at petal base	L 3–5mm; **clasping stem**	**English Stonecrop**
	pale yellow scales at petal base	L 6–12mm; **not clasping stem**	**White Stonecrop**
Yellow	PETALS: 5	peppery taste	**Biting Stonecrop**
	PETALS: 6–8	tasteless	**Reflexed Stonecrop**

ⓟ English Stonecrop *Sedum anglicum*

H to 50mm. **Form** dwarf; mat-forming. **Fls** D 6–12mm; PET 5; white; **pink scales** at base of petals; INFL two main branches. **Lvs** L 3–5mm; **clasping stem**. **Hab** rocks, dunes, shingle, walls. **SS** Thick-leaved Stonecrop *S. dasyphyllum* (N/I) [LVS opposite; pubescent; bluish-green (can be tinged purple)].

J F M A M J J A S O N D

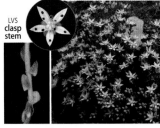

ⓟ White Stonecrop *Sedum album*

H to 150mm. **Form** mat-forming. **Fls** D 6–9mm; PET 5; white; **pale yellow scales** at base of petals; INFL many-branched. **Lvs** L 6–12mm; **not clasping stem**. **Hab** rocks, dunes, shingle, brownfield.

J F M A M J J A S O N D

ⓟ Biting Stonecrop *Sedum acre*

H to 100mm. **Form** low; mat-forming. **Fls** D 10–12mm; PET 5; bright yellow. **Lvs** L 3–5mm, overlapping; rounded in section; broadest near base; peppery taste to most people. **Hab** dry, grassland, sand dunes, shingle, brownfield.

J F M A M J J A S O N D

ⓟ Reflexed Stonecrop *Petrosedum rupestre*

H to 30cm. **Form** mat-forming with erect flower stem. **Fls** D 10–12mm; PET 6–8; yellow; in terminal cluster. **Lvs** L to 20mm; rounded in section; tapering to fine point. **Hab** walls, rocks, dry banks. **SS** Rock Stonecrop *P. forsterianum* (N/I) [LVS flat; leaf-tips abruptly pointed].

J F M A M J J A S O N D

10 Haloragaceae | Water-milfoil family | 2 spp. | 4 spp. B&I

Form aquatic perennials of still and slow-moving waters and adjacent mud. **Lvs** submerged; feathery; finely pinnate; in whorls. **Fls** tiny; (D ± 4 mm); unisexual; PETALS 4; SEPALS 4; ♂ with 8 stamens; in spikes held above the water.

♂ FL

♀ FL

SPIKED

IDENTIFY BY: ► inflorescence form ► leaf count and detail

SS Other **water-milfoils** (N/I) [LVS in whorls of 5]; **Water-violet** *Hottonia palustris* (N/I) [LVS in flattened rosettes; erect stem leafless].

ⓟ Spiked Water-milfoil
Myriophyllum spicatum AQ

H SPIKE L to 15 cm, erect; STEM L to 2·5 m. **Fls** pinkish, **in whorls**. **Lvs** whorls of 4; LFLTS **linear, 8–24 pairs**. **Hab** lowland; water pH neutral to alkaline; also brackish.

J F M A M J J A S O N D

LF >8 leaflet pairs

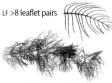

FL pinkish, in whorls

denser-looking; more 'rigid'

ⓟ Alternate Water-milfoil
Myriophyllum alterniflorum AQ

H SPIKE L to 5 cm, drooping; STEM L to 1·2 m. **Fls** yellowish, upper fls single or opposite. **Lvs** whorls of 4; LFLTS **linear, 3–10 pairs**. **Hab** lowland and upland; water pH slightly acid to alkaline; rarely brackish.

J F M A M J J A S O N D

LF ≤10 leaflet pairs

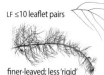

FL yellowish, single/opposite

finer-leaved; less 'rigid'

11 Fabaceae | Pea family | 28 spp. | 97 spp. B&I

Fls SYMMETRY bilateral; PETALS formed into standard, wings and keel (see *right*); SEPALS fused. **Fr** elongate pods in many. **Lvs** varied in form: spiny, trifoliate, pinnate or undivided; with distinctive stipules at base of stalk, and tendrils in some species.

SEPALS fused — STANDARD

WING

KEEL

IDENTIFY TO GROUP BY: ► woody or not ► inflorescence form ► leaf details

Pea groups + leaf types

Woody plants	Non-woody plants		
	FLS few or loose	INFL spiked or grouped	INFL rounded cluster
GORSES, BROOMS, RESTHARROWS LEAF TYPES IN GROUP spiny; trifoliate	**VETCHES, TARES** LEAF TYPES IN GROUP pinnate; pinnate + tendril	**BIRD'S-FOOT-TREFOILS, VETCHES, MELILOTS** LEAF TYPES IN GROUP pinnate; pinnate + tendril; trifoliate	**CLOVERS, TREFOILS MEDICKS, Kidney Vetch** LEAF TYPES IN GROUP trifoliate; pinnate
pp.81–82	*pp.82–83*	*pp.83–85*	*pp.86–87*

Pea family leaf types

SPINY

PINNATE

PINNATE + TENDRIL

TRIFOLIATE

STIPULES at the base of the leaves can be useful in identification

STIPULES

● **Woody peas – gorses, brooms, laburnums and restharrows**

Trees and shrubs with yellow flowers plus the much smaller, pink-flowered restharrows which are perennials with a woody base, rather than shrubs as normally recognized.

FORM spiny shrubs, H ≤ 200 cm; FLOWERS yellow

SS All gorses share the same broad form – green-stemmed and spiny; very young plants have trifoliate leaves. **Dwarf Gorse** *U. minor* (N/I) – most like **Western Gorse** – [FORM less robust; SPINES weaker; SEPAL L ≈ petal L; BRACT W < fl-stalk W].

Gorse identification		Gorse	Western Gorse
In flower		January–June (–December)	July–November (–February)
Flower	COLOUR/LENGTH	golden-yellow; L **10–20**mm	clear yellow; L **10–12**mm
	SEPALS	±⅔ petal length; HAIRS **spreading**	≤⅔ petal length; HAIRS **appressed**
	BRACTS	**2–5**mm; W >2× fl-stalk W	<0·8mm; W **1–2**× fl-stalk W
Stem	SPINES	L to 25mm; rigid; **deeply furrowed**	L to 25mm; rigid; **weakly furrowed**
Fruit	POD	L **12–19**mm, very hairy	L **8–14**mm, very hairy

Ⓟ Gorse *Ulex europaeus* **WS**

H to 200 cm. **Hab** moorland, heathland, grassland on acid soils.

can smell like coconut
SEPAL hairs spreading

BRACT large; W > 2× stalk W
SPINE deeply furrowed

POD 12–19mm

Ⓟ Western Gorse **WS**
Ulex gallii

H to 150 cm. **Hab** heathland, clifftop grassland.

more weakly scented
SEPAL hairs appressed

BRACT small; W 1–2× stalk W
SPINE weakly furrowed

POD 8–14mm

GORSE   WESTERN

J F M A M J J A S O N D   J F M A M J J A S O N D

FORM spineless shrubs, H ≤ 200 cm; FLOWERS yellow

Ⓟ Broom *Cytisus scoparius* **WS**

J F M A M J J A S O N D

H to 200 cm. **Form** tall, erect; STEM green, angled. **Fls** L 16–18mm; golden-yellow; stalked; in loose spikes. **Lvs** undivided or trifoliate; (found on young branches only). **Fr** pod L to 40mm; black; hairy. **Hab** acid grassland, heathland, dry scrub.

SS Spanish Broom *Spartium junceum* (N/I) [shrub (H to 3 m); STEM rush-like; FL larger (L to 28mm)].

STEM angled

LF (if present) trifoliate

POD black; hairy

81

FORM **deciduous spineless trees** H ≥ 200 cm; FLOWERS **yellow**

Ⓟ **Laburnum** *Laburnum anagyroides* WT

H to 8 m. **Form** BARK smooth; pale brown. **Fls** L 18–30 mm; **golden-yellow; in drooping clusters** (L to 20 cm). **Lvs** trifoliate; LFLTS entire, silvery beneath. **Fr** pod; L to 60 mm; hairy when young, blackish-brown when ripe. **Hab** woods, scrub, road verges, railway embankments.

FR young

LFLTS smooth-edged

J F M A M J J A S O N D

FORM **woody-based perennials with or without spines** H < 50 cm; FLOWERS **pink**

Ⓟ **Common Restharrow** *Ononis repens*

H to 30 cm. **Form** prostrate to erect; **woody base**; can be spiny. **Fls** L 10–20 mm; **pink**; wing = keel; in leafy spikes. **Lvs** LFLTS 1 or 3; broadly oval; toothed; covered in **sticky hairs; smell of sweaty armpits** when crushed; STIPULES as lflts. **Fr** pod (L < sepals), with 1–2 seeds. **Hab** dry grassland, usually calcareous. **SS Spiny Restharrow** *O. spinosa* (N/i) [typically has spines, LFLTS L >3× W; mainly coastal].

LFLTS toothed, with sticky hairs

J F M A M J J A S O N D

Peas with single flowers or few-flowered in loose inflorescences
IDENTIFY BY: ▶ inflorescence form　▶ leaf details　▶ stipule shape　▶ pod details

Ⓐ **Common Vetch** *Vicia sativa* ssp. *nigra*

H to 40 cm. **Form** erect, scrambling. **Fls** L 10–30 mm; purple-pink (keel can be whitish); 1–3 at base of lf-stalks. **Lvs** pinnate; LFLTS 4–8 pairs, widest at middle or above; TENDRILS branched or not; STIPULES toothed, 'half-arrow'; usually with dark spot. **Fr** pod; L to 80 mm; usually blackish; 4–12-seeded. **Hab** grassland, hedgebanks, scrub edges. **SS Spring Vetch** *V. lathyroides* (N/i) [smaller; TENDRILS unbranched/absent]. The widespread Common Vetch sspp. differ as follows:

J F M A M J J A S O N D

ssp. *nigra*

ssp. *segetalis*

STIPULE typically with **dark spot**

SSP. *NIGRA*

ssp. *nigra*
UPPER LVS

ssp. *segetalis*

LOWER LVS
[both sspp. similar]

subspecies	FL COLOUR	UPPER LFLTS
ssp. *nigra*	standard same as wings	much narrower than lower
ssp. *segetalis*	standard paler than wings	slightly narrower than lower

● **Tares** | **Form** small-flowered scrambling peas. **Lvs** pinnate with tendrils.

Ⓐ Smooth Tare *Ervum tetraspermum*

H to 60 cm. **Fls** D 4–8 mm; deep lilac; 1–2 in a head. **Lvs** LFLTS 3–6 pairs, tip pointed to round; TENDRILS unbranched; STIPULES 'cow's-horn'. **Fr** pod; L 12–16 mm; usually **4-seeded**; **brown** when ripe. **Hab** rough grassland, scrub edge.

J F M A M J J A S O N D

TENDRILS unbranched

POD **4-seeded**, **smooth**

TIP roundish

INFL 1–2 fls

LFLTS 3–6 pairs

STIPULE 'cow's-horn'

Ⓐ Hairy Tare *Ervilia hirsuta*

H to 80 cm. **Fls** D 3–5 mm; pale lilac; 2–7 in loose spike. **Lvs** LFLTS 6–8 pairs; tip squarish; TENDRILS branched; STIPULES 'antlers'. **Fr** pod; L 6–11 mm; downy; usually **2-seeded**; **black** when ripe. **Hab** rough grassland, scrub.

J F M A M J J A S O N D

TENDRILS branched

POD **2-seeded**, **downy**

TIP squarish

INFL 2–7 fls

LFLTS 6–8 pairs

STIPULE 'antlers'

Peas with multiple flowers in heads or spike-like inflorescences 1/2

IDENTIFY BY: ► flower details ► inflorescence form ► leaf details ► pod details

● **Melilots** | **Form** tall; erect (H to 150 cm). **Fls** yellow or white; in long spikes. **Lvs** trifoliate; LFLTS narrowly oval, toothed. **Fr** short pods. **Hab** disturbed grassland, roadsides and brownfield.

SS All melilots are similar in form; **White Melilot** *M. albus* (N/I) – most like **Tall Melilot** – [FLS white; PODS brown].

Melilot identification		Tall Melilot	Ribbed Melilot
Flower	COLOUR/LENGTH	golden-yellow; L 5–7 mm	lemon-yellow; L 5–7 mm
	PETALS	all ± equal length	L keel < L both standard and wings
Fruit	POD	downy; net-veined; black when ripe	hairless; ridged; brown when ripe

Ⓑⓟ Tall Melilot
Melilotus altissimus

J F M A M J J A S O N D

KEEL
WING

FL golden-yellow

LFLTS toothed

FR downy; net-veined; black

Ⓑ Ribbed Melilot
Melilotus officinalis

J F M A M J J A S O N D

KEEL
WING

FL lemon-yellow

LFLTS toothed

FR hairless; ridged; brown

RIBBED MELILOT

Peas with multiple flowers in heads or spike-like inflorescences

FLOWERS **yellow**

Meadow Vetchling *Lathyrus pratensis*

J F M A M J J A S O N D

L to 150 cm. **Form** scrambling; STEM angled but not winged. **Fls** L 12–18 mm; yellow; up to 12 on long stalk. **Lvs** LFLTS **1 pair**, L to 30 mm; TENDRIL ± branched; STIPULES long (± L lflts); arrow-shaped. **Fr** pod; L to 35 mm; flattened; black when ripe. **Hab** rough grassland, hedgebanks, meadows.

LFLTS 2, with tendrils

STIPULES long, arrow-shaped

Kidney Vetch *Anthyllis vulneraria*

J F M A M J J A S O N D

H to 30 cm. **Form** sprawling to erect. **Fls** L 12–15 mm; pale to orange-yellow (some sspp. pink/red); **heads rounded**, dense, surrounded by **ruff of leafy bracts**; SEP woolly. **Lvs** pinnate; STEM to 15 lflts; LOWER to 7 lflts, large terminal lflt only in many. **Fr** pod; L to 4 mm; within inflated sepal-tube. **Hab** grassland, especially calcareous; also sand dunes, cliff tops, mountain ledges.

INFL distinctive **woolly sepals**

POD is enclosed within sepals

● **Bird's-foot-trefoils** | **Fls** yellow; in heads. **Lvs** with 5 leaflets. **Fr pods splayed horizontally** – like a bird's foot.

SS All 5 bird's-foot-trefoil species are similar in form; **Narrow-leaved Bird's-foot-trefoil** L. *tenuis* (N/I) – most like **Common** – [LFLTS narrower (L >3× W); FLS fewer, smaller; mostly coastal].

GREATER

COMMON   GREATER

J F M A M J J A S O N D   J F M A M J J A S O N D

FR pods similar in both species

LF shape similar in both species; lower two leaflets stipule-like

Common Bird's-foot-trefoil *Lotus corniculatus*

H to 40 cm. **Form** ± hairless; **patch-forming**. **Fls** L 10–15 mm; yellow (can be flushed orange or red, esp. in bud); heads of 2–8. **Lvs** LFLTS 5–10 mm; L <3× W. **Fr** pod; L to 30 mm. **Hab** meadows, pastures.

FORM patch-forming

SEPALS ± **hairless**; ANGLE OF GAP AT BASE **obtuse**

Greater Bird's-foot-trefoil *Lotus pedunculatus*

H to 75 cm. **Form** usually very hairy; erect. **Fls** L 10–18 mm; paler yellow; heads of 5–12. **Lvs** LFLTS L to 15 mm. **Fr** pod; L to 35 mm. **Hab** marshes, fens, damp grassland and woodland rides.

FORM erect

SEPALS **very hairy**; ANGLE OF GAP AT BASE **acute**

FLOWERS mauve to reddish-purple [white]; LEAVES pinnate

🅿 Goat's-rue *Galega officinalis*

H to 150 cm. **Form** upright; hairless. **Fls** L 12–15 mm; mauve (or white) in **long loose spikes** (L ≥ leaf); SEPAL-TEETH bristle-like. **Lvs** LFLTS 4–8 pairs, broadly oval; STIPULES 'arrowhead'-shaped. **Fr** pod; L 20–30 mm; smooth; constricted between seeds.

Hab road verges, disturbed areas, brownfield. **SS** Sainfoin *Onobrychis viciifolia* (N/I) [FLS pink with red veins, INFL more compact].

SEPALS teeth bristle-like

LFLTS broad

STIPULES 'arrowhead'

POD constricted between seeds

🅿 Bush Vetch *Vicia sepium*

H to 60 cm. **Form** scrambling. **Fls** L 12–15 mm; **dusky purple**; 2–6 in **stalkless infl**. **Lvs** LFLTS 5–9 pairs, **broadly oval**, widest near base (can be notched at tip); TENDRILS branched; STIPULES 'spearhead'-shaped. **Fr** pod; L 20–35 mm; smooth; up to 10 seeds. **Hab** hedgebanks, woodland clearings/edges, rough grassland.

LF tendrils branched

LFLTS widest near base

STIPULES 'spearhead'

POD smooth

🅿 Tufted Vetch *Vicia cracca*

H to 200 cm. **Form** scrambling. **Fls** L 10–12 mm; purple-blue; 10–40 in **long-stalked, 1-sided infl**. (L 10–12 cm). **Lvs** LFLTS 8–12 pairs, **narrowly oval**; softly hairy; TENDRILS branched; STIPULES 'L'-shaped. **Fr** pod; L to 25 mm; hairless. **Hab** rough grassland, hedgebanks, scrub edges. **SS** Fodder Vetch *V. villosa* (N/I) [LFLTS 4–12 pairs; FL bicolored, keel and wings whitish].

LF tendrils branched

LFLTS narrow

STIPULES 'L'-shaped

POD smooth

🅿 Bitter-vetch *Lathyrus linifolius*

H to 40 cm. **Form** erect; STEM winged. **Fls** L 10–16 mm; **reddish-purple fading to bluish**; 2–6 in long-stalked heads. **Lvs** LFLTS 2–4 pairs, narrowly oval; LF terminal point; STIPULES 'half-arrowhead'-shaped. **Fr** pod; L 45 mm; reddish-brown. **Hab** wood margins, scrub, heathland.

STEM winged

LF terminal point

LFLTS narrow

STIPULES 'half-arrowhead'

POD long, reddish-brown

● **Clovers** | **Lvs** trifoliate. **Fls** usually numerous; in dense, rounded heads. **Fr** short, rounded pods.

SS White and Alsike Clovers can be confused; as can Zigzag and Red Clovers; leaves – wood-sorrels (*p.114*) [LFLTS held < horizontal].

Ⓟ **White Clover** *Trifolium repens*

H to 30 cm. **Form** patch-forming, rooting at nodes. **Fls** L 7–12 mm; white (pale pink). **Fl-hds** D 15–25 mm; on stalk (L to 20 cm). **Lvs** LFLTS rounded; STIPULES narrow, pointed, green, can be red-veined. **Hab** wide range of grassland *e.g.* meadows, lawns, verges.

J F M A M J J A S O N D

STIPULE
short point
LFLTS whitish chevron mark (usually)

more rounded

Ⓟ **Alsike Clover** *Trifolium hybridum*

H to 40 cm. **Fls** L 8–10 mm; TOP FLS white, BASE FLS **flushed pink**. **Fl-hds** D 15–22 mm, on stalk (L to 15 cm). **Form** erect, can be patch-forming, **not rooting at nodes**. **Lvs** LFLTS oval; STIPULES oval, long tapering point. **Hab** agricultural grassland, brownfield.

J F M A M J J A S O N D

STIPULE
long point
LFLTS plain; no chevron mark

more oval

Ⓟ **Zigzag Clover** *Trifolium medium*

H to 50 cm. **Form** creeping to erect. **Fls** L 12–20 mm; pinky-blue. **Fl-hds** D 25–35 mm. **Lvs** almost opposite pair below infl; LFLTS L to 15 mm, slightly hairy, narrowish, pale patch slight or absent; STIPULES narrow, tapering, green. **Hab** upland or clay grassland.

J F M A M J J A S O N D

STIPULE
pointed, green
LFLTS plain or with slight pale chevron

narrowish

Ⓟ **Red Clover** *Trifolium pratense*

H to 60 cm. **Form** erect; hairy. **Fls** L 12–18 mm; pinky-purple. **Fl-hds** L 20–40 mm; taller than wide. **Lvs** those below infl opposite; LFLTS L to 30 mm (L > W), hairy, usually with whitish chevron in centre; STIPULES triangular, bristle-tipped, purple-veined. **Hab** grassland.

J F M A M J J A S O N D

STIPULE
bristle tip; purple-veined
LFLTS usually with whitish chevron

Ⓐ **Hare's-foot Clover** *Trifolium arvense*

H to 20 cm. **Form** erect; hairy. **Fls** L 4–6 mm; white or pink; largely hidden within **abundant soft, whitish hairs**. **Fl-hds** L 10–20 mm; rounded to elongated. **Lvs** LFLTS ± narrowly oval, L to 15 mm × W 4 mm; LF-STALK L < lflt; STIPULES long, fine points. **Hab** dry sandy grassland, lowland heath, sand dunes.

J F M A M J J A S O N D

FLS hidden in long hairs

● **Trefoils** and **Medicks** | **Lvs** trifoliate. **Fls** usually numerous, in dense, rounded heads. **Fr** pods usually curved or coiled, longer than those of clovers.

SS All yellow trefoils and medicks are somewhat similar; **Lesser Trefoil** and **Black Medick** are perhaps the most frequently confused.

Ⓐ Slender Trefoil *Trifolium micranthum*

L

H to 15 cm. **Form** prostrate to sprawling; hairless. **Fls** L 2–3 mm; orange-yellow; standard **deeply notched**; 2–6 in small head. **Lvs** LFLTS L to 5 mm; notched at tip; stalks very short; **lengths equal**. **Hab** dry grassland, bare sandy soils.

J F M A M J J A S O N D

FLS 2–6

LFLTS stalks short; lengths ± equal

Ⓐ Hop Trefoil *Trifolium campestre*

L

H to 30 cm. **Form** erect; downy. **Fls** L 4–7 mm; pale yellow; standard **broad, can be pleated**; 20–30 in globular head (D to 15 mm). **Lvs** LFLTS L to 10 mm; central lflt longer-stalked. **Fr** pods; 'Hop'-like; dead standard lies flat over pod. **Hab** dry, especially calcareous grassland.

J F M A M J J A S O N D

FLS 20–30

LFLTS central stalk longer

STANDARD broad over brown POD

Ⓐ Lesser Trefoil *Trifolium dubium*

L

H to 25 cm. **Form** sprawling. **Fls** L 3–4 mm; deep yellow; standard **folded**; up to 20 in head (D to 9 mm); SEP hairless. **Lvs** LFLTS L to 10 mm; central lflt longer stalked. **Fr** pods; hairless; brown and drooping when ripe; covered by dead standard. **Hab** dry grassland, bare sandy soils.

J F M A M J J A S O N D

FLS up to 20

LFLTS central stalk longer

STANDARD folded over brown POD

Ⓢ Ⓟ Black Medick *Medicago lupulina*

L

H to 40 cm. **Form** sprawling or erect. **Fls** L 2–3 mm; bright yellow; **up to 50** in rounded head (D to 8 mm); SEP hairy. **Lvs** LFLTS L to 20 mm; downy; toothed; minute point at tip. **Fr** pods; curved (one coil, at most), **black when ripe** with net-like surface wrinkles. **Hab** dry grassland.

J F M A M J J A S O N D

FLS up to 50

LFLTS typically with minute point

PODS **black**

Ⓐ Spotted Medick *Medicago arabica*

L

H to 60 cm. **Form** sprawling; hairless when mature. **Fls** D 4–6 mm; bright yellow; **up to 8** in loose head. **Lvs** LFLTS L to 25 mm; usually with variable dark blotch in centre of each. **Fr** pods; green; in tight (up to 5) coils with numerous **hooked spines**. **Hab** grassland, especially coastal.

J F M A M J J A S O N D

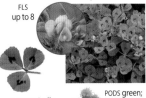

FLS up to 8

LFLTS **typically with dark blotch**

PODS green; spiny; coiled

12 Polygalaceae | **Milkwort** family

2 spp. | 4 spp. B&I COMMON

Form small perennials; herbaceous, although many with a woody base.
Fls unique structure: SEPALS 5, free, the inner 2 much larger and more petal-like than the outer 3; PETALS 3, fused, the lowest of which is divided into a **white fringe of narrow lobes**; STAMENS 8, fused into a tube attached to the petals.

IDENTIFY BY: ► **lower + upper leaf arrangement** ► **sepal details**

SS All milkworts are similar. Two rarer milkworts are distinguished from the common species shown here by their lower leaves being longer than the upper, and congested into a rosette.

In both these species LVS at the stem-base are smaller than those further up the stem; FLS typically blue, but can be purple, pink or white.

P **Common Milkwort** *Polygala vulgaris*

H to 30 cm. **Form** trailing to erect; base woody.
Fls L 5–8 mm; OUTER SEPALS pointed at apex. **Infl** usually **>10 flowers on main spikes**. **Lvs all alternate**. **Hab** calcareous to acid grassland, heathland.

J F M A M J J A S O N D

P **Heath Milkwort** *Polygala serpyllifolia*

H to 25 cm. **Form** usually trailing; base scarcely woody.
Fls L 5–6 mm; OUTER SEPALS rounded at apex. **Infl** usually **<10 flowers on main spikes**. **Lvs at least some lower leaves (if present) opposite**. **Hab** acid and upland grassland, heathland.

J F M A M J J A S O N D

typically more flowers on main spikes than Heath Milkwort

flowers can also be pink, purple or white

ALL LVS
alternate

LOWER LVS
opposite

stem-base leaves often lost; if they are absent, look at leaf scars to establish leaf arrangement

The typical blue-and-white flowers of milkworts stand out from the surrounding vegetation.

13 Rosaceae | **Rose** family

43 spp. | 178 spp. B&I

Form a large and diverse family, from trees and shrubs to tiny annuals. **Fls** typically PETALS 5, SEPALS 5, STAMENS numerous; some also have an outer ring of sepal-like structures below the true sepals (epicalyx). **Lvs** alternate, typically with stipules. **Fr** structures particularly diverse, ranging from the familiar fleshy apples and plums, to berries, hips and haws, and dry fruits.

IDENTIFY TO GROUP BY: ▶ inflorescence ▶ flower details ▶ leaf shape ▶ fruit

Rose family groups identification

TREES/SHRUBS/SCRAMBLERS | PETALS white to pink | FLOWER D > 10mm | FLOWER D < 10mm

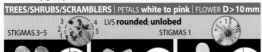

STIGMAS 3–5 ³ ⁴ **LVS rounded; unloaded**
² ¹ ⁵ **STIGMAS 1** · ¹

FLS numerous; in flat-topped clusters

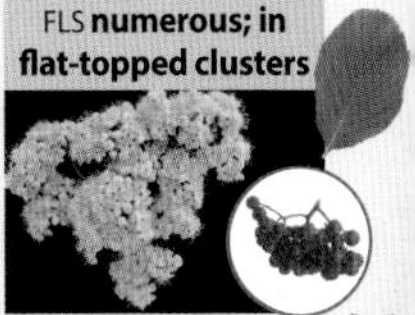

APPLES + PEARS
FRUIT **apple, pear**
pp. 90–91

BLACKTHORN/PLUM
FRUIT **sloe, plum**
p. 92

CHERRIES
FRUIT **cherry**
p. 93

ROWAN/WHITEBEAMS
LVS rounded to pinnate;
FRUIT **red berry**
p. 91

LVS **with separate leaflets** (trifoliate or pinnate)

LVS lobed

FLS clusters of up to 3

STIP large; joined to lf-stalk for much of length

STIP narrow; attached to lf-stalk only at base

FLS some spp.

ROSES
FRUIT **hip**
pp. 94–95

BRAMBLES
FRUIT **blackberry, raspberry**
p. 96

HAWTHORN
FRUIT **haw**
p. 94

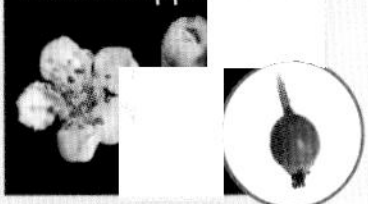

COTONEASTERS
LVS **rounded**
FRUIT **haw**
p. 90

HERBACEOUS PLANTS

FLOWERS **solitary; showy; colourful (white, yellow, pink)**

STRAWBERRIES, CINQUEFOILS, TORMENTILS and AVENS
LVS lobed to pinnate; FRUIT various: 'strawberry' to dry clusters of seeds
pp. 97–99

FLOWERS **in showy heads or spikes**

FLOWERS tiny (D 1–2mm); green; PET absent

FLS white to cream | INFL spike-like | INFL dense, rounded | FLS yellow-green

♀ ♂

FLS
STIPULES

'MEADOWSWEETS'
LVS **pinnate**
p. 100

AGRIMONIES
LVS **pinnate**
p. 99

BURNETS
LVS **pinnate**
p. 100

LADY'S-MANTLES
LVS **lobed**
p. 101

PARSLEY-PIERTS
LVS **lobed**
p. 102

Trees or large shrubs; FLS D < 10mm; in clusters of up to 3

● **Cotoneasters** | As a group, cotoneasters are frequently encountered, as berries are eaten and spread by birds. However, more than 90 species are established in B&I and identification is not straightforward, even using specialist keys. The ones below appear to be the most common though other species may prove to be more widespread in the future. The one native species is confined to the Great Orme, Conwy.

FLS some spp. open; some bud-like

ⓟ Himalayan Cotoneaster *Cotoneaster simonsii* **WS**

H to 3 m. **Form** erect semi-evergreen shrub; branches in more than one plane. **Fls** D 5–8 mm; groups of 2–4; PET **pinkish-white**, erect. **Lvs** L 10–25 mm; oval; shiny green above; light green, with some soft hairs below; alternate. **Fr** globular to pear-shaped; D 8–11 mm; **orange-red**. **Hab** rocks, walls, quarries, scrub on chalk.

J F M A M J J A S O N D

LF UP shiny

LF UND soft hairs

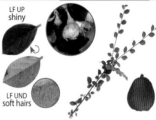

ⓟ Wall Cotoneaster *Cotoneaster horizontalis* **WS**

H to 1m. **Form** low shrub with regular **'herringbone' branching in one plane**; arching or appressed to walls and rocks. **Fls** D ± 5 mm; groups of 2–3; PET **pink**, erect. **Lvs** L 6–12 mm; oval; shiny green above; matt green below. **Fr** almost globular; D 4–6 mm; orange-red. **Hab** rocks, walls, quarries, scrub on chalk. **SS** Entire-leaved Cotoneaster *C. integrifolius* (N/I) [FLS white, PET spreading; BRANCHES not all in one plane].

J F M A M J J A S O N D

PETALS pink; erect

LF UP shiny

LF UND matt

in fruit

Trees or large shrubs; FLS D > 10mm; LVS unlobed; STIGMAS 3–5

● **Apples** | **Form** trees; H to 10m. **Fls** D 25–50mm; white flushed with pink, especially in bud; ANTHERS yellow. **Lvs** oval, pointed. **Fr** globular, yellow-green apple.

SS Wild Pear *Pyrus pyraster* (N/I) [usually spiny; FR often pear-shaped; flesh texture gritty]; **'Plums'** (p. 92) [OVARY superior].

BEWARE: The native Crab Apple and introduced Apple readily hybridize producing individuals with a full range of intermediate characters. Because of this, it is difficult to identify Crab Apple with confidence. Fruit size alone is not enough: even small-fruited, hairless examples can only be regarded as 'likely Crab Apples' without molecular analysis.

Apple ID		**ⓟ Crab Apple** *Malus sylvestris* **WT**	**ⓟ Apple** *Malus domestica* **WT**
Branches		can be **spiny**	never spiny
Flower	STALKS	**hairless** when mature	**at least a few hairs on lower stalk**
	SEPALS	hairless	hairy
Leaves		L to 5 cm; **hairless when mature**	L to 15 cm; underside hairy
Fruit ('apple')		D 2–3 cm, sharp-tasting	D up to 12 cm, often sweet-tasting
Habitat		woods and hedgerows	hedges, scrub, waysides as relic of cultivation or from picnic throwaways

Trees or large shrubs; FLS D < 10 mm; numerous in flat-topped clusters

● **Whitebeams (including Rowan and service-trees)** |
Form deciduous, small to medium-sized trees. **Fls** D 6–8 mm;
creamy-white; PETALS 5. **Lvs** variable between species; ranging
from oval to pinnate. **Fr** berry-like; shades of red; L to 10 mm;
slightly longer than broad.

COMMON
WHITEBEAM

SS The whitebeams and service-trees are an aggregate of more
than 50 species with a range of leaf shapes from oval to pinnate.
Some species are very rare (just a handful of individuals). Many
are a challenge and identification requires experience and
specialist knowledge beyond the scope of this book.

IDENTIFY BY: ► leaf shape details ► fruit details (some species)

Ⓟ **Rowan** *Sorbus aucuparia* WT

H to 18 m (typically much less).
Lvs L to 25 cm; **pinnate**; LFLTS
5–8 pairs; toothed. **Fr** bright
red; almost globular; L to
9 mm. **Hab** woods, heaths,
moors, mountains.

J F M A M J J A S O N D

LF pinnate

Ⓟ **Swedish Whitebeam** *Sorbus intermedia* WT

H to 10 m. **Lvs** L 6–10 cm; oval;
lower part divided into lobes
(≤⅓ to midrib); UNDERSIDE
felted yellowish. **Fr** orange-red;
L to 15 mm; L > W. **Hab** woods,
hedges, urban plantings.
SS Common Whitebeam
S. aria (N/i) [LVS unlobed;
underside felted whitish].

J F M A M J J A S O N D

LF lobed at base

LF UND
felted
yellowish

CRAB APPLE

LF-STALK
hairless

LF UND
hairless

SEPALS
hairless

The fruit and leaves of Crab
Apple are generally much
smaller than those of Apple

APPLE

LF-STALK
hairy

LF UND
hairy

SEPALS
hairy

CRAB APPLE

J F M A M J J A S O N D

Trees or large shrubs; **LVS unlobed; FLS D > 10 mm; STIGMAS 1**

● **Prunus (Blackthorn and Plums)** | **Form** suckering.
Fls short-stalked; solitary or in clusters of up to 3; PETALS 5,
white to pinkish. **Lvs** broadly oval. **Fr** 'sloes' or 'plums'.

SS Blackthorn and **Wild Plum** subspecies and forms
frequently hybridize; giving rise to a wide range of variants.

IDENTIFY BY: ► flowering time, size and sepals ► fruit
► twig colour

P ● Cherry Plum *Prunus cerasifera* **WT**

J F M A M J J A S O N D

H to 8 m. **Form** shrub,
not usually very spiny,
young TWIGS **green, can
be red-tinged**. **Fls** white
to pinkish; D **15–22 mm**;
appear with opening
leaves; SEP **bent back**.
Lvs L to 70 mm;
broadly oval; ± glossy.
Fr globular; D 20–30 mm; ripen **red** or
yellow; RIPE TASTE slightly sweet. **Hab**
hedges; much planted in urban areas.

LF ±
glossy

FR red or yellow;
slightly sweet

FLS as leaves
open

SEPALS bent back

TWIG green when young　　no spines

P ● Blackthorn *Prunus spinosa* **WT**

J F M A M J J A S O N D

H to 5 m. **Form** very
spiny shrub; prostrate on
shingle; TWIGS **blackish-
grey with sharp spines**.
Fls white; D **12–16 mm**;
appear before leaves;
SEP erect. **Lvs** L to 30 mm;
broadly oval; dull.
Fr ± globular; D 8–15 mm;
blue-black with greyish waxy coating;
RIPE TASTE bitter. **Hab** hedges, scrub,
woodland edges; shingle.

LF
small;
dull

FR blue-black;
± globular;
bitter

FLS before
leaves

FLS much smaller
than in Cherry Plum　　SEPALS erect

TWIG blackish-grey

sharp spines

P ● Wild Plum *Prunus domestica* **WT**

D J F M A M J J A S O N D

H to 8 m. **Form** large
shrub/small tree; some
forms with sparse spines.
TWIGS grey-brown.
Fls white; D **15–25 mm**;
in clusters of 2–3; appear
with leaves; SEP erect
(to bent back). **Lvs** L to
80mm; ± broadly-oval;
dull. **Fr** variable in size/shape (L>20mm;
usually L > W) and colour (green, yellow,
red, purplish); RIPE TASTE sweet.
Hab hedges, scrub, gardens.

LF
dull

FR green, yellow, red
or purple; sweet

FLS with
leaves

SEPALS usually erect

TWIG grey-brown; ± sparse spines

● **Prunus (Cherries)** | **Form** mostly suckering. **Fls** in loose clusters of up to 6 or cylindrical spikes of up to 40; PETALS 5, white to pinkish. **Lvs** broadly oval. **Fr** 'cherries'.

SS Numerous species and varieties of cherry are grown in gardens and also planted as street trees. Many have pink-flushed flowers and/or >5 petals; these individuals are usually grafted onto Wild Cherry rootstock, so basal shoots may produce 'wild-type' features from below the graft.

Deciduous; LF-STALKS with red glands		**Evergreen;** LF-STALKS no glands

Ⓟ Wild Cherry WT
Prunus avium

H to 30 m.
Form suckering tree; BARK with **strong horizontal lines**. **Fls** white; D 15–30 mm; **long-stalked; in clusters** of 2–6. **Lvs** L to 15 cm; pointed tip; finely toothed. **Fr** globular; D 20–30 mm; red to almost black when ripe, (varies in cultivated examples). **Hab** woods, especially on slopes with springs; hedges, cultivated orchards and gardens.

Ⓟ Bird Cherry WT
Prunus padus

H to 15 m.
Form suckering shrub/small tree; BARK peeling. **Fls** white; D 10–20 mm; up to 40 in **elongate erect or drooping cylindrical spikes**. **Lvs** L to 10 cm; pointed tip; finely, regularly toothed. **Fr** globular; D 6–8 mm shiny black when ripe. **Hab** woods, especially upland; much planted in gardens and parks.

Ⓟ Cherry Laurel WT
Prunus laurocerasus

H to 10 m.
Form shrub/small tree.
Fls white; D 8–12 mm; in cylindrical spikes (L to 13 cm; < leaf L). **Lvs** L to 15 cm; ± rounded; shiny, leathery; LF-STALKS + YOUNG STEM green. **Fr** ± globular; D 10–12 mm; purplish-black when ripe. **Hab** widely planted; naturalized in woodland.
SS Portugal Laurel *P. lusitanica* (N/I) [LF-STALKS + YOUNG STEM red-purple; infl L > leaf L]; leaves – **Rhododendron** (*p. 159*) [CRUSHED LVS scentless].

FLS long-stalked; in clusters

FLS in erect or drooping spikes

LF shiny; whiff of bitter almonds (cyanide) when crushed

FR sweet, edible

FR bitter ✗

FR do not eat ✗

93

Trees or large shrubs; LVS lobed

● **Hawthorns** | **Form** H to 12 m; twigs with rigid spines (L to 25 mm). **Fls** D <15 mm; white, in flat clusters. **Lvs** broadly oval in outline; lobed. **Fr** red, tough-skinned berry (haw); L to 10 mm; slightly longer than broad; up to 3 seeds.

SS In woodlands, the two hawthorns readily hybridize, giving a full spectrum of intermediates; hybrids are often planted; cultivated/naturalized hawthorns with pink, often double, flowers are also generally of hybrid origin.

ⓟ **Hawthorn** WT
Crataegus monogyna

Fls STYLES 1. **Lvs** divided >⅔ way to midrib into **5–7 lobes**. **Fr** SEEDS 1. **Hab** hedges, scrub, woodland edges.

LF-LOBES 5–7

STYLES 1   SEEDS 1

J F M A M J J A S O N D

ⓟ **Midland Hawthorn** WT
Crataegus laevigata

Fls STYLES 2–3. **Lvs** divided <⅔ way to midrib into **±3 lobes**. **Fr** SEEDS 2–3. **Hab** woodland interiors; planted in hedges and gardens.

FLS fewer and often slightly larger

LF-LOBES ±3

LVS often more **glossy** and leathery

STYLES 2–3   SEEDS 2–3

J F M A M J J A S O N D

Small shrubs/woody-based scramblers; LVS with separate leaflets   1/2

● **Roses** | **Form** long; usually arching; STEM prickly. **Fls** showy; PETALS 5; SEPALS typically entire. **Lvs** pinnate. **Fr** fleshy (hip).

IDENTIFY BY: ► flower, sepal details + scent ► stem + prickle form ► leaves + glands ► fruit details ► crushed leaf scent

SS All roses are readily recognizable as such, but numerous additional scarcer species and hybrids make specific identification often very difficult.

Hip blackish when ripe; **Hab** dry sandy, heathy or limestone grassland

ⓟ **Burnet Rose** *Rosa spinosissima* WS

H to 50 cm. **Form** patch-forming; PRICKLES **dense, slender. Fls** white; D 20–40 mm; BRACTS absent. **Lvs** LFLTS 3–5 pairs; L 10–25 mm. **Fr** globular L 5–15 mm; **blackish when ripe**.

RIPE HIP blackish

LFLTS more rounded than those of other roses

STEM prickles slender; dense

J F M A M J J A S O N D

Small shrubs/woody-based scramblers; LVS with separate leaflets 2/2

Hip red when ripe; **L ≥ W; Hab** hedgerows, scrub open woodland

ⓅDog-rose *Rosa canina* **WS**

H to 3 m. **Form** STEM arching; PRICKLES stout, strongly curved. **Fls** pink to white; D 30–60 mm; STYLES **not fused into a column**; SEP fall before fruiting; STALK ± smooth. **Lvs** LFLTS 2–3 pairs; L to 25 mm; hairless; GLANDS absent; MARGIN teeth uneven.
Fr ± egg-shaped; L to 25 mm; L > W.

FR
STYLES not fused
STALK smooth
LF UND no glands
TEETH uneven
STEM prickles stout, curved

ⓅField-rose *Rosa arvensis* **WS**

H to 1·5 m. **Form** STEM sprawling; PRICKLES hooked, all ± equal size. **Fls** white; D 30–50 mm; STYLES **fused into a column**; SEP can have a few lobes, falling before fruiting; STALK typically with **stalked glands**. **Lvs** LFLTS 2–3 pairs; L to 35 mm; hairless; GLANDS absent; MARGIN teeth ± even. **Fr** globular to egg-shaped; L 12–15 mm.

FR
STYLES fused
STALK stalked glands
LF UND no glands
TEETH ± even
STEM prickles hooked

ⓅSherard's Downy-rose *Rosa sherardii* **WS**

H to 1·5 m. **Form** compact; STEM slender, arching; PRICKLES slender, straight to curved. **Fls** usually deep pink; D 25–40 mm; STIGMAS form **domed head**; SEP lobed; remain until fruit ripe; STALK with glands + stiff hairs. **Lvs** LFLTS 2–3 pairs; L to 40 mm; softly hairy; GLANDS on underside, **resin-scented**; MARGIN double-toothed. **Fr** globular to egg-shaped; L 10–20 mm. **SS Sweet-briar** *R. rubiginosa* (N/I) [LVS apple-scented].

FR
STIGMAS domed head
STALK glands + hairs
LF UND glands
TEETH uneven
STEM prickles slender

Hip red when ripe; **L < W; Hab** brownfield, sand dunes, dry grassland

ⓅJapanese Rose *Rosa rugosa* **WS**

H to 1·5 m. **Form** suckering; PRICKLES **dense, variously sized**. **Fls** white to red; D 60–90 mm; strong, sweet scent; SEP tip expanded, remain at fruiting. **Lvs** LFLTS 2–4 pairs; L 20–50 mm; surface raised between sunken veins. **Fr** L to 25 mm (W > L).

RIPE FRUIT large; wide
LF distinctly wrinkled upperside
STEM prickles dense

Small shrubs/woody-based scramblers; OVARY **superior**;
FRUIT **an aggregation of small thin-skinned fleshy fruits**

● **'Brambles'** | **Form** STEM with spines or prickles. **Fls** PETALS 5; white
to deep pink; numerous ♂ and ♀ parts on conical base (receptacle).
Lvs trifoliate, palmate or pinnate; LFLTS broadly oval, 2–3 pairs.
Fr aggregation (L 1–2 cm) of many small thin-skinned fleshy fruits
(drupelets), each containing a seed.

SS The genus *Rubus* includes a few well-defined species such as **Raspberry**
and **Dewberry**, but the bulk of the 300+ recognized types are 'bramble' microspecies
that set seed without fertilization. **Although differences in these are constant, they are
often minor and require specialist keys for identification.** However **Bramble** (or Blackberry)
as an aggregate of these microspecies is well-known and very common.
The one sexual microspecies, '**Elm-leaved Blackberry**' *R. ulmifolius*, is widespread and
relatively straightforward to identify and so is included here.

ℙ Raspberry *Rubus idaeus* WS

J F M A M J J A S O N D

FR

H to 150 cm. **Form** erect, patch-
forming; STEM **prickles weak,
straight**. **Fls** D 10–12 mm; PET
white, narrow; in loose, few-
flowered clusters. **Lvs** pinnate;
LFLTS 3, 5 or 7 (L to 12 cm), **whitish-
green below**. **Fr** L 5–15 mm; up to
100+ drupelets; **red when ripe**;
downy. RIPE TASTE 'raspberry'.
Hab woodland rides and clearings, heaths; also
escapes from cultivation.

LF UND
whitish-green

STEM prickles weak

ℙ Dewberry *Rubus caesius* WS

J F M A M J J A S O N D

FR

H to 50 cm. **Form** prostrate,
arching; STEM **prickles few**, short
and slender. **Fls** D 20–30 mm; PET
white (always). **Lvs** trifoliate;
LFLTS 3 (L to 14 cm), basal pair
almost stalkless. **Fr** L 5–15 mm;
<20 drupelets; **black with greyish
surface coating**; RIPE TASTE more
tart than most brambles.
Hab woods, scrub, rough grassland, dunes.

flowers earlier; fruit aggregation smaller but
drupelets larger and fewer than brambles

ℙ Bramble *Rubus fruticosus* agg. WS

J F M A M J J A S O N D

H to 300 cm. **Form** patch-forming;
STEM with **hooked spines**.
Fls D 20–30 mm; PET white to
deep pink. **Lvs** trifoliate or palmate;
LFLTS 3, 5 or 7 (L to 10 cm). **Fr**
L 10–20 mm; 25–100 drupelets;
purplish-black when ripe; RIPE
TASTE usually sweet. **Hab** woods,
hedges, scrub, brownfield.

FR

LF UND
green

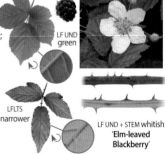

'**Elm-leaved Blackberry**' *R. ulmifolius* LFLTS smaller
and usually with curled-down margins, STEM + LF
UNDER more whitish, FLS deeper pink and later in
the season than many of the other microspecies.

LFLTS
narrower

LF UND + STEM whitish
'**Elm-leaved
Blackberry**'

Herbaceous (can have woody rootstock); INFLORESCENCE **showy, solitary**   1/2

FLOWERS **white**

● **Strawberries** | **Form** prostrate; with surface runners. **Fls** PETALS 5.
Lvs trifoliate; lobes with toothed margins. **Fr** tiny and dry; or seeds on the
outside of a rounded or conical fleshy base. **Hab** woods, hedgerows, scrub.

P Barren Strawberry *Potentilla sterilis*

L stem to 15 cm. **Fls** D 8–15 mm;
PET **separated so that green
sepals visible** between them;
tip notched. **Lvs** LFLTS 3, L to
25 mm, grey-green, **end tooth <
adjacents**. **Fr** dry.

J F M A M J J A S O N D

PETAL L ≈ SEPAL L

FR

LF

P Wild Strawberry *Fragaria vesca*

L stem to 20 cm. **Fls** D 10–
20 mm; PET **L > sepal L; tip
not notched**. **Lvs** LFLTS 3, L to
60 mm, glossy green above, **end
tooth > adjacents**. **Fr** fleshy,
red; D < 10 mm; RIPE TASTE
'strawberry', can be intensely so.
SS introduced/garden varieties
(N/I) [FLS + FR larger].

J F M A M J J A S O N D

PETAL L > SEPAL L

FR

LF

FLOWERS **purplish or orange-pink**

P Marsh Cinquefoil *Comarum palustre*

H to 50 cm. **Form** erect.
Fls D 20–30 mm; PET + SEP
purple. **Lvs** pinnate; LFLTS 2 or
3 pairs, L to 60 mm, end lflt
same size as others. **Hab**
marshes and other wetland
habitats.

J F M A M J J A S O N D

BASAL LVS pinnate;
leaflets equal

FLS
'flat'

P Water Avens *Geum rivale*

H to 60 cm. **Form** erect with
nodding fls. **Fls** D 15–20 mm;
PET **orange-pink**; SEP purple.
Lvs STEM-LVS usually 3-lobed;
BASAL LVS pinnate, LFLTS 3–6
pairs, L to 20 mm, end lflt larger
(L 20–50 mm) and broader
than the others. **Fr** bur-like;
styles hooked. **Hab** marshes,
damp woodland, banks of watercourses.

J F M A M J J A S O N D

BASAL LVS pinnate;
end leaflet larger

FLS
nodding

FR bur-like

SS Hybrids between **Water Avens** and **Wood Avens** (*p. 99*), with mixed features, are frequent.

Herbaceous (can have woody rootstock); INFLORESCENCE **showy, solitary**

FLOWERS **yellow**

● **Tormentils and Creeping Cinquefoil** | **Form** three very similar species, all of which can be prostrate and trailing. **Fls** PETALS 4 or 5; yellow. **Lvs** trifoliate or palmate.

IDENTIFY BY: ► seed count ► flower, leaf, leaf-stalk and stipule details as below

Flower	Length	Form		Flower		
				DIAMETER	PETAL COUNT	SEEDS
Tormentil	to 45 cm	erect to trailing; not rooting at nodes		7–12mm	a mix of 4	very few 5	<20
Trailing Tormentil	to 80 cm	prostrate; rooting at nodes	from late July	12–18mm	and 5	± 25% 5	<20–50
Creeping Cinquefoil	to 100 cm		always	15–25mm	all 5		60–120

Leaf	Stipules	Stem-leaves		Lower leaves	
		LFLT COUNT	LF-STALKS	LFLT COUNT	LF-STALKS
Tormentil	lobed; like the leaves	3	very short (L<5mm)	most with 3 (a few with 4 or 5)	long
Trailing Tormentil	unlobed; unlike the leaves	3	upper L<10mm; lower L 10–20mm	±50% with 5; the rest a mix of 3 and 4	L 10–20mm
Creeping Cinquefoil		palmate, with 5 leaflets; lf-stalks L >10mm; all same length			

SS Hybrids (sterile) between **Tormentil/Trailing Tormentil** and **Creeping Cinquefoil** are very frequent and virtually identical to **Trailing** [but STEM LF-STALKS *all same length* and L >10mm].

ℙ Tormentil
Potentilla erecta

Hab dry grassland, heathland, mountains.

ROOTS AT NODES
absent

STIPULES leaf-like

STEM-LF ± stalkless

LOWER LVS mostly 3-lobed

SEEDS <20

ℙ Trailing Tormentil
Potentilla anglica

Hab dry grassland, heathland.

ROOTS AT NODES
from July

STIPULES unlobed

STEM-LVS stalked

LOWER LVS ± 50% 5-lobed

SEEDS <20–50

ℙ Creeping Cinquefoil
Potentilla reptans

Hab open grassland, tracks and disturbed habitats, sand dunes.

ROOTS AT NODES
present

STIPULES unlobed

STEM + LOWER LVS

SEEDS 60–120

P Wood Avens *Geum urbanum*

H to 70 cm. **Form** sprawling to erect. **Fls** D 10–15 mm; PETALS 5. **Lvs** STEM 3-lobed; ROSETTE pinnate; 2–3 pairs of variable lflts with large, often 3-lobed, end lflt. **Fr** globular, **bur-like**. **Hab** woods, hedges, gardens.

J F M A M J J A S O N D

FR bur-like

LF

P Silverweed *Potentilla anserina*

L to 80 cm. **Form** prostrate; RUNNERS long, often reddish. **Fls** D ±10 mm; PETALS 5; solitary. **Lvs** pinnate; silvery-hairy underneath. **Hab** open grassland, tracks and disturbed habitats, sand dunes, upper saltmarshes.

J F M A M J J A S O N D

LF 7–12 pairs of toothed main leaflets, alternating with small ones

---

Herbaceous (can have woody rootstock); INFLORESCENCE **showy heads or spikes**   1/2

● **Agrimonies** | Two very similar species. **Form** erect. **Fls** yellow; D 5–8 mm; in a **long spike**. **Lvs** pinnate; LFLTS 3–6 pairs of main leaflets alternating with variable smaller ones; ± aromatic when bruised. **Hab** hedges, banks, rough grassland.

Agrimony Identification		**P Agrimony** *Agrimonia eupatoria*	**P Fragrant Agrimony** *Agrimonia procera*
Form		H to 80 cm	H to 100 cm; **more robust**
Stem hairs		scattered long over short; often tinged red	scattered long hairs with glandular hairs
Lvs	GLANDS (UNDERSIDE)	**absent**, or very few	**numerous; shiny, stalkless**
	AROMA	**none** to faintly 'apricot'	**strong citrus**
Ripe fruit	GROOVES	**deeper**; extending from apex to base	shallower; **not** extending all the way to base
	HOOKED BRISTLES	outermost **spreading to forward-pointing**	outermost **bent back**

AGRIMONY   FRAGRANT

J F M A M J J A S O N D   J F M A M J J A S O N D

AGRIMONY   FRAGRANT AGRIMONY

GLANDS ± absent   GLANDS **present**

FR grooves ± reach base   FR grooves **well short of base**

STEM   STEM

AGRIMONY

Herbaceous (can have woody rootstock); INFLORESCENCE **showy heads or spikes**

● **'Meadowsweets'** | Two somewhat similar species of differing habitats.
Form erect. **Infl** an irregular terminal loose branched cluster. **Lvs** pinnate with main leaflets alternating irregularly with smaller ones; STIPULES large, toothed.

SS Both species could conceivably be confused with members of the carrot family (p.222) [FLS in umbels].

Ⓟ **Meadowsweet** *Filipendula ulmaria*

H to 120cm. **Fls** D2–5mm; PET **usually 5**, white; sweetly scented. **Lvs** pinnate; MAIN LFLTS **2–5 pairs; broad**; L20–80mm. **Fr** twisted. **Hab** marshes and banks of watercourses.

J F M A M J J A S O N D

Ⓟ **Dropwort** *Filipendula vulgaris*

H to 120cm. **Fls** D5–9mm; PET **usually 6**, creamy-white (may have pink tinge); unscented. **Lvs** pinnate; MAIN LFLTS **8–20 pairs; narrow**; L5–15mm. **Fr** splayed. **Hab** calcareous grassland.

J F M A M J J A S O N D

FL 5 petals — FR twisted — LFLTS 2–5 pairs

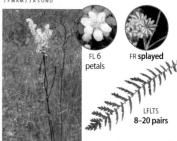

FL 6 petals — FR splayed — LFLTS 8–20 pairs

● **Burnets** | **Form** erect grassland species. **Fls** in dense terminal spike; PETALS absent; SEPALS 4. **Basal lvs** pinnate; LFLTS toothed.

Ⓟ **Salad Burnet** *Poterium sanguisorba*

H to 50cm. **Infl** compact, **globular**; L7–15mm. **Fls** green; UPPER ♀ (2 reddish, **feathery stigmas**); LOWER ♂ (up to ± 12 stamens) or bisexual. **Lflts** 4–12 pairs; rounded; L to 20mm; AROMA cucumber when bruised. **Hab** dry grassland, especially calcareous.

J F M A M J J A S O N D

LFLTS shortly stalked

FLS ± unisexual

Ⓟ **Great Burnet** *Sanguisorba officinalis*

H to 120cm. **Infl** rounded-elongate; L10–30mm. **Fls** purplish; all bisexual (1 undivided stigma, 4 stamens). **Lflts** 3–7 pairs; broadly oval; L to 40mm. **Hab** damp grassland, fens.

J F M A M J J A S O N D

LFLTS distinctly stalked

FLS bisexual

● Lady's-mantles | **Form** distinctive perennials; STEM ascending (H to 50 cm). **Fls** small (D 2–4 mm); yellow-green; PETALS absent; SEPALS 4; STAMENS 4. **Infl** terminal cluster. **Lvs** palmately lobed.

SS The lady's-mantles produce seeds without fertilization (apomictic) and so differ only in minor characters such as the size and shape of the leaf-teeth. Identification features of the four most widespread species are detailed below. **Confident identification especially of the rarer forms is possible only with experience and specialist keys, although geographic distribution can narrow the choice.**

IDENTIFY BY: ► epicalyx and sepal length ► leaf and leaf-lobe shape ► leaf-margin teeth ► type and location of leaf-hairs ► other specific features

EPICALYX segments **as long as** sepals ('8-pointed star')

❶ Soft Lady's-mantle
Alchemilla mollis

Lf-stalk + Fl-stalk HAIRS spreading. **Lf** LOBES 9–11; TEETH all ± same size; slightly incurved. **Hab** gardens; colonist of grassland and wood edge habitats.

JFMAMJJASOND

EPICALYX shape diagnostic
LF-STALK | hairs spreading
LF UP | softly hairy
LF UND | softly hairy
TEETH incurved
LF-STALK hairs spreading

EPICALYX segments **shorter than** sepals

❷ Smooth Lady's-mantle
Alchemilla glabra

Lf-stalk + Fl-stalk hairless or with a few appressed hairs. **Lf** LOBES 7–9; TEETH teeth at tip of lobe broad, curved and larger than the others. **Hab** damp grassland, woodland rides, mountain rocks.

JFMAMJJASOND

LF-STALK | ± hairless
LF UP | hairless
LF UND | ± hairless
can be a few hairs on veins near lobe tip
LF 'leathery'
TEETH at lobe tip broader than others
LF-STALK ± hairless

❸ Slender Lady's-mantle
Alchemilla filicaulis

Lf-stalk + Fl-stalk HAIRS dense, spreading; PLANT BASE wine-red (diagnostic). **Lf** LOBES 5–9; TEETH triangular, incurved. **Hab** damp grassland, woodland rides, mountain pastures.

JFMAMJJASOND

PLANT BASE diagnostic
LF-STALK | hairs spreading
LF UP | softly hairy
LF UND | softly hairy
PLANT BASE wine-red
TEETH incurved
LF-STALK hairs spreading

❹ Pale Lady's-mantle
Alchemilla xanthochlora

Lf-stalk + Fl-stalk HAIRS dense, spreading; PLANT BASE **brown**. **Lf** LOBES 7–11; TEETH large, ± straight, sharply pointed. **Hab** damp grassland, road verges.

JFMAMJJASOND

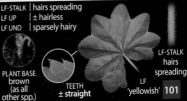

LF-STALK | hairs spreading
LF UP | ± hairless
LF UND | sparsely hairy
PLANT BASE brown (as all other spp.)
TEETH ± straight
LF-STALK hairs spreading
LF 'yellowish' 101

Herbaceous (can have woody rootstock); INFLORESCENCE **tiny green flowers**

● **Parsley-pierts** | Two very similar, but distinctive species. **Form** sprawling to shortly erect (H to 10 cm). **Infl** dense cluster. **Fls** tiny (D 1–2 mm) located opposite upper leaves; PETALS absent; SEPALS 4; STAMENS 1 (usually). **Lvs** L < 10 mm; deeply palmately lobed; STIPULES of upper lvs fused into a leafy cup which surrounds the fruiting inflorescence.

SS Both species very similar and only safely identified when in fruit using the details below.

Parsley-piert Identification	Ⓐ **Parsley-piert** *Aphanes arvensis*	Ⓐ **Slender Parsley-piert** *Aphanes australis*
Plant colour	more green to yellow-green	more greyish-green
Stipule lobes	approx. ¼ of stipule-cup height	approx. ½ of stipule-cup height
Sepals in fruit	erect; L 0.6–0.8 mm	converging; L 0.3–0.5 mm;
Fruit length	reaches tip of stipules	does not reach tip of stipules
Fruit shape	constricted at base of sepals	not constricted
Habitat	cultivated land; dry bare patches in grassland	
	on a wide range of soils	mainly on sandy soils

PARSLEY-PIERT

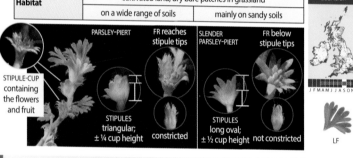

PARSLEY-PIERT · FR reaches stipule tips · SLENDER PARSLEY-PIERT · FR below stipule tips

STIPULE-CUP containing the flowers and fruit

STIPULES triangular; ± ¼ cup height · constricted

STIPULES long oval; ± ½ cup height · not constricted

J F M A M J J A S O N D

SLENDER

J F M A M J J A S O N D

LF

---

14 Elaeagnaceae | **Sea-buckthorn** family   **1 sp.** | 4 spp. B&I

Form deciduous shrub. **Fls** ♂ + ♀ on separate plants; petal-less; SEPALS 2. **Lvs** linear; with silvery scales. Not remotely similar to the two true buckthorns (*opposite*).

Ⓟ **Sea-buckthorn** *Hippophaë rhamnoides* WS

H to 3m; can be larger. **Form** spreading; suckering widely; thorny. **Fls** appear before leaves; tiny (D 3–4 mm); ♂ 2 stamens. **Lvs** linear to narrowly oval; L to 80 × 15 mm.
Fr orange berries (D 6–8 mm); in dense clusters on older twigs. **Hab** sand dunes and other coastal habitats; also planted inland.

J F M A M J J A S O N D

♂ FLS

♀ FL

LF

FR

Most obvious when in fruit

15 Rhamnaceae | **Buckthorn** family   `2 spp.` | 2 spp. B&I

Form deciduous shrubs or small trees. **Fls** greenish; PETALS + SEPALS 4 or 5.
Lvs broadly oval; ± pointed; with stipules. **Fr** berries; black when ripe.

IDENTIFY BY: ► leaf shape and veining   ► inner bark colour

SS Both buckthorns, particularly Buckthorn, are similar to **Dogwood** (*p. 155*) [FLS white in flat-topped clusters; LVS entire and 4-veined].

Ⓟ **Buckthorn** *Rhamnus cathartica* **WS**

H to 7 m. **Form** usually
spiny. **Fls** D 4–5 mm; PET 4 +
SEP 4 (both rarely 5); **green**;
loose clusters in leaf-axils on
last year's wood. **Lvs** L 40–
60 mm; **toothed**; turn yellow
in autumn. **Fr** 3–4-seeded
berries (D 6–10 mm); green,
ripening black. **Hab** hedges,
scrub, open woodland; usually calcareous;
also planted.

J F M A M J J A S O N D

BARK
orange
beneath

LF-VEINS
2–4

FLS + FR clustered

Ⓟ **Alder Buckthorn** *Frangula alnus* **WS**

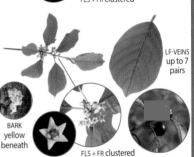

H to 7 m. **Form** not spiny.
Fls D 3–4 mm; PET 5,
greenish-white; SEP 5, green;
in loose clusters in leaf-axils
on new twigs. **Lvs** L up to
60 mm; **entire**; turn red in
autumn. **Fr** 2–3-seeded
berries; green, **ripening red
then black**. **Hab** damp scrub,
bogs, open woodland.

J F M A M J J A S O N D

BARK
yellow
beneath

LF-VEINS
up to 7
pairs

FLS + FR clustered

16 Cannabaceae | **Hop** family   `1 sp.` | 2 spp. B&I

Form rampant **climber**; twining clockwise. **Fls** ♂ (small (±4 mm); INFL branched); ♀ (larger; INFL compact); on separate plants. **Lvs** palmately lobed; rough. **Fr** 'pine-cone'-like papery clusters.

Ⓟ **Hop** *Humulus lupulus* **WC**

L to 8 m. **Form** roughly
hairy. **Fls** clustered in bases
of upper leaves; ♂ clusters
rather loose; ♀ compact;
'pine cone'-like. **Lvs** opposite;
rough; **3–5-lobed**; divided
¾ of the way to base; LOBES
coarsely toothed. **Fr pale
green cone-like** fruiting
structures (L to 5 cm) with overlapping
papery bracts. **Hab** hedges, scrub and fen
woodland; also cultivated.

J F M A M J J A S O N D

IN FRUIT

♂ FLS

♂ FL   ♀

FR

LF

17 Ulmaceae | **Elm** family

2 spp. | 2 spp. B&I   ENGLISH ELM (mature tree)

Form deciduous trees, often suckering. **Fls** tiny; **purplish**; in showy **rounded clusters**; appearing **before leaves** emerge. **Lvs** ± broadly oval; pointed; toothed; with **asymmetric base**. **Fr** seed within a notched disc formed by 2 broad wings. **Hab** hedges and woods; river margins (Wych Elm).

SS A complex of numerous clonal microspecies meaning identification is difficult. Dutch Elm Disease, caused by the fungus *Ophiostoma novo-ulmi*, has eliminated most mature trees (*right*). Look for distinctive bark-beetle galleries, the dispersal agent for fungal spores, under the bark of dead trunks. Elms are also recognizable as a group by the distinctively shaped leaf-holes produced by Elm Zigzag Sawfly *Aproceros leucopoda* larvae, first found in B&I in 2017 and now spreading rapidly.

IDENTIFY BY: ▶ Sep–Feb bud-hairs; ▶ Feb–Apr seed position; ▶ Jul–Sep leaf shape (mature shoots growing in full sun)

Elm in flower

WYCH ELM
FLS + FR before leaves

Elm in fruit
WYCH ELM

FIELD ELM

Elm in leaf

Elm Identification	Form			Leaves		Fruit
	HEIGHT	SUCKERS	MATURE OUTLINE	SHAPE	UPPERSIDE	SEED
Ⓟ Wych Elm **WT** *Ulmus glabra*	H to 30 m (large trees outside of woodland rare)	does not sucker freely	CROWN rounded; TRUNK usually divided low down	L to 100 mm; UPPERSIDE **very rough**		located **centrally**
Ⓟ 'Field Elm' **WT** *Ulmus minor* agg.	H to 30 m (large trees rare)	suckers freely (H to 5 m)	TRUNK usually divided	CROWN spreading	L to 90 mm; L>¾ W; UPPERSIDE **smooth**	located **towards tip**
English Elm *Ulmus minor* 'Atinia' or *Ulmus procera*				CROWN rounded to oblong	L to 90 mm; L<¾ W; UPPERSIDE **rough**	

WYCH ELM
FR seed located centrally

'FIELD ELMS'
FR seed located near tip

WYCH

ALL TWIGS not corky

BUDS hairs obvious

2ND-YEAR TWIGS **usually corky**

BUDS ± hairless

J F M A M J J A S O N D

LF-STALK usually L<3 mm, mostly covered by leaf-base

WYCH ELM

LF-STALK usually L>3 mm; at most partly covered by side of leaf-blade

'FIELD ELMS'   ENGLISH ELM

'FIELD'

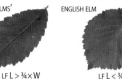

LF L > ¾ × W   LF L < ¾ × W

J F M A M J J A S O N D

18 Urticaceae | **Nettle** family

Form usually widely creeping; some with **stinging hairs**. **Fls** usually inconspicuous; unisexual; 1 whorl of 4 segments; STIGMA densely branched; some species with ♂ + ♀ on separate plants.

● Nettles | **Lvs** margins toothed.

SS Leaf shape similar to *e.g.* **hemp-nettles**, **dead-nettles** and **woundworts** (Lamiaceae *pp. 184–186*) [all have showy flowers and lack stinging hairs].

Ⓟ **Common Nettle** *Urtica dioica*

J F M A M J J A S O N D

H to 200 cm. **Form** erect; ♂ + ♀ **plants separate**. **Fls** loose, green, drooping clusters from base of upper lf-stalks. **Lvs** heart-shaped to triangular; L to 100 mm; deeply toothed; hairy; typically with abundant stinging hairs. **Hab** nutrient-enriched areas – *e.g.* waysides, woods, fens, gardens, pastures.

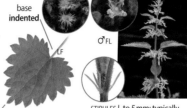

♀FLS
base indented
LF
tooth at tip **longer** than adjacents

STIPULES L to 5 mm; typically pointed; can be toothed and leafy

♂FL

Ⓐ **Small Nettle** *Urtica urens*

J F M A M J J A S O N D

H to 60 cm. **Form** erect; **unisexual**. **Fls** clusters with many ♀ and a few ♂ fls. **Lvs** L to 40 mm; broadly oval; deeply toothed; hairless apart from stinging hairs. **Hab** cultivated ground, gardens, especially on sandy soils.

base **not indented**
LF
tooth at tip **similar size** to adjacents

♂ + ♀FLS

STIPULES always toothed and leafy

● Other nettle family members | **Lvs** margins without teeth.

Ⓟ **Mind-your-own-business** *Soleirolia soleirolii*

J F M A M J J A S O N D

L stem to 20 cm. **Form** prostrate; **mat-forming**; rooting at nodes. **Fls** tiny; **solitary**; pink; at base of leaves. **Lvs** L to 6 mm; round to broadly oval; sparsely hairy. **Hab** shady walls, rocks and paving. **SS** Corsican Mint *Mentha requienii* (N/i) [strong minty scent; LVS opposite].

Ⓟ **Pellitory-of-the-wall** *Parietaria judaica*

J F M A M J J A S O N

H to 50 cm. **Form** prostrate to erect; STEM usually reddish. **Fls** D 2–3 mm; segments green, tinged with red; clusters of ♂, ♀ and bisexual fls at base of leaves. **Lvs** L to 40 mm; oval; pointed; hairy; entire. **Hab** walls, rocks, cliffs, hedgebanks.

INFL **clusters in leaf-axils**

LF

♀FL ♂FL

LVS round to broadly oval; alternate

STEM **reddish**

19 Fagaceae | **Beech + Oak** family `7 spp.` | 8 spp. B&I

Form trees. **Fls** ♂ and ♀ in separate clusters on same plant; ♂ in dangling catkins (long in some species). **Fr** nuts; at least partly enveloped in enlarged scales from around the ♀ flowers.

IDENTIFY BY: ► leaf shape and details ► flowers ► nut details

● **Beech** | **Fls** small; ♂ with 8–16 stamens, in heads on long stalks; ♀ on shorter stalks.

Ⓟ **European Beech** *Fagus sylvatica* `WT`

H to 40m. **Form** deciduous. **Lvs** L to 8 cm; oval; pointed; W ≈ L; margins fringed with silky hairs. **Fr** 1–2 sharply angled nuts in capsule with pointed scales. **Hab** woodland, plantations, parks and gardens especially on calcareous soils. **SS Hornbeam** (*p. 110*) [BUDS appressed to the twig; LF-MARGIN toothed].

J F M A M J J A S O N D

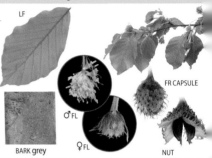

LF

♂ FL

♀ FL

BARK grey

FR CAPSULE

NUT

● **Sweet Chestnut** | **Fls** tiny; ♂ with 10–20 stamens, in long, stiff catkins; ♀ in groups of 3 at base of catkins.

Ⓟ **Sweet Chestnut** *Castanea sativa* `WT`

H to 35m. **Form** deciduous. **Lvs** L 10–25 cm; oblong to narrowly oval; pointed; teeth regular, sharp. **Fr** shining brown roundly angled nut (L to 35 mm) enclosed within softly spiny case until maturity. **Hab** woodland, parks.

J F M A M J J A S O N D

NOTE: not related to, or even similar to, Horse-chestnut (maple family – *p. 129*).

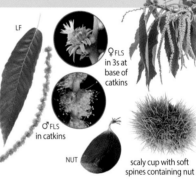

LF

♀ FLS in 3s at base of catkins

♂ FLS in catkins

NUT

scaly cup with soft spines containing nut

● **Evergreen oaks** | **Fls + Fr** as deciduous oaks (opposite).

Ⓟ **Evergreen Oak** *Quercus ilex* `WT`

H to 30m. **Form** evergreen. **Lvs** L 4–7 cm; broadly oval; leathery; dark green above; grey-woolly below; entire or with small sharp teeth. **Fr** small acorn; cup scales appressed; STALK L <20 mm. **Hab** parks, planted windbreaks, coastal scrub; self-seeds. **SS Holly** (*p. 169*) [BARK smooth grey; LVS spines sharper].

J F M A M J J A S O N D

LF holly-like when young

ACORN small

BARK brown; scaled

LF UND grey-woolly

● **Deciduous oaks** | **Fls** ♂ in drooping catkins; ♀ in fewer-flowered, stiff clusters; appearing before the leaves. **Lvs** with rounded to quite pointed lobes. **Fr** acorns in scaly cups.

SS Deciduous oaks are somewhat similar; particularly Pedunculate and Sessile Oaks, which hybridize readily; the hybrids with mixed and/or intermediate characters.

♀ FL
♂ FL

PEDUNCULATE OAK

Ⓟ Pedunculate Oak *Quercus robur* **WT**

H to 37 m. **Form** CROWN broad, spreading. **Lvs** L 8–12 cm; LOBES 3–6 pairs, divided < ½ way to midrib; LF-STALK **very short**, L < 10 mm; LF-BASE **rounded to heart-shaped; often overlapping stalk. Fr** acorn; STALK L 20–90 mm. **Hab** woodland and hedgerows; parks.

J F M A M J J A S O N D

STALKS: LF **very short**; ACORN **long**

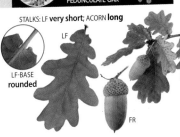

LF

LF-BASE **rounded**

FR

Ⓟ Sessile Oak *Quercus petraea* **WT**

H to 40 m. **Form** CROWN narrower than in Pedunculate Oak. **Lvs** L 7–12 cm; LOBES 3–6 pairs, shallow; LF-STALK L 10–25 mm; LF-BASE usually **wedge-shaped, tapering into stalk. Fr** acorn; STALK absent or short (L < 20 mm). **Hab** woodland and hedgerows, especially on acidic soils.

J F M A M J J A S O N D

STALKS: LF **long**; ACORN **absent/very short**

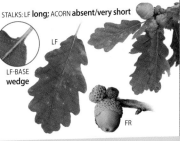

LF

LF-BASE **wedge**

FR

● **Non-native deciduous oaks.**

Ⓟ Turkey Oak *Quercus cerris* **WT**

H to 40 m. **Lvs** L 6–12 cm; LF-STALK L to 25 mm; LOBES 7–8 pairs, divided almost ½ way to midrib; ± pointed at tip. **Fr** acorn; cup scales long, spreading; STALK L to 20 mm. **Hab** parks and gardens.

J F M A M J J A S O N D

BUDS with linear scales

LVS slightly deeper lobes than in Pedunculate or Sessile Oaks

ACORN cup with long, spreading scales

Ⓟ Red Oak *Quercus rubra* **WT**

H to 30 m. **Lvs** L to 25 cm; LOBES **sharply pointed**, divided ½ way to midrib; LF-STALK L to 50 mm. **Fr** acorn; STALK L < 20 mm. **Hab** plantations, parks and gardens. **SS Scarlet Oak** *Q. coccinea* [LF longest lobe longer than width of leaf at its narrowest point].

J F M A M J J A S O N D

RED OAK

ACORN short-stalked

SCARLET OAK

LVS usually turn **bright red in autumn**

20 Myricaceae | **Bog-myrtle** family

1 sp. B&I

Form deciduous shrubs. **Infl** in stiff catkins at base of bracts; ♂ + ♀ typically on separate plants. **Lvs** strongly aromatic.

ⓟ Bog-myrtle *Myrica gale* **WS**

H to 150 cm. **Form** erect, suckering. **Fls** tiny; appear before leaves. **Infl** stiff catkins; ♂ stiff reddish spike, L to 15 mm; ♀ globular, D to 5 mm, with red styles. **Lvs** L to 60 mm; narrow oval; usually grey-green, downy beneath; AROMA **sweetly resinous**.
Fr 2-winged seed. **Hab** bogs, wet moorland and heathland. **SS** when not in flower – **Creeping Willow** (*p.113*) [LVS not aromatic].

21 Juglandaceae | **Walnut** family

1 sp. | 3 spp. B&I

Form deciduous trees. **Fls** appearing before leaves; ♂ in drooping catkins; ♀ with 2 styles and branched stigmas in few-flowered stiff clusters. **Lvs** pinnate.

ⓟ Walnut *Juglans regia* **WT**

H to 25 m. **Form** deciduous tree. **Lvs** L to 45 cm; pinnate; LFLTS usually 3–5 pairs + end lflt; **sweetly aromatic** when rubbed. **Fr** D to 5 cm; rounded, smooth green capsule enclosing a **strongly wrinkled, hard-shelled seed**.
Hab parks and gardens; sometimes self-seeds
SS other walnuts (N/I) [LF + FR details].

the edible 'nut' is not a botanical nut, but actually the inner part of the stony seed

22 Betulaceae | **Birch** family

7 spp. | 14 spp. B&I

Form deciduous trees. **Fls** ♂ in dangling catkins; ♀ in separate catkins or erect small groups. **Lvs** unlobed.

Key to Birch family in flower

♂ Fls	♀ Fls	SPECIES
dangling catkins	tiny bud-like structures	**Hazel** *opposite*
	stalked, cone-like structures	**Alders** *opposite*
	single ± erect spike	**Birches** *p.110*
	dangling catkins	**Hornbeam** *p.110*

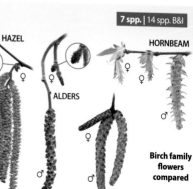

Birch family flowers compared

108

Ⓟ Hazel *Corylus avellana* **WT**

H to 8m. **Form** deciduous shrub or small tree; multi-stemmed. **Fls** appear before leaves; ♂ catkins, L **to 80mm**, yellow when ripe; ♀ tiny bud-like clusters, with **star-like red styles** when ripe. **Lvs** L to 12cm; broadly oval to heart-shaped; pointed; MARGIN sharply double-toothed. **Fr** oval, pointed nut (L to 20mm) surrounded by a **ruff of deeply lobed bracts**. **Hab** woods, hedges, scrub.

J F M A M J J A S O N D

LF-TIP pointed

LF-MARGIN double-toothed

NUT surrounded by deeply lobed bracts

Ⓐ **Alders** | **Form** deciduous trees; all broadly similar. **Infl** unisexual catkins; opening before leaves emerge; ♂ dangling; ♀ cone-like structures; green at first; woody when ripe/open. **Lvs** ± broadly oval. **Fr** cone-like structures; persisting all year.

IDENTIFY BY: ▶ bark colour; then ▶ leaf shape and details ▶ ♀ 'cone' size

---

BARK **dark brown; deeply fissured;** LF-TIP **blunt or notched**

Ⓟ Alder *Alnus glutinosa* **WT**

H to 25m. **Lvs** broadly oval; L to 100mm; TIP **blunt or notched**; VEINS 4–8 pairs. **Fr** up to 8 'cones'; L to 20mm; stalked. **Hab** damp woods; lake and river margins.

J F M A M J J A S O N D

LF-BASE wedge-shaped
LF-VEINS 4–8 pairs

FR 'cones' stalked

In all alders old 'cones' persist and are reminiscent of some conifers, particularly cypresses (*p. 295*)

young   old

BARK **grey; smooth to lightly fissured;** LF-TIP **usually pointed**

Ⓟ Grey Alder *Alnus incana* **WT**

H to 20m. **Lvs** broadly oval; L to 110mm; VEINS **7–15 pairs**. **Fr** up to 8 'cones'; L to 18mm; **short stalks if any**. **Hab** shelterbelts, gardens; may self-seed.

J F M A M J J A S O N D

LF-BASE wedge-shaped

FR 'cones' shortly stalked at most

ITALIAN ALDER

LF-VEINS 7–15 pairs

FR (old)

♀ FLS

Ⓟ Italian Alder *Alnus cordata* **WT**

H to 20m. **Lvs** ± heart-shaped; L to 110mm; TIP usually pointed; VEINS 8–9 pairs. **Fr** up to 3 'cones'; L to 30mm; stalked. **Hab** shelterbelts, gardens; often self-seeds.

J F M A M J J A S O N D

LF ± heart-shaped

♂ FLS

Flowering alder with the previous year's fruit still present

LF marginal teeth regular but shallow

FR 'cones' much larger than other alders

109

Ⓟ **Hornbeam** *Carpinus betulus* **WT**

H to 32 m. **Form** deciduous tree; BARK smooth, **trunk fluted** like 'rippling muscles'. **Fls** ♂+♀ in unisexual catkins. **Lvs** ± narrowly oval; L to 100 mm; pointed; **pleated along veins**; MARGIN toothed. **Fr** in drooping catkins; L to 50 mm; each seed with large 3-lobed bract. **Hab** woods, hedges, especially on clay soils. **SS** Beech (*p. 106*) [winter buds splayed; LF-MARGIN not toothed].

J F M A M J J A S O N D

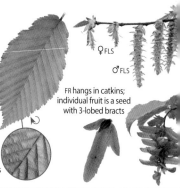

♀FLS

♂FLS

FR hangs in catkins; individual fruit is a seed with 3-lobed bracts

LF pleated along veins

● **Birches** | Deciduous trees; **H** to 28 m. **Infl** unisexual catkins; opening as leaves emerge; ♂ dangling, L to 50 mm; ♀ ± erect. **Lvs** triangular; tip pointed; margins toothed. **Fr** drooping catkins; BRACT 3-lobed; SEED with papery wings.

Ⓟ **Downy Birch** *Betula pubescens* **WT**

Form BARK often brownish; can be white; most with black **horizontal fissures** near trunk base; BRANCHES not drooping; young twigs hairy. **Lvs** L to 50 mm; toothed; pointed tip short. **Fr** BRACTS 3-lobed, **side lobes directed backwards**; SEED wings up to 2× width of seed, not extending forward of stigmas. **Hab** woods, heaths, especially on damper sites, including on peaty soils.

J F M A M J J A S O N D

SS Birches hybridize readily; hybrids show a full range of intermediate characters and consequently many individuals cannot be assigned to species.

♀FLS

BRANCHES not drooping

♂FLS

Both birches look similar in flower

BARK horizontal fissures

LF typically with shorter point

YOUNG TWIGS hairy

BRACT lobes directed downwards

SEED wings ± W seed; not extending beyond stigma

Ⓟ **Silver Birch** *Betula pendula* **WT**

Form BARK almost always white; typically with black **diamond-shaped fissures** near trunk base; BRANCHES typically drooping; young twigs hairless, red, with white dots. **Lvs** L to 60 mm; irregularly double-toothed; pointed **tip drawn out** to up to ¼ of leaf length. **Fr** BRACTS 3-lobed, **side lobes directed ± sideways**; SEED papery wings ≥ 2× width of seed, extending forward of stigmas. **Hab** woods, heaths, especially on acid soils.

J F M A M J J A S O N D

BRANCHES typically drooping

BARK 'diamond' fissures

LF typically with longer point

BRACT lobes 'bird of prey'

SEED wings ≥2× W seed; extending beyond stigma

YOUNG TWIGS hairless; white-dotted

23 Salicaceae | **Willow** family

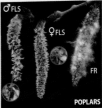

WILLOWS ♂ FLS ♂ FLS ♀ FLS FR

♂ FLS ♀ FLS FR

POPLARS

Form deciduous trees and shrubs.
Infl catkins; ♂ + ♀ on separate plants.
Fr capsule with cottony, plumed
seeds.

11 spp. | 28 spp. B&I

SS The many species, cultivars and
hybrids make identification difficult.

● **Poplars** | **Form** deciduous trees. **Infl** drooping catkins. **Lvs** broad
(often very); long-stalked; lobed in some species.

BARK smooth, grey, with diamond-shaped fissures

Ⓟ Aspen *Populus tremula* **WT**

YOUNG TWIGS brown

H to 31 m. **Infl** ♂ L to
100 mm, pinkish; ♀ L to
125 mm, reddish to green.
Lvs broadly oval; **shallowly
lobed**; L to 80 mm; hairless
(longer and hairy on
suckers); flutter in the
lightest breeze. **Hab** damp
woods and hedges.

♀ FLS

J F M A M J J A S O N D

Ⓟ White Poplar *Populus alba* **WT**

H to 28 m. **Infl** ♂ (very rare)
L to 75 mm, reddish; ♀ L
to 50 mm, yellowish. **Lvs
palmately 3–5 lobed**; L to
100 mm; hairless (longer
and hairy on suckers).
Hab plantations and
shelterbelts; woodland
and scrub; sand dunes.

J F M A M J J A S O N D

ASPEN

GREY POPLAR

WHITE POPLAR

most
trees
are ♀

♀ FLS

YOUNG TWIGS
softly felted
greyish-white

LF palmately
3–5 lobed

SS Grey Poplar *P.* × *canescens* [LVS scarcely
palmately lobed; trees predominantly ♂].

SS Aspen, Grey and White Poplars readily hybridize, producing a range of
trees with intermediate characters.

BARK dark with regular fissures

Ⓟ Hybrid Black-poplar **WT**
Populus deltoides × *nigra* = *P.* × *canadensis*

J F M A M J J A S O N D

H to 40 m. **Form** MAIN
BRANCHES ± upswept. **Infl**
L to 60 mm; ♂ reddish;
♀ yellowish-green. **Lvs**
broadly triangular with
drawn-out tip. **Hab** parks,
gardens, plantations. **SS**
Black-poplar *P. nigra* (N/I)
[BARK more rugged; MAIN
BRANCHES down-arching; LF-STALKS many
with spiral galls]; many hybrids and cultivars.

LVS broadly
triangular with
drawn-out tip

♂ FLS

♀ FLS

● **Willows** | **Form** trees or shrubs. **Infl** catkins; ♂ + ♀ on separate plants; generally erect and, except where stated, expanding before leaves. **Lvs** broadly oval to almost linear; unlobed; short-stalked.

IDENTIFY BY: ▶ leaf and stipule details ▶ catkins ▶ twigs

Trees or large shrubs, LEAVES > (often very much more than) 3× as long as broad

Ⓟ Crack-willow *Salix × fragilis* WT

H to 29 m. **Form** variable; (variants and cultivars). **Catkin** fl at same time as lvs; ♂ L to 50 mm, yellow; ♀ L to 80 mm, green. **Lvs** L to 16 cm, L 5–9 × W. **Hab** banks of streams, rivers and ponds.

J F M A M J J A S O N D

LF-MARGIN **coarsely toothed**

LF **almost hairless when mature**

TWIG tends to **break when snapped**

Ⓟ White Willow *Salix alba* WT

H to 33 m. **Catkin** fl at same time as lvs, ♂ L to 55 mm, yellow; ♀ L to 50 mm, green. **Lvs** L to 10 cm; L 6–8 × W; **silky hairy at first. Hab** marshes; banks of streams, rivers and ponds.

J F M A M J J A S O N D

LF-MARGIN **finely toothed**

LF hairs on underside, at least when mature

TWIG glossy brown; yellow in some

Ⓟ Purple Willow *Salix purpurea* WT

H to 5 m. **Form** upright shrub. **Catkin** L to 40 mm; ♂ with **purple-tipped stamens** and scales. **Lvs** L to 80 mm; L >2 × W; hairless, **purplish-green**, waxy. **Hab** marshes, damp woodland.

J F M A M J J A S O N D

FL **stamens fused into 1**

LVS mostly **opposite** (unlike all other willows)

TWIG yellowish to purple-brown

Ⓟ Osier *Salix viminalis* WT

H to 6 m. **Form** small tree or upright shrub. **Catkin** fl before lvs ♂ L to 18 mm; ♀ L to 30 mm. **Lvs** L to 18 cm; L 7–8 × W; underside silky hairy. **Hab** marshes; often in plantations. **SS Olive Willow** *S. eleagnos* (N/I), often planted [LF-HAIRS matted, not silky].

J F M A M J J A S O N D

LF-MARGIN **downrolled**

LF silky hairs on underside

TWIG dull yellow-brown

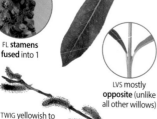

Small trees/large shrubs, LEAVES usually ≤ 3× as long as broad

Three species that can be difficult to identify. They are all shrubby (**Grey** and **Goat** also occur as small trees). **Fls** catkins; ♂ silky-grey, bursting yellow with pollen, ♀ green when ripe. Hybrids, with intermediate features, are frequent.

GREY   GOAT   EARED

J F M A M J J A S O N D   J F M A M J J A S O N D   J F M A M J J A S O N D

♂ CATKIN

♀ CATKIN

LEAVES **margin not (or not very) wavy**

LEAVES **margin wavy**

℗ Grey Willow WT
Salix cinerea ssp. *oleifolia*

H to 10 m. **Catkin** egg-shaped; L to 35 mm. **Lvs** L to 70 mm; L 2–4 × W. **Hab** marshes, damp woodland.

LF-MARGIN **not wavy**; underside many with **rusty hairs** among dense greyish hairs

2ND YR TWIG longitudinal ridges under bark

℗ Goat Willow WT
Salix caprea

H to 6 m. **Catkin** egg-shaped; L to 35 mm. **Lvs** L to 80 mm; L 1–2 × W. **Hab** marshes, open woodland, hedges.

LF-MARGIN **slightly wavy**; underside hairs greyish

2ND YR TWIG **smooth** under bark

℗ Eared Willow WT
Salix aurita

H to 3 m. **Catkin** egg-shaped; L to 20 mm. **Lvs** L to 70 mm; L 1·5–2·5 × W. **Hab** heathland; acid, damp uplands.

STIPULES **large, ear-shaped**, do not fall early

LF-MARGIN **wavy**; underside hairs greyish

2ND YR TWIG dark red; longitudinal ridges under bark

Shrub, H ≤ 150 cm; often creeping

℗ Creeping Willow *Salix repens* WS

J F M A M J J A S O N D

H to 150 cm, **often shorter. Form** spreading to erect shrub. **Fls** L to 25 mm; egg-shaped. **Lvs small** (L to 35 mm); narrowly to broadly oval; UNDERSIDE densely silky-hairy. **Hab** 3 forms: heaths, moors (var. *repens*); fens (var. *fusca*) and dune-slacks (var. *argentea*). **SS** Bog-myrtle (*p. 108*) [LVS aromatic; FLS tiny].

♀ FLS variable; red to green

LF vary in shape (can be narrower); hairs on underside

Var. repens *in wet lowland heath*

♂ FLS

24 Cucurbitaceae | **White Bryony** [Gourd] family

1 sp. | 3 spp. B&I

Form non-woody climber with coiled tendrils. **Fls** ♂ + ♀ on separate plants; PETALS 5, fused at base; ♂ with distinctive stamens and pollen sacs.

℗ **White Bryony** *Bryonia dioica* ✕ **C**

H to 5m. **Fls pale green with darker green veins**; in loose clusters; ♂ D 12–18mm stalked; ♀ D 10–12mm; stalkless, with 3 hairy stigmas. **Lvs** deeply palmately lobed, roughly hairy; **spirally-coiled tendrils** arising from leaf-axils. **Fr** berry; D to 9mm; **red when ripe**. **Hab** hedges, scrub, especially on calcareous soils.

LF **palmate**; coiled tendrils (*below*) distinctive

♂ FL 5 stamens; 2 pairs partly fused

♂

♀ FL 3 hairy stigmas, each divided into 2 parts

FR red when ripe

NOTE: wholly unrelated to Black Bryony (*p. 236*); their only common feature being their climbing habit

25 Oxalidaceae | **Wood-sorrel** family

2 spp. | 10 spp. B&I

Form low-growing; patch-forming; spreading by runners. **Fls** PETALS 5; STAMENS 10. **Lvs** trifoliate; leaflets equal-sized; taste sour.

Wood- and Yellow-sorrels have a distinctive leaf shape

℗ **Wood-sorrel**
Oxalis acetosella

Form bulbous, spreading by runners. **H** to 10cm. **Fls** D 10–20mm; cup-shaped; solitary on long stalks; PET **white with mauve veins**. **Lvs** trifoliate; sparsely hairy; often purplish below; lflts droop below horizontal. **Fr** 4-angled capsule (L to 4mm). **Hab** woods, hedgebanks and other shady habitats. **SS** leaves – **clovers** (*p. 86*) [leaflets held ≥ horizontal].

℗ **Procumbent Yellow-sorrel**
Oxalis corniculata

H usually <50mm; SPREAD to 50cm. **Form** creeping; rooting at nodes. **Fls** D 4–7mm; up to 8 in loose clusters; PET **yellow**. **Lvs** trifoliate; often **strongly tinged purple**. **Fr** capsule (L 8–20mm), usually bent back. **Hab** gardens, paving, bare sandy ground. **SS** several other scarcer, **non-native yellow-sorrels** (N/i) [INFL details].

FL white with mauve veins

26 Celastraceae | Spindle family   `1 sp.` | 3 spp. B&I

Form deciduous shrubs. **Fls** SEPALS, PETALS + STAMENS 4 or 5; with a nectar-secreting disc at the flower base. **Fr** orange seeds within a pink fleshy coat.

P Spindle *Euonymus europaeus* `WS`

J F M A M J J A S O N D

H to 6m. **Form** TWIGS green, 4-angled.
Fls D 8–10mm; PET 4, **greenish-white**, alternating with 4 short stamens; up to 10 in loose stalked clusters. **Lvs** L to 8cm; opposite; broadly oval; pointed. **Fr** D to 15mm; lobed; outer fleshy coat ripens to coral-pink.
Hab hedges and scrub on base-rich soils.

LF

FLS distinctive

FR **pink** when ripe

TWIGS green; 4-angled

The pink fruits are highly distinctive, even at distance.

27 Linaceae | Flax family   `2 spp.` | 5 spp. B&I

Form erect; STEMS wiry. **Fls** typically blue or white in open, branched clusters; SEPALS, PETALS + STAMENS 5. **Lvs** lack stalks and stipules. **Fr** rounded capsule with 10 seeds.

A Flax *Linum usitatissimum*

J F M A M J J A S O N D

H to 80 cm. **Fls** D 20–25mm; **blue**.
Lvs narrowly **oval to linear;** W 1·5–3·0mm; alternate. **Fr** capsule; L 6–9mm. **Hab** cultivated; arable field margins and fallow areas. **SS** two native blue flaxes (N/I) [FL size + SEPAL details; much scarcer; permanent grassland].

FR

LVS arranged **alternately**

LVS 3-veined

A Fairy Flax *Linum catharticum*

J F M A M J J A S O N D

H to 25 cm. **Fls** D 4–6mm; **white**; INFL widely forked. **Lvs** L 2–3mm; narrowly oval; 1-veined; opposite. **Fr** capsule; L 2–3mm. **Hab** calcareous and sandy grassland, moorland.

LVS arranged **oppositely**

28 Euphorbiaceae | Spurge family   **7 spp. | 20 spp. B&I**

Disparate genera with few common characteristics other than the presence of acrid sap (watery or milky). **Spurges** have a unique condensed inflorescence structure (cyathium); **mercuries** (*opposite*) have inconspicuous 3-segmented flowers.

CYATHIUM

GLANDS

● **Spurges** | **Form** erect; **STEM** with milky sap. **Infl** compound umbels. **Fls** cyathium comprising a ♀ and numerous 1-stamened ♂'s in a group; surrounded by glands and enclosed within bracts.

SS Spurges are easily recognizable as a group; once the group's features are known only a few need detailed examination for confident identification.

SPURGES   **PETTY SPURGE**

IDENTIFY BY: ▶ leaf shape/toothing ▶ bracts ▶ ray quantity ▶ 'flower' + fruit details

Small, flaccid annuals

Ⓐ Petty Spurge *Euphorbia peplus*

J F M A M J J A S O N D

H to 30 cm. **Infl** 3 rays. **Fls** GLANDS crescent-shaped, extended into long, slender horns. **Lvs** L to 20 mm; broadly oval, stalked, untoothed. **Fr** capsule (L 2 mm); smooth with 2 raised lines on each valve. **Hab** gardens, cultivated ground, brownfield.

UMBEL 3-rayed

GLANDS 'horned'

LF (+ BRACTS) untoothed

Ⓐ Sun Spurge *Euphorbia helioscopia*

J F M A M J J A S O N D

H to 50 cm. **Infl** 5 rays, flat-topped each with a large, round, toothed bract. **Fls** GLANDS rounded on outer edge. **Lvs** L to 30 mm; broadly oval; MARGIN finely toothed. **Fr** capsule (L 3–5 mm); smooth. **Hab** arable, gardens, brownfield.

UMBEL 5-rayed

GLANDS outer edge rounded

LF (+ BRACTS) finely toothed

Ⓐ Dwarf Spurge *Euphorbia exigua*

J F M A M J J A S O N D

H to 20 cm. **Infl** 3 rays. **Fls** GLANDS crescent-shaped, extended into long, slender horns; BRACTS triangular. **Lvs** L to 30 mm; grey-green; linear, unstalked. **Fr** smooth capsule (L 2 mm) with 1 raised line on each valve. **Hab** gardens, cultivated ground, brownfield.

BRACTS triangular

GLANDS 'horned'

LVS linear

Larger, robust perennials; STEM almost woody at base

🅟 Wood Spurge *Euphorbia amygdaloides*

H to 90 cm. **Form** erect, patch-forming. **Infl** 5–10 rays. **Fls** GLANDS crescent-shaped, **horn tips converging**; BRACTS yellowish, fused below flower. **Lvs** L to 80 mm; narrowly oval; softly hairy. **Fr** capsule (L to 4 mm); minutely rough. **Hab** woods, hedges. **SS** ssp. *robbiae* (N/I), a frequent garden escape [LVS leathery, hairless, shiny].

GLANDS 'horns' converging

STEM-LVS narrow; rosette-like arrangement

🅟 Caper Spurge *Euphorbia lathyris*

H to 200 cm. **Form** erect. **Infl** 2–6 rays. **Fls** GLANDS crescent-shaped, with **blunt horns**; BRACTS broadly triangular. **Lvs** L to 150 mm; narrowly triangular; grey-green; opposite; each pair at right-angles to the pair above and below. **Fr** capsule (L to 20 mm); smooth; **3-angled**. **Hab** shady hedges, wood edges, brownfield, gardens.

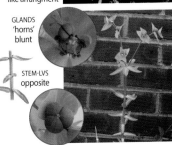

GLANDS 'horns' blunt

STEM-LVS opposite

FR ✖ do not mistake for the edible buds of capers *Capparis* spp.

🔵 **Mercuries** | **Form** erect; STEM with watery sap. **Fls** inconspicuous (D 4–5 mm); green; 3-segmented: ♂ + ♀ on separate plants – ♂ in spikes, ♀ in smaller clusters.

SS Mercuries are easily recognizable as a group and are separated from one another by the features shown below.

Mercury ID	**🅟 Dog's Mercury** *Mercurialis perennis*	**🅐 Annual Mercury** *Mercurialis annua*
Form	H to 40 cm; **short hairs, unbranched**	H to 50 cm; **hairless, branched**
Flowers ♂	in spikes	
Flowers ♀	in clusters; **stalked**	in clusters; **stalkless**
Leaves	L to 80 mm; hairy, short-toothed	L to 50 mm; hairless, toothed
Fruit	W 6–8 mm; hairy	W 3–4 mm; shiny
Habitat	woods, hedgerows, shady places	gardens and cultivated ground

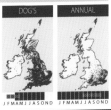

DOG'S | ANNUAL

ANNUAL MERCURY

♂ ♀

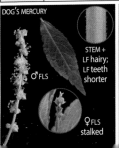

DOG'S MERCURY

STEM + LF hairy; LF teeth shorter

♂ FLS

♀ FLS stalked

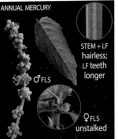

ANNUAL MERCURY

STEM + LF hairless; LF teeth longer

♂ FLS

♀ FLS unstalked

117

29 Violaceae | **Violet** family

8 spp. | 15 spp. B&I

Form low-growing. **Fls** solitary; with a distinctive 'face-like' appearance; PETALS 5, the lowest with an elongated spur; SEPALS with backward-pointing appendages. **Lvs** ± heart-shaped to round kidney-shaped; STIPULES present.

● **Dog-violets** | **Fls** ± violet; STYLE tip not swollen. **Lvs** from aerial stem; STIPULES undivided.

SS A difficult group to identify. Some characters are variable and overlap between species. They also hybridize freely producing plants with intermediate characters. The table below covers the features that need to be assessed in combination to confirm a 'good', rather than hybrid, individual. It is often prudent to look at a range of plants in any population.

Dog-violet ID		**Ⓟ Early Dog-violet** *Viola reichenbachiana*	**Ⓟ Common Dog-violet** *Viola riviniana*	**Ⓟ Heath Dog-violet** *Viola canina*
Form		H to 20 cm; basal leaf-rosette and lateral flower-stalk; somewhat **hairy**	H to 20 cm; tufted; **slightly hairy** at most	H to 40 cm; spreading to erect; ± hairless
Flowers	SHAPE	H 12–18mm; H ≥ W	H 14–25mm; usually H ± W	H 10–18mm; H ≥ W
	COLOUR	violet; centre usually darker	deep bluish-violet	bright blue or violet
	SPUR	**dark; slender; straight; not notched**	**whitish or pale; stout; many curved and notched**	spur whitish or greenish-yellow; can be notched
	SEPALS	appendages **very short**	appendages square	appendages conspicuous
Leaves	FORM	broadly heart-shaped; L to 80mm; L ± W	broadly heart-shaped; L to 80mm; L ≥ W	triangular to heart-shaped; L to 40mm;
	STIPULES	**fringed**	narrow; with hair-like teeth	untoothed or toothed towards tip
Habitat		open woodland, woodland edges, scrub, shady hedgebanks; mainly calcareous	deciduous woodland, heaths, old pastures, chalk downs	open woods, grassy heaths, sandy commons, fens, dunes; often acid

EARLY

SPUR **dark**

LF

STIPULE **fringed**

COMMON

SPUR whitish

LF

STIPULE **hair-like teeth**

HEATH

SPUR whitish or greenish-yellow

STIPULE ± untoothed

LF

● **Violets** | **Fls** ± violet; STYLE tip not swollen. **Lvs** all basal; STIPULES undivided.

ⓟ Sweet Violet *Viola odorata*

H to 20 cm. **Form** tufted; with surface runners. **Fls** H ± 15 mm; deep violet to white; AROMA **sweetly scented**. **Lvs** L to 80 mm; broadly heart-shaped; tip **rounded**. **Fr** capsule; hairy. **Hab** open woodland and edges, scrub, shady hedgebanks; mainly calcareous. **SS** cultivars [vary in form + colour].

J F M A M J J A S O N D

our only fragrant violet

LF-TIP rounded

LF-STALK **hairs angled downwards**

ⓟ Hairy Violet *Viola hirta*

H to 15 cm. **Form** tufted; with short underground runners. **Fls** H ± 15 mm; pale blue-violet. **Lvs** L to 80 mm; narrowly heart-shaped; tip pointed. **Fr** capsule; hairy. **Hab** open woodland, woodland edges, scrub, shady hedgebanks; mainly calcareous.

J F M A M J J A S O N D

LF-TIP pointed

LF-STALK **hairs spreading**

ⓟ Marsh Violet *Viola palustris*

H to 15 cm. **Form** wide-spreading by long underground runners. **Fls** H 10–15 mm; stemless; lilac, with darker veins. **Lvs** L to 40 mm; rounded kidney-shaped. **Fr** capsule; hairless. **Hab** bogs, fens, marshes, wet heaths, wet woods.

J F M A M J J A S O N D

LF ± round

FL contrasting, conspicuous dark veins

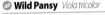

● **Pansies** | **Fls** yellow to purple; STYLE tip thickened. **Lvs** stipules leaf-like. **Lvs** stalks hairy.

Ⓐ Field Pansy *Viola arvensis*

H to 40 cm. **Form** sprawling to erect; well-branched. **Fls** H 8–20 mm; **yellow or cream**; can have violet blotches; SPUR L 2–4 mm. **Lvs** L to 30 mm; narrowly oval; bluntly toothed. **Hab** cultivated ground, brownfield. **SS Dwarf Pansy** *V. kitaibeliana* (N/I) [tiny; FL H <8mm, SPUR L 1–2mm].

J F M A M J J A S O N D

SEP L > PET L

STIPULE middle lobe oval

ⓐⓟ Wild Pansy *Viola tricolor*

H to 40 cm. **Form** erect; well-branched. **Fls** H 10–35 mm; **variable** (purple-violet, blotched with yellow, or all yellow). **Lvs** L to 30 mm; narrowly oval; bluntly toothed. **Hab** cultivated ground, brownfield; some forms/ssp. in specialised habitats such as sand dunes.

J F M A M J J A S O N D

SEP L < PET L

L

STIPULE middle lobe usually narrow

30 Hypericaceae | St John's-wort family `7 spp.` | 15 spp. B&I

A clearly defined, readily recognizable family. Fls PETALS 5, yellow; STAMENS numerous, partially fused into 3 or 5 bundles; STYLES 3 (in the spp. covered here). In most species, LEAVES, SEPALS and/or PETAL-MARGINS have stalked black glands and/or black or translucent dots (see *below*).

SS Other rarer **St John's-worts** [various characters as in *identify by* below]; other 5-petalled yellow flowers (see **Gallery 11** (*p. 55*)).

IDENTIFY BY: ► leaf (translucent dots) ► leaf (at stem)
► stem details ► gland details

LEAF translucent dots (T-DOTS), appear as perforations, especially if the leaf is backlit (hand-lens may be needed)

Plants not woody; STAMENS **in 3 bundles;** FRUIT **dry capsule**

Plants hairless; STEM **round**

Ⓟ Slender St John's-wort
Hypericum pulchrum

J F M A M J J A S O N D

H to 60 cm. **Form** erect. **Fls** D 12–18 mm; bright yellow; **tinged reddish underneath;** PET + SEP with **stalked black glands. Lvs** L to 10 mm; triangular; almost clasping stem; T-DOTS present. **Hab** heathlands and dry, open woodland on acid soils.

LF translucent dots

LVS ± **clasping stem**

FL red-tinged underside (most obvious in bud)

Plants hairless; STEM **with 2 raised ridges**

Ⓟ Perforate St John's-wort
Hypericum perforatum

J F M A M J J A S O N D

H to 80 cm. **Form** erect. **Fls** D 15–25 mm; bright yellow; PET + SEP with a few black dots. **Lvs** L to 20 mm; T-DOTS **numerous. Hab** dry grassland, open woodland.

FLS a few black dots

LVS not clasping stem

Ⓟ Trailing St John's-wort
Hypericum humifusum

J F M A M J J A S O N D

H to 20 cm. **Form** usually **trailing.** D 8–12 mm; bright yellow; PET + SEP with a few dots and/or stalked black glands. **Lvs** L to 10 mm; ± narrowly oval; T- DOTS present. **Hab** open woodland, dry heaths.

FLS ≤12 mm; a few black dots

SEPALS **unequal,** 3 longer and broader than the other 2

LVS ± narrowly oval

Plants hairless; STEM square (unwinged or winged)

P Imperforate St John's-wort
Hypericum maculatum

H to 60 cm. **Form** erect.
Fls D 15–25 mm; bright
yellow; PET + SEP with black
dots. STEM **unwinged**.
Lvs L to 20 mm; with black
dots; T-DOTS **absent**.
Hab grassland, usually on
the damp side.

J F M A M J J A S O N D

LF black dots

FLS black dots

STEM
unwinged

P Square-stalked St John's-wort
Hypericum tetrapterum

H to 60 cm. **Form** erect.
Fls D 9–13 mm; pale yellow;
PET + SEP may have black
dots; PETAL L <2x SEPAL L.
STEM wings > 0·25 mm.
Lvs L to 20 mm; with black
dots; T-DOTS present. **Hab**
damp meadows, marshes.

J F M A M J J A S O N D

LF black +
translucent
dots

FLS may have
black dots

STEM **winged**

Plants hairy; STEM round

P Hairy St John's-wort *Hypericum hirsutum*

H to 100 cm. **Form** erect.
Fls D 15–22 mm; pale
yellow; PET + SEP with
stalked black glands. **Lvs**
L to 50 mm; **hairy**; T-DOTS
present. **Hab** grassland and
open woodland, especially
on calcareous soils.

J F M A M J J A S O N D

LF hairy +
translucent dots

FLS stalked
black glands

STEM **hairy**

Woody-stemmed shrub; STAMENS in 5 bundles; FRUIT fleshy

P Tutsan *Hypericum androsaemum* WS

H to 80 cm. **Form** branched,
semi-evergreen **shrub**;
STEM woody, 2-ridged.
Fls D 15–25 mm; pale yellow.
Lvs L to 100 mm; with black
dots; T-DOTS absent.
Fr fleshy berry, red, ripening
black. **Hab** damp woodland
and shady hedgebanks.

J F M A M J J A S O N D

SS several similar garden species and
hybrids may be found wild [FLS D > 25 mm].

STAMENS in
5 bundles

FR red,
ripening
black

121

31 Geraniaceae | **Crane's-bill** family | **10 spp. | 22 spp. B&I**

A well-defined family. **Fls** showy; PETALS 5; SYMMETRY radial or almost so; usually pink or blue. **Fr** 5-seeded; each with a long beak that coils when ripe to aid dispersal. **Lvs** palmate, palmately lobed or pinnate.

IDENTIFY BY: ► basal leaf shape; then by
► type/extent of lobe division

FR distinctive

● Crane's-bills | **Fls** PETALS notched at tip in most spp.; STAMENS 10, in 2 whorls – not always present at the same time; 1 whorl can lack anthers. **Lvs** palmate or palmately lobed.

LEAVES **deeply divided;** OUTLINE **palmate to rounded**

LEAVES **typically D 100–200 mm; with diamond-shaped spiky lobes**

ⓟ Meadow Crane's-bill *Geranium pratense*

H to 80 cm. **Form** erect; tufted; hairy. **Fls** D 25–45 mm; **bright violet-blue. Fr** STALKS spreading or bent back when immature. **Hab** mainly **lowland** meadows and verges.

J F M A M J J A S O N D

LF-LOBES deeply cut and well-separated; **basal lobes close to or overlapping stalk**

ⓟ Wood Crane's-bill *Geranium sylvaticum*

H to 70 cm. **Form** erect; tufted; hairy. **Fls** D 22–33 mm; **pinkish-mauve with white centre. Fr** STALKS upright when immature. **Hab** mainly **upland** meadows and verges.

J F M A M J J A S O N D

LF-LOBES moderately deeply cut and separated; **basal lobes forming distinct angle with stalk**

LEAVES **typically D 20–70 mm; narrow linear segments with pinnate lobes**

ⓟ Bloody Crane's-bill *Geranium sanguineum*

H to 40 cm. **Form** spreading; branched, hairy. **Fls** D 25–30 mm; **bright purplish-crimson; usually solitary. Lvs** D to 60 mm. **Hab** grassland, woodland, sandy soils, limestone.

J F M A M J J A S O N D

FL-STALKS long, with **pair of reddish bracts** halfway

LF

Ⓐ Cut-leaved Crane's-bill *Geranium dissectum*

H to 60 cm. **Form** straggly; hairy. **Fls** D 8–10 mm; pink. **Lvs** D to 70 mm. **Hab** cultivated and brownfield land.

J F M A M J J A S O N D

FL-STALKS short (L <15 mm); SEP **densely hairy with bristle at tip**

LF

LEAVES divided to ± half depth; OUTLINE palmate to rounded

LEAVES smaller; typically D 10–50mm

SS These two spp. of dry grasslands, lawns and cultivated ground are notoriously confusing. It is best to examine hairs on FL-STALKS and LF-STALKS. STAMEN count is not straightforward as the two whorls ripen sequentially and can appear to be a single whorl of 5 depending on development stage (5 may not yet be out, or may be over).

🔵 Dove's-foot Crane's-bill
Geranium molle

H to 40 cm. **Form** short; densely hairy. **Fls** D 6–10mm; pinkish-purple, can be very pale; PET **deeply notched**.

LF

STAMENS **all 10 fertile**

FL-STALKS with glandular and non-glandular short hairs and non-glandular long hairs

FR* **hairless**

🔵 Small-flowered Crane's-bill
Geranium pusillum

D

H to 40 cm. **Form** short, densely hairy. **Fls** D 4–6mm; pale lilac; PET shallowly notched.

STAMENS **inner 5 fertile; outer 5 infertile (missing anthers)**

LF

FL-STALKS with short, usually **non-glandular hairs only**

FR* **hairy**

* sepals pulled back to reveal fruit

LEAVES larger; typically D 50–80mm

🔵 Hedgerow Crane's-bill
Geranium pyrenaicum

H to 60 cm. **Form** erect, hairy (glandular and non-glandular). **Fls** D 14–18mm; purplish pink, in pairs. **Hab** meadows, hedgerows, field margins, brownfield.

J F M A M J J A S O N D

PET **deeply notched**

LF divided into 5–7 wedge-shaped lobes; LF-TEETH (along tip) blunt

🔵 Shining Crane's-bill
Geranium lucidum

D

H to 40 cm. **Form** erect, very sparsely hairy. **Fls** D 10–14mm; pink; usually in pairs. **Hab** shady places, walls, hedgebanks, calcareous soils.

J F M A M J J A S O N D

PET **rounded**; oval; distinctly narrowing towards base

LF **shiny green**; divided into 5 lobes; LF-TEETH rounded

123

LEAVES **divided;** OUTLINE ± **triangular**

ⓐⒷ **Herb-Robert** *Geranium robertianum*

H to 40 cm. **Form** hairy; may be **flushed red;** AROMA unpleasant when crushed. **Fls** D 14–18 mm; bright pink; can be white. **Lvs** trifoliate; LFLTS deeply lobed or pinnate. **Hab** shady places, woods, tracksides, limestone pavement, coastal shingle.

J F M A M J J A S O N D

PET rounded; distinctly narrowing towards base

LF

LEAVES **pinnate**

● **Stork's-bills** | **Fls** PETALS rounded; STAMENS 10, in 2 whorls, 1 whorl lacks anthers. **Lvs** pinnate.

Ⓐ **Common Stork's-bill** *Erodium cicutarium*

H to 60 cm; usually much less. **Form** prostrate or straggly; with basal rosette; hairy; very variable. **Fls** D 7–18 mm; pink, in loose umbel of up to 12. **Lvs** pinnate or 2-pinnately lobed; LFLTS divided almost to midrib. **Fr** hairy. **Hab** dry grassland, disturbed ground, especially coastal. **SS** Musk Stork's-bill *E. moschatum* (N/I) [LFLTS divided < half-way to midrib].

J F M A M J J A S O N D

PET rounded at tip; STAMENS 5 fertile, 5 infertile (reduced)

LF

FR long-beaked; hairy

32 Lythraceae | **Purple-loosestrife** family

2 spp. | 4 spp. B&I

Fls PETALS 6; SEPALS 6 plus 6 outer sepals (epicalyx) attached on the rim of a cup that surrounds the ovary – a combination of features diagnostic of this family.

Ⓐ **Water-purslane** *Lythrum portula*

L to 25 cm. **Form** trailing; rooting at nodes. **Fls** D ± 2 mm; singly in leaf-axils; purple; PET 0–6; STAMENS usually 6. **Lvs** rounded with pronounced stalk ('**spoon-shaped**'); opposite. **Hab** in or on mud beside shallow water; damp trackways.

J F M A M J J A S O N D

FL inconspicuous at base of leaves

Ⓟ **Purple-loosestrife** *Lythrum salicaria*

H to 120 cm. **Form** erect. **Fls** D 10–15 mm; **purple;** in whorls; STAMENS 6; STYLES 3 forms (long, medium and short). **Lvs** very narrowly triangular; opposite or whorled. **Hab** marshes, fens, river and lake margins.

J F M A M J J A S O N D

FL **6-petalled**

INFL distinctive spike of flowers in whorls

33 Onagraceae | **Willowherb** family

A distinctive family: all clearly with OVARY inferior. **Fls** typically 4-parted; STIGMA club-shaped or 4-lobed; OVARY elongated in many. **Fr** SEEDS with plumed hairs in most species.

Willowherb groups and species identification

FORM **various**; FLS **not pink**; FRUITS **various**; SEEDS **lack a hairy plume**

EVENING-PRIMROSES; Fuchsia; Enchanter's-nightshade
below

FORM **'typical' willowherb**; FLS **pink**; SEPALS, PETALS + STAMENS **4**; FRUIT **a long ± cylindrical pod**; SEEDS **with a hairy plume**

FL SYMMETRY **radial**; LVS **some/all opposite**

'TYPICAL' WILLOWHERBS
p. 126

FL SYMMETRY **bilateral**; LVS **all alternate**

Rosebay Willowherb
p. 128

P Fuchsia WS
Fuchsia magellanica

H to 3 m. **Form** deciduous shrub. **Fls** L 30–40 mm; **hanging. Lvs** opposite or whorled; broadly oval; pointed; toothed. **Fr** black berry; D to 20 mm. **SS** several other garden species and cultivars may be found wild (N/I) [FL size and colour].

P Enchanter's-nightshade
Circaea lutetiana

H to 70 cm; usually much less. **Form** erect. **Fls** H to 5 mm; white or pale pink; well-spaced in elongated infl. **Lvs** L to 100 mm; opposite; rounded at base; obscurely toothed; softly hairy. **Fr** with hooked bristles. **Hab** woods, hedges, gardens.

aB Large-flowered Evening-primrose *Oenothera glazioviana*

H to 180 cm. **Form** erect; hairy. **Fls** 30–50 mm (L<W); **yellow**; open; fragrant in evening. **Lvs** oval; pointed; **wavy margins. Fr** long ± cylindrical pod; seeds lack hairy plume. **Hab** sand dunes, brownfield. **SS** several other **evening-primroses** (N/I) [identification difficult and not wholly resolved].

FLS distinctive

FLS distinctive

PETALS **purple**; SEPALS + FL-TUBE **red**; STIGMA club-shaped, protruding from fl

PETALS **2, deeply divided**; SEPALS **2, fall when fl opens**; STIGMA 2-lobed

SEPALS **red-striped**

STYLE L > STAMEN L

STEM + LF-HAIRS **long with red, bulbous base**

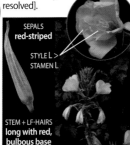

● **'Typical' willowherbs** | **Fls** symmetry radial. **Lvs** some/all opposite.

STIGMA **club-shaped, unlobed; small, creeping plants, flowers solitary in leaf-axils**

ℙ New Zealand Willowherb *Epilobium brunnescens*

L to 20 cm. **Form prostrate,** creeping or trailing; rooting at the nodes; **mat-forming. Fls** D 4–6 mm; white or pale pink. **Lvs** L to 10 mm; broadly oval; STALK L 0·5–3·0 mm. **Fr** L 25–45 mm; on stalk (L to 75 mm). **Hab** upland moist stony areas, stream banks, quarries.

FL small; white or pale pink

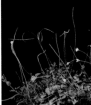

J F M A M J J A S O N D

LF small; oval

SS The four widespread willowherbs (*below*) with a club-shaped stigma present considerable problems as a suite of somewhat subjective and overlapping characters is needed for confident identification. In addition, hybridization within this group is frequent.

IDENTIFY BY: ► leaf arrangement + leaf-margin + stem details

SHORT-FRUITED SQ-STALKED AMERICAN MARSH

FR distinctive

SEEDS with hairy plume

J F M A M J J A S O N D   J F M A M J J A S O N D   J F M A M J J A S O N D   J F M A M J J A S O N D

NOTE: As for all willowherbs, the distinctive slender fruiting pods vary in size and there is much overlap between species. Consequently they are of little help in identification.

SQUARE-STALKED WILLOWHERB

STIGMA **club-shaped, unlobed;** FORM **erect with terminal clusters of flowers and leaf-like bracts**

UPPER + LOWER LEAVES **opposite**

ℙ Short-fruited Willowherb	**ℙ Square-stalked Willowherb**
Epilobium obscurum	*Epilobium tetragonum*

H to 60 cm. **Form** spreading by late-season runners. **Fls** D 7–9 mm; **Lvs** L to 70 mm; narrowly triangular; UPPER alternate; LOWER opposite. **Fr** L 40–60 mm. **Hab** damp woodland, streamsides, marshes.

H to 60 cm. **Form** spreading by late-season runners. **Fls** D 6–8 mm. **Lvs** L to 75 mm; very narrowly oval (can be ± parallel-sided); pointed. **Fr** L 65–100 mm. **Hab** damp woodland, streamsides, cultivated ground.

a) FL ± deep rose-pink; b) SEP-TUBE **glandular hairs;** c) STEM 4 raised lines

LF ± stalkless; **rounded base** runs down into lines on stem; LF-MARGIN a few irregular teeth

LF-BASE **rounded**

a) FL ± pale lilac; b) SEP-TUBE appressed hairs; c) STEM 4 raised lines

LF stalkless; **tapering base** runs down into lines on stem; LF-MARGIN strongly and irregularly toothed

LF-BASE **tapering**

STIGMA **4-lobed**

ⓟ **Broad-leaved W'herb**
Epilobium montanum

H to 60 cm.
Form erect;
almost hairless.
Fls D 6–9 mm;
rose-pink.
Lvs L to 70 mm;
narrowly oval;
with rounded
base; **stalked**
(L 2–6 mm). **Fr** L 40–80 mm.
Hab woods, hedges,
cultivated ground,
mountain screes.

ⓟ **Hoary Willowherb**
Epilobium parviflorum

H to 60 cm.
Form erect;
densely
hairy; with
leafy surface
runners.
Fls D 6–9 mm;
pale pink.
Lvs L to 70 mm;
narrow; pointed; stalkless; **not
clasping stem. Fr** L 35–65 mm.
Hab fens, marshes, river banks.

ⓟ **Great Willowherb**
Epilobium hirsutum

H to 180 cm.
Form erect;
densely hairy.
Fls D 12–25 mm;
deep pink. **Lvs**
L to 120 mm;
narrowly
oval; pointed;
stalkless;
clasping stem. **Fr** L 50–80 mm.
Hab damp woodland, streamsides, marshes.

FL small; pale pink

FL large; deep pink

STEM ±
hairless

LF
stalked

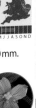

PLANT
densely
hairy

LF
unstalked

UPPER LEAVES **alternate**; LOWER LEAVES **opposite**

ⓟ **American Willowherb** *Epilobium ciliatum*

H to 90 cm. **Form** well-branched. **Fls** D
4–6 mm PET cleft to ½ depth. **Lvs** L to 100 mm;
narrowly triangular. **Fr** L 40–65 mm.
Hab brownfield, gardens, streamsides, damp
woodland rides.

ⓟ **Marsh Willowherb** *Epilobium palustre*

H to 60 cm. **Form** spreading by underground
runners with fleshy buds at the tip. **Fls** D 4–
6 mm. **Lvs** L to 70 mm; narrowly oval; pointed;
UPPER + LOWER opposite. **Fr** L 50–80 mm. **Hab**
marshes, fens, avoids calcareous sites.

a) | b) | c)

a) FL pale pink; b) **FR curled glandular
hairs**; STEM upper (c) with 4 raised lines;
basal part with 2 lines

LF stalkless; base
rounded to **short stalk**;
LF-MARGIN with **many
small, irregular teeth**

a) | b) | c)

a) FL pale pink; b) SEP-TUBE almost
hairless; c) STEM smooth; can have
curled hairs in 2 rows

LF ± stalkless; **wedge-
shaped base not running
down stem;** LF-MARGIN
virtually untoothed

127

● **Rosebay Willowherb** | **Fls** symmetry bilateral. **Lvs** all alternate.

℗ Rosebay Willowherb
Chamaenerion angustifolium

H to 150 cm. **Form** erect. **Fls** D 20–30 mm; rose-purple; in **tapering spike**. **Lvs** L to 15 cm; narrowly oval; tapered at both ends. **Fr** L 25–80 mm producing abundant plumed seeds. **Hab** woodland clearings, brownfield, scree slopes.

J F M A M J J A S O N D

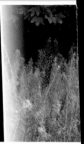

FL

PETS upper 2 **slightly smaller than lower 2**; STAMENS + STYLE curved downwards; STIGMA 4-lobed

34 Sapindaceae | **Maple** family   **4 spp.** | 10 spp. B&I

Form deciduous trees; several divergent groups. **Fls** in branched clusters at the end of shoots. **Lvs** palmate or palmately lobed in those covered here.

LVS **palmately lobed**; INFL **loose clusters**; FLS **yellow-green**; FR **winged seed pairs**

SS Other cultivated *Acer* spp., sometimes found growing wild, are all somewhat similar [separated by various characters as per *identify by* below].

IDENTIFY BY: ► leaf shape ► inflorescence form ► seed wing details

℗ Field Maple *Acer campestre* WT

H to 20 m. **Bark** fissured; can be corky. **Infl** as leaves emerge. **Lvs** L to 7 cm; 3–5 lobes with wide teeth; LOBES + TEETH rounded. **Fr** wings diverge at almost 180°. **Hab** woods, parks, gardens.

J F M A M J J A S O N D

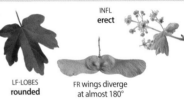

INFL **erect**

LF-LOBES **rounded**

FR wings diverge at almost 180°

℗ Norway Maple *Acer pseudoplatanus* WT

H to 20 m. **Bark** fissured; not flaking. **Infl before leaves emerge**. **Lvs** L to 15 cm; 5–7 lobes with a few teeth; LOBES + TEETH with long, sharp points. **Fr** wings diverge at almost 180°. **Hab** woods, parks, gardens.

J F M A M J J A S O N D

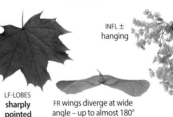

INFL ± hanging

LF-LOBES **sharply pointed**

FR wings diverge at wide angle – up to almost 180°

℗ Sycamore *Acer platanoides* WT

H to 40 m. **Bark** smooth; flaking with age. **Infl** as leaves emerge. **Lvs** L to 15 cm; 5 toothed lobes; tips pointed. **Fr** wings diverge at an acute angle. **Hab** woods, parks, gardens. **SS Guelder-rose** (*p. 217*) [shrubby; FLS showy white; FR red berries].

J F M A M J J A S O N D

LF-LOBES **pointed**

INFL hanging

FR wings diverge at **acute angle**

LVS **palmate**; INFL **dense conical cluster**; FLS **bilateral**; FR **rounded capsules, spiny in most**

℗ Horse-chestnut *Aesculus hippocastanum* WT

FL

H to 40 m. Fls PET white with yellow to pink blotch at base; STAMENS 5–7, long. **Lvs** palmate; LFLTS 5–7; L to 25 cm; stalkless. **Fr** spiked capsule; D 50–80 mm; containing 1–3 nuts (conkers). **Hab** parks, large gardens, scrub. **SS** other planted horse-chestnuts [FR details].

J F M A M J J A S O N D

LF

FR CAPSULE

NUT (conker)

NOTE: unrelated to Sweet Chestnut (p. 106); the apparent similarity of fruits within a spiny capsule is superficial

---

35 Malvaceae | **Mallow + Lime** family   **5 spp.** | 19 spp. B&I

A family with very divergent subfamilies that have only been classified as related as a result of molecular studies. However, perhaps nature knows best: the hemipteran Firebug *Pyrrhocoris apterus*, seemingly colonizing Britain, feeds on the seeds of both lime trees and mallows, and little else.

● **Limes** | **Form** deciduous trees. **Fls** with distinctive, long papery bracts.

℗ Lime WT
Tilia cordata × platyphyllos = T. × europaea

H to 45 m. Fls in drooping clusters of **4–10**, hidden among leaves; BRACTS L 80–110 mm. **Lvs** L 60–150 mm; UPPERSIDE shiny or dull green. **Twigs** reddish. **Fr** almost globular; D 7–8 mm. **Hab** woods (rare as a native), parks, churchyards, gardens. **SS** Large-leaved Lime *T. platyphyllos* (N/i) [FL clusters of 3–5].

J F M A M J J A S O N D

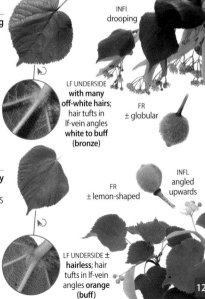

INFL drooping

LF UNDERSIDE **with many off-white hairs**; hair tufts in lf-vein angles **white to buff (bronze)**

FR ± globular

℗ Small-leaved Lime *Tilia cordata* WT

H to 30 m. Fls in obliquely erect clusters of **5–11**, held above leaves; BRACTS L 35–60 mm. **Lvs** L 25–80 mm; UPPERSIDE dull green. **Twigs** greenish to reddish. **Fr** lemon-shaped; D 6–8 mm. **Hab** ancient woods, often coppiced, plantations, parks.

J F M A M J J A S O N D

FR ± lemon-shaped

INFL angled upwards

LF UNDERSIDE ± hairless; hair tufts in lf-vein angles **orange (buff)**

129

● **Mallows** | **Form** typically herbaceous; some woody at base. **Lvs** palmately lobed to round. **Fls** STAMENS united into a tube; BRACTS 3; forming a whorl (epicalyx) just below the flower.

℗ **Common Mallow** *Malva sylvestris*

H to 100 cm. **Form** erect or spreading; sparsely to densely hairy. **Fls** D 25–40 mm; 1–several in lf-axils. **Lvs** D 50–150 mm; can be folded; can have dark base. **Hab** roadsides, hedgerows, rough grassland. **SS** Smaller Tree-mallow *M. multiflora* (N/i) [BRACTS fused at their base].

J F M A M J J A S O N D

FL **pink-purple with darker veins**

LVS all palmately lobed

℗ **Musk-mallow** *Malva moschata*

H to 80 cm. **Form** erect or spreading; almost hairless. **Fls** 30–60 mm; in terminal cluster + solitary at base of upper leaves. **Lvs** D 50–80 mm; variable. **Hab** roadsides, hedgerows and rough grassland on well-drained soils.

J F M A M J J A S O N D

UPPER LVS usually divided

FL ± **unmarked; rose-pink or white**

LOWER LVS usually lobed

Ⓐ **Dwarf Mallow** *Malva neglecta*

H to 60 cm. **Form** sprawling or erect; hairy; can be densely so. **Fls** 18–25 mm; in clusters of 2–5 on stalks in leaf-axils. **Lvs** D 40–70 mm; variable. **Hab** field margins, tracksides, rough grassland and coastal habitats, usually on dry soils.

J F M A M J J A S O N D

FL **very pale pink or lilac with darker veins**

LVS all ± round, with 5–7 shallow lobes

---

36 Cistaceae | **Rock-rose** family

1 sp. | 4 spp. B&I

Form woody-based perennials. **Fls** PETALS 5, **delicately crumpled;** SEPALS 5, 3 of which are distinctly larger than the others; STAMENS numerous. **Lvs** surfaces with star-shaped hairs.

℗ **Common Rock-rose** *Helianthemum nummularium*

H to 20 cm. **Form** woody based; STEM (L to 50 cm) creeping, wiry. **Fls** D 20–25 mm; **yellow;** in terminal clusters; STIGMA 3-lobed. **Lvs** narrowly oval; 1-veined; sparsely covered (at least) in star-shaped hairs; UPPERSIDE green; UNDERSIDE whitish; MARGIN somewhat inrolled; STIPULES small + linear (L > LF-STALKS).

LF UP

Fr 3-valved capsule. **Hab** calcareous grassland. **SS** other **rock-roses** (N/i) [FL + LF details]; other 5-petalled yellow flowers – see **Gallery 11** (p.55).

J F M A M J J A S O N D

LF UND

37 Resedaceae | **Mignonette** family

2 spp. | 3 spp. B&I

A distinctive family in which the ovary is open at the top. **Fls** short-stalked; in ± elongate spikes; UPPER + SIDE PETALS deeply lobed; BOTTOM PETALS undivided; STAMENS arising from a nectar-secreting disc.

Ⓑ Weld *Reseda lutea*

H to 150 cm. **Form** erect; hairless; little-branched. **Fls** D 4–5 mm; yellowish; numerous in **elongate spikes**; PET 4; SEP 4; STAMENS 20–25. **Lvs** BASAL rosette; STEM **entire; linear; margins wavy; midrib pale**. **Fr** capsule (L 5–6 mm), ± globular; 3 apical points; crowded up stem. **Hab** disturbed ground, brownfield, especially on calcareous soils.

STAMENS ≥ 20

LF undivided

FR ± globular

ⓑⓟ Wild Mignonette *Reseda luteola*

H to 75 cm. **Form** erect; hairless; well-branched. **Fls** D 6 mm; yellowish; numerous in **conical spikes**; PET 6; SEP 6; STAMENS 12–20. **Lvs** BASAL rosette (withers before flowering); STEM **deeply pinnately lobed**, LOBES linear. **Fr** capsule (L 10–20 mm), ± **oblong**; 3 short apical points. **Hab** disturbed ground, brownfield, especially on calcareous soils.

STAMENS ≤ 20

LF divided

FR ± oblong

38 Santalaceae | **Mistletoe** family

1 sp. | 2 spp. B&I

The very different-looking members of this family, including the familiar evergreen Mistletoe, are all semiparasitic (photosynthetic but taking water and mineral nutrients from their host).

Ⓟ Mistletoe *Viscum album*

D to 200 cm. **Form** wide; repeated branching; older plants with circular outline. **Fls** ♂ + ♀ on separate plants; D 3–5 mm; yellow-green; inconspicuous; stalkless; PET 4; sepal-like; in clusters of 3–5.

Lvs opposite; L to 80 mm; narrowly oval; evergreen; leathery; yellowish-green; **widest near the rounded tips**. **Fr** white berry; D to 10 mm. **Hab** on branches of various trees, particularly apples (*p. 90*), poplars (*p. 111*), limes (*p. 129*) and hawthorns (*p. 94*). **SS** Witches'-broom twig galls, caused by e.g. *Taphrina* fungi can look similar at a distance (INSET).

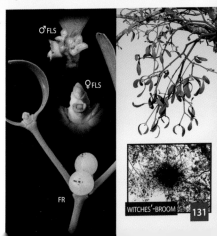

♂ FLS

♀ FLS

FR

WITCHES'-BROOM

131

39 Brassicaceae | Cabbage family

28 spp. | 103 spp. B&i

Fls flower structure diagnostic: PETALS 4 and **SEPALS 4; arranged cross-wise** (hence the former family name Cruciferae); **STAMENS** typically 6, the outer 2 shorter than the inner 4. **Fr** pods or capsules; often **highly distinctive in form and alignment,** the details of which can be a crucial part of the identification process.

IDENTIFY TO GROUP BY: ► **flower colour** ► **fruit shape, alignment and details**

FLOWERS yellow

FR BEAK L ≥ 4 mm	FR BEAK L < 4 mm

FLOWERS purple, mauve or pink

Charlock; 'CABBAGES'
below and opposite

YELLOW 'MUSTARDS'; YELLOW 'CRESSES'
p. 134

Radish; Honesty, Dame's-violet, Aubretia
p. 135

Wild Radish (p. 135) can be yellow or white [pods constricted]; Honesty [pods large, round] and Dame's-violet [pods long, upcurved] can be white

FLOWERS white or absent

FRUIT relatively short (L < 4× W)	FRUIT relatively long (L > 4× W)

WHITLOWGRASSES; SCURVY-GRASSES; PENNY-CRESSES; SWINE-CRESSES + others
p. 136–138

BITTER-CRESSES; WATER-CRESSES; WHITE 'MUSTARDS'; WHITE 'CRESSES' + others
p. 139

FLOWERS yellow

1/3

BEAK OF FRUIT **long (≥ 4 mm)**; FRUIT VALVES **at least 3-veined**

A Charlock *Sinapis arvensis*

H to 100 cm. **Form** erect; often well-branched. **Fls** D 15–30 mm in dense infl. **Lvs** variable; broadly oval to pinnately lobed; smooth to roughly hairy. **Fr** L 20–55 mm; BEAK conical (L to 16 mm). **Hab** arable and other disturbed areas; brownfield. **SS** White Mustard *S. alba* (N/i) [BEAK flattened].

J F M A M J J A S O N D

SEP narrow; inrolled can be spreading or bent back

INFL

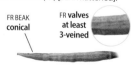

FR BEAK conical

FR valves at least 3-veined

LVS variable; if lobed then end lobe much larger than the others

BEAK OF FRUIT long (≥ 4 mm); FRUIT VALVES **1 main vein**

FR 1 main vein

● **Brassicas (Cabbages)** | An important genus of edible plants, leaves and roots used as vegetables and seeds as source of oil and mustard; many forms are found as escapes from or relics of cultivation. All have yellow flowers; fruiting characters offer the most reliable separation.

AB **Rape** *Brassica napus*

J F M A M J J A S O N D

H to 120 cm.
Form erect.
Fls D 18–30 mm in flat-topped head. **Lvs** UPPER clasp stem; LOWER blue-grey, slightly hairy at most.
Fr L 50–100 mm; BEAK conical (L to 16 mm).
Hab brownfield and disturbed ground, arable margins, road verges.

BUDS **slightly overtop open fls**

UPPER LVS clasp stem

LOWER LVS blue-grey; slightly hairy at most

FR not appressed; FR BEAK can contain 1 seed

AB **Turnip** *Brassica rapa*

J F M A M J J A S O N D

H to 120 cm.
Form erect.
Fls D 15–20 mm in flat-topped head. **Lvs** UPPER clasp stem; LOWER green, roughly hairy.
Fr L 30–65 mm; BEAK conical (L to 16 mm).
Hab arable and other disturbed areas; brownfield. **SS Cabbage** *B. oleracea* (N/i) [mainly coastal; BUDS greatly overtop open fls].

BUDS **slightly overtopped** by open fls

UPPER LVS clasp stem

LOWER LVS green; roughly hairy

FR not appressed; FR BEAK seedless

A **Black Mustard** *Brassica nigra*

J F M A M J J A S O N D

H to 200 cm.
Form erect. **Fls** D 10–15 mm in branched infl.
Lvs UPPER not clasping stem; LOWER bristly, bright green.
Fr L 8–25 mm; 4-angled; beak L ± 5 mm. **Hab** brownfield, rough grassland, river banks, sea cliff slopes. **SS Hoary Mustard** *Hirschfeldia incana* ((N/i) [larger flowers, 1–2 seeds in beak].

UPPER LVS not clasping stem

LOWER LVS bright green; bristly

FR **appressed** to stem; BEAK L ± 5 mm; seedless

133

BEAK OF FRUIT **short (< 4 mm)**; FRUIT **short + slender (L < 5×W)**

Ⓐ Marsh Yellow-cress *Rorippa palustris*

H to 60 cm. **Form** erect.
Fls D 3–5 mm; PET L = SEP L.
Lvs STEM almost pinnate; LOBES
2–6 pairs; narrow; shallow-
toothed. **Fr** L 5–10 mm; gently
curved, 1–2× as long as stalk.
Hab marshes, wetland fringes,
often colonising bare mud.
SS Creeping Yellow-cress *R.*
sylvestris (N/I) [FR longer, more slender;
PET L 1·5–2·0× SEP L; patch-forming].

PETALS + SEPALS
equal in length

FR **gently
curved**

LF 2–6 pairs of leaflets

J F M A M J J A S O N D

BEAK OF FRUIT **short (< 4 mm)**; FRUIT **long + slender (L > 5×W)**

ⒶⒷ Hedge Mustard *Sisymbrium officinale*

H to 100 cm. **Form** erect,
wiry. **Infl** widely branched
elongating in fruit.
Fls D 4–6 mm. **Lvs** deeply
lobed; BASAL in rosette.
Fr L to 20 mm; appressed to
stem. **Hab** brownfield,
hedgebanks, road verges.
SS other *Sisymbrium* spp.
(N/I), especially in urban areas [FR longer
and spreading more widely from stem].

FR beakless;
**appressed
to stem**

STEM-LF

BASAL LF

J F M A M J J A S O N D

ⒷⓅ Winter-cress *Barbarea vulgaris*

H to 80 cm. **Form** erect,
branched. **Infl** elongating
in fruit. **Fls** D 7–10 mm. **Lvs**
BASAL pinnate, in rosette; STEM
shining green, shallowly lobed,
clasping stem. **Fr** L 15–30 mm;
not appressed to stem. **Hab**
ditch and streamsides; damp
road verges. **SS** other winter-
cresses. [FR + LF details].

FR **erect**
but not
appressed
to stem; beak
L to 3·5 mm

STEM-LF

J F M A M J J A S O N D

Ⓟ Wallflower *Erysimum cheiri*

H to 60 cm. **Form** erect, base
woody. **Fls** D 20–35 mm; wild-
type orange-yellow, many
other colours from cultivation;
petal L > 2× sepal L; AROMA
strong, sweet. **Lvs** very
narrowly oval; entire. **Fr** L 30–
70 mm; flattened; ± appressed;
hairy; valves 1-veined. **Hab**
walls, rocks, cliffs, brownfield.

J F M A M J J A S O N D

FLOWERS typically purple or mauve

Colour variation: Cuckooflower (*p. 139*) is usually virtually white although can be pink to lilac; **Wild Radish** can be yellow or white; **Honesty** and **Dame's-violet** can be white.

B Honesty *Lunaria annua*

H to 100 cm. **Form** erect. **Fls** D 18–25 mm; purple, pink or white; may have darker veins. **Lvs** oval; sharply toothed; lower long-stalked. **Fr** L 25–60 mm; flattened; **round to oval**. **Hab** roadsides, brownfield, gardens, tips.

J F M A M J J A S O N D

FR pod translucent when ripe

LF sharp, irregular teeth

bP Dame's-violet *Hesperis matronalis*

H to 120 cm. **Form** erect; well-branched. **Fls** D 18–25 mm; purple, pink or white; may have darker veins. **Lvs** STEM narrowly oval with drawn-out point. **Fr** L to 100 mm; slender; cylindrical. **Hab** riverbanks, brownfield, hedgebanks, road verges, gardens.

J F M A M J J A S O N D

FR pod gently upcurved

LVS close to stem

LF ± stalkless; sharply and regularly toothed

A Wild Radish *Raphanus raphanistrum*

H to 60 cm. **Form** erect; bristly hairy. **Fls** D 12–20 mm; **yellow, mauve or white**; usually with darker veins. **Lvs** pinnate, with large, rounded end-lobe. **Fr** L 20–50 mm; BEAK L to 25 mm; pod deeply constricted between seeds. **Hab** field margins, other disturbed sites.

J F M A M J J A S O N D

SS Sea Radish ssp. *maritimus* (N/I) [coastal; FL usually yellow; LFLTS overlap].

FR constricted between seeds when ripe

FL

LF pinnate; LFLTS not overlapping

P Aubretia *Aubrieta deltoidea*

H to 20 cm. **Form** mat-forming; many non-flowering shoots. **Fls** D 18–25 mm; **purple to mauve**. **Lvs** diamond-shaped with star-shaped hairs and a few sharp teeth. **Fr** L to 100 mm; slightly compressed; L 2–3× W. **Hab** gardens, walls, rocks.

J F M A M J J A S O N D

FL

FL

LF

FLOWERS white

FORM **robust**; FRUIT **no mature fruits in B&I**

❷ Horse-radish
Armoracia rusticana

flowers often imperfect and rarely fruits in B&I

FL

H to 150 cm. **Form** erect; hairless. **Fls** D 8–9 mm; in branched infl. **Lvs** L to 100 mm; broadly oval; glossy; sharply toothed. **Hab** brownfield, riverbanks, road verges. **SS Dittander** *Lepidium latifolium*. (N/i) [FL smaller, in larger clusters; LVS less robust].

J F M A M J J A S O N D

LF broad, like a dock (*p. 141*), but sharply toothed

FRUIT **pods; relatively broad (L < 4×W)** | **strongly flattened front to back**

ⓑ Shepherd's-purse
Capsella bursa-pastoris

H to 40 cm. **Form** erect, rosette-forming. **Fls** D 2–3 mm in loose spike. **Lvs** BASAL lobed; STEM stalkless, clasping. **Fr** L 6–9 mm; ± heart-shaped on long stalk; raised ridge on broad face. **Hab** gardens, arable fields, disturbed ground.

J F M A M J J A S O N D

Ⓐ Field Penny-cress
Thlaspi arvense

H to 60 cm. **Form** erect, rosette-forming. **Fls** D 4–6 mm in loose spike. **Lvs** BASAL lobed; STEM stalkless. **Fr** D 10–22 mm; circular; 2-winged; on long stalk. **Hab** arable margins, brownfield, disturbed areas. **SS** other *Thlaspi* spp. (N/i) [FL, LF and FR details].

J F M A M J J A S O N D

Ⓐ Common Whitlowgrass
Erophila verna

H to 20 cm. **Form** erect, rosette-forming. **Fls** D 3–6 mm on leafless stem; PET notched. **Lvs** all basal; coarsely toothed; short stalk. **Fr** L 4–9 mm; oval on long stalk. **Hab** open sandy ground, heaths, rocks, pavements. **SS** other *Erophila* spp. (N/i) [LF, STEM + SEED details].

J F M A M J J A S O N D

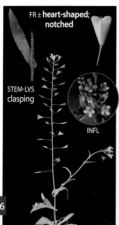

FR ± heart-shaped; notched

STEM-LVS clasping

INFL

FR winged; notched at tip; style short

STEM-LVS stalkless

INFL

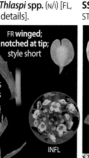

FR oval

STEM leafless

PET deeply cleft

FRUIT pods; relatively broad (L < 4×W) | not, or barely, flattened front to back   1/2

INFL at apex of shoots; LVS ± entire to palmately lobed

🅐🅑 Danish Scurvygrass
Cochlearia danica

H to 25 cm.
Form patch-forming.
Fls D 4–5 mm (can have a hint of mauve). **Lvs** BASAL ± heart-shaped (can be purplish); STEM palmately 3–5-lobed.
Fr L 3–5 mm. **Hab** coastal habitats, **salted road verges**.

🅑🅟 Common Scurvygrass
Cochlearia officinalis

H to 40 cm.
Form fleshy; spreading to erect.
Fls D 4–5 mm.
Lvs BASAL long-stalked, usually heart-shaped; STEM clasping. **Fr** L 3–7 mm; ± globular. **Hab** saltmarsh, cliffs, salted verges. **SS** English Scurvygrass *C. anglica* (N/I) [FR compressed; LF base tapers to stalk].

🅟 Hoary Cress
Lepidium draba

H to 60 cm.
Form erect; patch-forming.
Fls D 5–6 mm in dense flat-topped heads; STYLE projecting.
Lvs greyish-green, entire to shallowly toothed. **Fr** L 3–5 mm; W ≥ L; somewhat inflated; ridged. **Hab** brownfield, road verges, railway banks, sea walls.

BASAL LVS ±
heart-shaped   FR

STEM-LVS
3–5-lobed;
not clasping

BASAL LVS base
± heart-shaped   FR

STEM-LVS
clasping

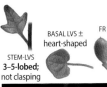

distinctive
flat-topped
form

LF **clasping stem**

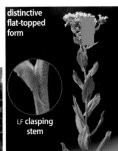

INFL

INFL

🔴 Swine-cresses | **Form** largely prostrate. **Fls** in dense clusters.
Hab bare, trampled and disturbed areas, arable margins.

INFL attached opposite leaf-stalks; LVS irregularly deeply lobed

🅐 Swine-cress
Lepidium coronopus

W to 30 cm. **Fls** D 2·5 mm; STAMENS **6**; PETAL L > SEPAL L. **Lvs** AROMA faint at most. **Fr** coarsely ridged; L 2·3–3·5 mm; L > stalk.

STAMENS
6   FR ridged;
L > stalk

🅐 Lesser Swine-cress
Lepidium didymum

W to 40 cm. **Fls** D < 2 mm; STAMENS **2**; PET tiny/absent. **Lvs** AROMA **unpleasant** when crushed. **Fr** smooth; L < stalk. .

STAMENS
usually
2

FR smooth;
L < stalk

SWINE-CRESS   LESSER

LESSER SWINE-CRESS

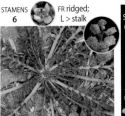

137

FLOWERS **white**

FRUIT pods; relatively long (L > 4×W) | not, or barely, flattened front to back

LVS **not pinnate**

Ⓑ Garlic Mustard *Alliaria petiolata*

H to 120 cm. **Form** erect; AROMA **garlic when crushed. Fls** D 4–6 mm in rounded heads. **Lvs** round, pointed, regularly toothed. **Fr** L 35–75 mm; **slender**, held **semi-erect and curving upwards. Hab** woods, hedges, shady places.

J F M A M J J A S O N D

INFL

LF regularly and deeply toothed

ⒶⒷ Thale Cress *Arabidopsis thaliana*

H to 50 cm. **Form** erect, rosette-forming. **Fls** D 3 mm in elongating, branched spike. **Lvs** mostly basal, toothed, hairy. **Fr** L 10–18 mm; **slender. Hab** gardens, walls and rocks, open sandy and stony ground. **SS Hairy Rock-cress** *Arabis hirsuta* (N/I) [taller; FR pressed closely to stem].

J F M A M J J A S O N D

INFL

The rosette is similar to that of the whitlowgrasses, (*p. 136*)– which have rounded fruits

STEM-LF

INFL

leaves on stem

BASAL LF

LVS **pinnate**   1/2

Ⓟ Water-cress *Nasturtium officinale*

H to 60 cm. **Form** prostrate to erect, patch-forming, evergreen. **Fls** D 4–6 mm in loose heads; ANTHERS yellow. **Lvs** LFLTS basal lvs rounder than stem-lvs. **Fr** L 13–18 mm; cylindrical. **Hab** in and around shallow still and slow-moving waterbodies. **SS Narrow-fruited Water-cress** *N. microphyllum* (INSET) – probably overlooked [seeds in single rows, LVS turn purple in autumn]. **Fool's Water-cress** and **Lesser Water-parsnip** (*p. 225*) [LVS with basal sheath].

J F M A M J J A S O N D

INFL

BASAL LF

LVS no basal sheath

WATER-CRESS
FR 2 rows of seeds

NARROW-FRUITED WATER-CRESS
FR 1 row of seeds

	2/2	
	2/2	
LVS **pinnate**	2/2	FR

● Wavy and Hairy Bitter-cresses | Fl D ≤ 4 mm; Fr L 12–25 mm.

ⓐⓑ **Wavy Bitter-cress**
Cardamine flexuosa

H to 50 cm. **Form** erect; STEM very wavy. **Fls** D 2·5–4·0 mm in elongating, branched spike; STAMENS 6 (4 longer; 2 shorter). **Lvs** lower form loose rosette. **Fr** slender; **barely overtopping flowers. Hab** damp woodland, gardens, shady places.

ⓐⓑ **Hairy Bitter-cress**
Cardamine hirsuta

H to 30 cm. **Form** erect; STEM wavy or straight. **Fls** D 2·0–3·5 mm in elongating, branched spike; STAMENS 4. **Lvs** lower form compact rosette. **Fr** slender; **usually overtopping flowers. Hab** gardens, walls, disturbed ground.

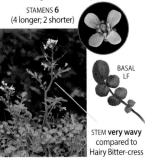

STAMENS **6** (4 longer; 2 shorter)

BASAL LF

STEM **very wavy** compared to Hairy Bitter-cress

STAMENS typically **4**

BASAL LF

ROSETTE more compact than that of Wavy Bitter-cress

HAIRY BITTER-CRESS

● Large Bitter-cress and Cuckooflower | Fl D < 10 mm; Fr L 20–40 mm.

ⓟ **Large Bitter-cress** *Cardamine amara*

H to 60 cm. **Form** erect. **Fls** D 10–14 mm in loose spike; ANTHERS **violet**. **Lvs** pale green. **Fr** L 20–40 mm. **Hab** damp woodland, streamsides.

ANTHERS **violet**

LVS noticeably pale green

ⓟ **Cuckooflower** *Cardamine pratensis*

H to 60 cm. **Form** erect. **Fls** D 12–18 mm in rounded heads; ANTHERS **yellow**. **Lvs** LFLTS basal lvs rounder than stem-lvs. **Fr** L 25–30 mm. **Hab** damp meadows, marshes, open woodland.

ANTHERS **yellow**

STEM-LVS

LFLTS: basal lvs rounder than those on stem-lvs

BASAL LVS

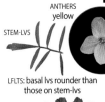

139

40 Plumbaginaceae | **Thrift + Sea-lavender** family   `1 sp.` | `13 spp. B&I`

Infl branched or compact + rounded. **Fls** pink to purple; 5-parted; SEPALS distinctively papery brown, fused at their base. **Hab** largely coastal and mountain species.

Ⓟ Thrift *Armeria maritima*

H to 30 cm. **Form** evergreen; cushion-like; rootstock woody. **Fls** D 8 mm; pink or white; in tight ±globular, **terminal head** on stiff, leafless stalks. **Lvs** L to 10 cm; linear, slightly succulent, dark green. **Fr** dry, papery, single-seeded capsule. **Hab** coastal cliffs, shingle, saltmarsh; salted road verges; montane grassland and rocks.

J F M A M J J A S O N D

● **Sea-lavenders** *Limonium* spp. | fall into two groups: **those of saltmarshes** [LVS with pinnate veins] and **those of sea-cliff or shingle sites**, 'Rock Sea-lavenders' [LVS with veins all arising from base]. The numerous rock sea-lavenders are all very similar, and typically have very restricted ranges. The two saltmarsh species (*right*) are variable, and best separated by a combination of branching pattern, anther colour and degree of flower-cluster overlap.

LAX-FLOWERED

ANTHERS purple

COMMON

ANTHERS yellow

41 Polygonaceae | **Dock + Knotweed** family   `15 spp.` | `45 spp. B&I`

Fls inconspicuous; massed to showy effect in many; tepals, usually brown or greenish, remain when fruiting, can have distinctive teeth, bumps or wings. **Fr** a single dry fruit; 3-angled in many. **Lvs** distinctive papery stipules (ochrea) fused around leaf-base and stalk.

IDENTIFY TO GROUP BY: ▶ tepal number + shape   ▶ leaf shape   ▶ inflorescence form

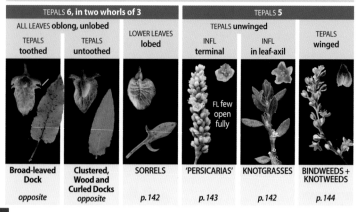

TEPALS 6, in two whorls of 3			TEPALS 5		
ALL LEAVES oblong, unlobed		LOWER LEAVES lobed	TEPALS unwinged		TEPALS winged
TEPALS toothed	TEPALS untoothed		INFL terminal	INFL in leaf-axil	
			FL few open fully		
Broad-leaved Dock	**Clustered, Wood and Curled Docks**	**SORRELS**	**'PERSICARIAS'**	**KNOTGRASSES**	**BINDWEEDS + KNOTWEEDS**
opposite	*opposite*	*p. 142*	*p. 143*	*p. 142*	*p. 144*

TEPALS 6, in two whorls of 3

LEAVES ± narrowly triangular; unlobed | TEPALS toothed

Broad-leaved Dock *Rumex obtusifolius*

H to 120 cm. **Form** erect; well-branched. **Fls** INNER TEPALS L 5–6 mm; strongly toothed in fruit. **Lvs** L to 50 cm; rounded at base. **Fr** usually just 1 tepal with developed swelling. **Hab** rough grassland, road verges, riverbanks, brownfield.

TEPALS strongly toothed in fruit

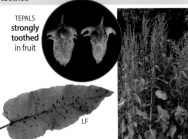

LF

LEAVES broadly oval; unlobed | TEPALS untoothed

SEPARATE BY: ▶ inflorescence structure + fruit details

Curled Dock *Rumex crispus*

Clustered Dock
Rumex conglomeratus

H to 80 cm. **Form** erect. **Infl** leafy almost to the top. **Lvs** L to 30 cm; broadly oval. **Fr** all 3 inner tepals with an egg-shaped swelling. **Hab** damp hedgebanks, grassland, marshes and riverbanks.

Wood Dock
Rumex sanguineus

H to 80 cm. **Form** erect. **Infl** leafy only at base. **Lvs** L to 30 cm; broadly oval. **Fr** only one inner tepal with a globular swelling. **Hab** woods, hedges, riverbanks, brownfield.

H to 120 cm. **Form** erect; BRANCHES short. **Infl** leafy almost to the top. **Fls** INNER TEPALS L 4–5 mm. **Lvs** L to 35 cm; narrowly oval; tapering to stalk; MARGIN **very wavy**. **Fr** 1–3 inner tepals with developed swelling. **Hab** rough grassland, marshes, brownfield.

TEPALS all with egg-shaped swelling (fr)

NOTE: fruit swellings start **white** and ripen **red**

LVS similar in Clustered and Wood Docks

TEPALS just one with globular swelling (fr)

LF-MARGIN wavy

TEPALS **untoothed**; 1–3 with swelling

INFL leafy near top

FORM most branches at >30° from stem

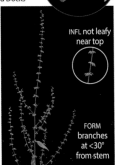

INFL not leafy near top

FORM branches at <30° from stem

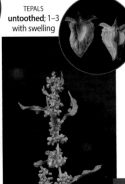

TEPALS **6, in two whorls of 3** | 2/2

LEAVES **lower with basal lobes**

℗ **Sheep's Sorrel** *Rumex acetosella*

H to 30 cm. **Form** erect.
Fls green, becoming **bright reddish-brown**; TEPALS L 1·5–2·0 mm; small swelling at base. **Lvs** L to 4 cm; UPPER LVS **stalked**. **Fr** seed as long as tepals. **Hab** dry, acid grassland and heathland.

J F M A M J J A S O N D

℗ **Common Sorrel** *Rumex acetosa*

H to 60 cm. **Form** erect.
Fls green, becoming reddish; TEPALS L 2·5–4·0 mm; small swelling at base. **Lvs** L to 10 cm; UPPER LVS **clasping**. **Fr** seed much shorter than tepals. **Hab** grassland.

J F M A M J J A S O N D

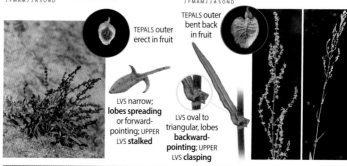

TEPALS outer **erect in fruit**

TEPALS outer **bent back in fruit**

LVS narrow; **lobes spreading** or forward-pointing; UPPER LVS **stalked**

LVS oval to triangular, lobes **backward-pointing**; UPPER LVS **clasping**

TEPALS **5** | 1/3

TEPALS **unwinged** | INFL **in leaf-axil, with < 6 flowers**; TEPALS L **1·5–3·0 mm**

● **Knotgrasses** | **Form** largely prostrate patch-forming. **Fls** TEPALS green with white lobes. **Hab** arable margins, paths, brownfield, seashores.

SS The common knotgrasses are both similar to other rarer species [SEED + OCHREA details].

Ⓐ **Knotgrass** *Polygonum aviculare*

Spread to 200 cm. **Lvs** variably oval; shorter on branches than on main stem. **Fr** L 2·5–3·5 mm.

J F M A M J J A S O N D

TEPALS lobes **fused at the base**

OCHREA silvery; fringed

SEED 3-angled; **all faces concave**

Ⓐ **Equal-leaved Knotgrass** *Polygonum arenastrum*

Spread to 50 cm. **Lvs** narrowly oval, **all equal in length**; overlapping on shoots. **Fr** L <2·5 mm.

J F M A M J J A S O N D

TEPALS lobes **fused >⅓ of their length**

OCHREA silvery; fringed

SEED 3-angled; **1 face concave; 2 faces convex**

Ⓐ Water-pepper *Persicaria hydropiper*

H to 70 cm. **Form** erect to spreading. **Fls** greenish to pale pink, in **lax, tapering, often drooping spike**. **Lvs** L to 100 mm; narrowly triangular. **Fr** L 2·5–3·8 mm; SEED **matt black**. **Hab** damp, shaded areas, woodland rides, marshes.

J F M A M J J A S O N D

SS Tasteless Water-pepper *P. mitis* (N/I) [no peppery taste].

LVS burning peppery taste

OCHREA slightly toothed

INFL with yellow dot-like glands

Ⓐ Pale Persicaria *Persicaria lapathifolia*

H to 70 cm. **Form** erect to spreading. **Fls** greenish to pink, in **dense spike**. **Lvs** L to 100 mm; narrowly oval; may have large blackish blotch. **Fr** L 2·0–3·3 mm; SEED shiny black. **Hab** arable margins, brownfield, often in damp areas.

J F M A M J J A S O N D

LVS usually plain

OCHREA fringed with short hairs

INFL with **yellow dot-like glands**

Ⓐ Redshank *Persicaria maculosa*

H to 70 cm. **Form** erect to spreading. **Fls** deep pink (occasionally whitish), in **dense spike**. **Lvs** L to 100 mm; narrowly oval; usually with large blackish blotch. **Fr** L 2·0–3·2 mm; SEED shiny black. **Hab** arable margins, brownfield, often in damp areas.

J F M A M J J A S O N D

LVS usually blotched

OCHREA fringed with long hairs

INFL **lacks** yellow dot-like glands

Ⓟ Amphibious Bistort *Persicaria amphibia* AQ

H to 60 cm. **Form** IN WATER floating leaves and erect, emergent FL spikes; ON LAND sprawling. **Fls** pink; in dense spike. **Lvs** L to 150 mm; IN WATER narrowly oval, long-stalked, hairless; ON LAND short-stalked at most, hairy, can have blackish blotch. **Fr** L 2–3 mm; SEED shiny

J F M A M J J A S O N D

black. **Hab** still waters, marshes, damp arable margins. **SS** pondweeds (*p. 233*).

FL **stamens protrude**

INFL **lacks** yellow dot-like glands

TERRESTRIAL LF

AQUATIC LF

TEPALS 5   3/3

TEPALS winged

Ⓐ Black-bindweed *Fallopia convolvulus* Ⓒ

L to 1·2m. **Form** scrambling or twining. **Fls** L 2–3mm; greenish-white; INFL long sparse spikes. **Lvs** L to 6cm; triangular, mealy underneath. **Fr** L 4–5mm; SEED dull black. **Hab** arable fields, gardens. **SS** leaves – **Hedge Bindweed** (p.162) [LVS not mealy textured].

J F M A M J J A S O N D

TEPALS greenish-white

LF

Ⓟ Russian-vine *Fallopia baldschuanica* WC

L to 10m+. **Form** woody-based scrambler/climber. **Fls** L 2–3mm; white; midrib green; INFL showy, branched. **Lvs** L to 15cm; broadly triangular to **heart-shaped**. **Fr** L 4–5mm; SEED shiny dark brown. **Hab** gardens, hedges, scrub, brownfield.

J F M A M J J A S O N D

TEPALS white with green midrib

LF-BASE curved in

Ⓟ Japanese Knotweed *Reynoutria japonica*

H to 2m. **Form** erect, **densely patch-forming**. **Fls** L 2–3mm; greenish-white; INFL showy, branched. **Lvs** L to 12cm; broadly triangular, pointed; BASE **squared-off**. **Fr** L 4mm; SEED glossy dark brown. **Hab** riverbanks, disturbed areas, brownfield.

J F M A M J J A S O N D

TEPALS greenish-white

LF-BASE squared-off

---

42 Balsaminaceae | **Balsam** family   1 sp. | 4 spp. B&I

Fls unique structure: PETALS 5, the lower 4 fused in left and right pairs; SEPALS 3, petal-like, the lowest forming a spurred pouch; STAMENS 5, fused at their tips so that the anthers are joined round the ovary. **Fr** touch-sensitive and explosive capsule.

Ⓐ Himalayan Balsam *Impatiens glandulifera*

H to 200cm. **Form** erect; patch-forming through seeding. **Fls** L 25–40mm; deep purplish-pink or white; lowest sepal with **sharply curved spur** (L 2–7mm). **Lvs** L to 15cm; opposite or in whorls of 3; narrowly oval; stalked; toothed. **Fr** an **explosive** 5-celled capsule. **Hab** river and canal banks; damp woodland and brownfield.

J F M A M J J A S O N D

FL lowest sepal with sharply curved spur

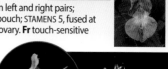

FR distinctive; explosive 5-celled capsule

43 Droseraceae | **Sundew** family

1 sp. | 3 spp. B&I

Form insectivorous; found in boggy places. **Fls** in a cluster atop a long stalk arising from the rosette; often not opening and may self-pollinate in bud. **Lvs** unique; covered in sticky hairs that trap and digest insects.

SS Often found growing with Round-leaved Sundew are two rarer species – **Great Sundew** *D. anglica* (N/I) and **Oblong-leaved Sundew** *D. intermedia* (N/I), [LF shape + LF-STALK length].

P Round-leaved Sundew   *Drosera rotundifolia*

J F M A M J J A S O N D

H to 150 mm. **Fls** D 4–6 mm; white; PET + SEP 5–8; STAMENS 5–8. **Form** rosette with erect flower spike arising from centre. **Lvs** LF-BLADE ± round (D 10 mm); LVS + STALKS L < 50 mm. **Hab** acid bogs and wet heaths, especially in bare patches.

FLS open only in bright sunshine

LF ± circular

44 Caryophyllaceae | **Campion** [Pink] family

24 spp. | 81 spp. B&I

A recognizable family, although difficult to define precisely. **Infl** a repeating pattern of a terminal flower on a stalk which itself has paired branches that each have a terminal flower (see *p. 17*). **Fls** SEPALS 5; PETALS 5, often notched or deeply divided into two. **Lvs** typically opposite. **Fr** a many-seeded capsule.

Pink groups and species identification

LEAVES **with stipules**	LEAVES **without stipules 1/2**		
		SEPALS **forming a tube**	SEP-TUBE
	PETALS **white**	STYLES **pink or red**	
SPURREYS SEPALS 5; PETALS 5; LVS **unlobed** *p. 146–147*	**CAMPIONS** *p. 146*	**Red Campion, Ragged-robin, Soapwort** *p. 147*	

LEAVES **without stipules 2/2**			
SEPALS **separate; not forming a tube**			SEPALS
PETALS **deeply split**		PETALS **entire or slightly notched**	PETALS **minute or absent**
STYLES **3**	STYLES **4 or 5**		
CHICKWEEDS + STITCHWORTS *p. 148*	**MOUSE-EARS** *p. 149*	**SANDWORTS + Knotted Pearlwort** *p. 150*	**PEARLWORTS** *p. 150*

LEAVES with stipules

🌸 **Lesser Sea-spurrey** *Spergularia marina*

Spread to 30 cm. **Form** prostrate; STEM can be stickily hairy. **Fls** D 5–8 mm; **deep pink. Lvs** L to 20 mm; linear; fleshy. **Fr** L 3–6 mm; SEEDS L 0·6–0·8 mm. **Hab** saltmarshes, sandy coastal habitats, salted road verges.

J F M A M J J A S O N D

🌸 **Greater Sea-spurrey** *Spergularia media*

Spread to 40 cm. **Form** prostrate; STEM can be stickily hairy. **Fls** D 10–12 mm; **pale pink. Lvs** L to 25 mm; linear; fleshy. **Fr** L 7–9 mm; SEEDS L 0·7–1·0 mm. **Hab** saltmarshes, sandy coastal habitats, salted road verges.

J F M A M J J A S O N D

PET L usually < sepals; STAMENS 2–7

STYLES 3

SEEDS can be winged or wingless

LF fleshy

PET L usually ≥ sepals; STAMENS 10

STYLES 3

SEEDS most distinctly winged

LF fleshy

LEAVES without stipules

SEPALS forming a tube; PETALS white; LVS narrowly oval

🌸 **White Campion** *Silene latifolia*

H to 100 cm. **Form** erect. **Fls** D 25–30 mm; SEPAL-TUBE ♂ L 15–22 mm, veins 10; ♀ L 20–30 mm, veins 20. **Lvs** clasping stem. **Fr** capsule; L 15–20 mm. **Hab** disturbed ground, hedgebanks, road verges.

J F M A M J J A S O N D

STYLES 5

♀ FL

♂ FL

SEP-TUBE narrowed at apex

♂ FL

🌸 **Bladder Campion** *Silene vulgaris*

H to 90 cm. **Form** erect. **Fls** D 18–20 mm; in drooping, forked infl; SEPAL-TUBE ribbed, conspicuously net-veined. **Lvs** pointed; greyish-green. **Fr** capsule; L 10–12 mm. **Hab** grassland, hedgebanks, road verges, especially on calcareous soils.

J F M A M J J A S O N D

STYLES 3

SEP-TUBE inflated

🌸 **Sea Campion** *Silene uniflora*

H to 25 cm. **Form** cushion-forming with erect flowering shoots. **Fls** D 20–25 mm; borne singly or in loose heads of 2–4. **Lvs** pointed; greyish-green; waxy. **Fr** capsule; L 17–20 mm. **Hab** coastal shingle, sea cliffs, mountain ledges and lake shores.

J F M A M J J A S O N D

STYLES 3

SEP-TUBE not narrowed at apex

⚘ Sand Spurrey *Spergularia rubra*

Spread to 25 cm. **Form** prostrate; STEM stickily hairy. **Fls** D 3–6 mm; **bright pink. Lvs** L to 25 mm; linear; pointed; **grey-green. Fr** L 3·5–5·0 mm; SEEDS L 0·7–1·0 mm, unwinged. **Hab** bare sandy ground in heathland and acid grassland.

PET L usually < sepals; STAMENS 5–10

STYLES 3

STIPULES **noticeably silvery**

LF not fleshy

Ⓐ Corn Spurrey *Spergula arvensis*

H to 50 cm. **Form** weakly erect. **Fls** D 4–7 mm; **white**, in open branched infl. **Lvs** L to 30 mm; linear; blunt; stickily hairy; STIPULES short, silvery. **Fr** L 5–8 mm; SEEDS L 1·0–1·5 mm; unwinged. **Hab** arable land on sandy soils; coastal grassland.

LVS **appear whorled**

STYLES 5

1/3

SEPALS **forming a tube**; PETALS **pink or red**; LVS ± **narrowly oval**

⚘ Red Campion *Silene dioica*

H to 100 cm. **Form** erect; hairy; can be sticky. **Fls** D 25–30 mm; SEPAL-TUBE L 10–15 mm; ♂ 10-veined, cylindrical; ♀ 20-veined, egg-shaped. **Lvs** LOWER with winged stalks. **Fr** capsule; L 10–20 mm; teeth bent back. **Hab** woods, hedges; open clifftop grassland, mountains.

PET pink to red; notched

Ⓟ Soapwort *Saponaria officinalis*

H to 90 cm. **Form** erect; hairless. **Fls** D 20–25 mm; pink; in **forked clusters**; SEPAL-TUBE L 20 mm. **Lvs** L to 60 mm; LOWER with winged stalks. **Fr** capsule; L 18–20 mm; 4 unequal teeth. **Hab** road verges, hedgebanks, especially near houses. **SS Phlox** *Phlox paniculata* (N/I) [INFL more compact; both PET + SEP fused into a tube].

PET pink; rounded

Ⓟ Ragged-Robin *Silene flos-cuculi*

H to 100 cm. **Form** erect; hairy; can be sticky. **Fls** D 25–30 mm; SEPAL-TUBE L 10–15 mm; ♂ 10-veined, cylindrical; ♀ 20-veined, egg-shaped. **Lvs** LOWER with winged stalks. **Fr** capsule; L 10–12 mm; teeth bent back. **Hab** woods, hedges and other shaded places; open clifftop grassland, mountains.

PET deep pinkish-red; **deeply 4-lobed**; segments narrow

147

LEAVES **without stipules**

PETALS **white, deeply split;** STYLES **3;** SEPALS **separate, not forming a tube**

PET L ≤ sepals; ANTHERS **reddish**	PET L ≤ sepals	PET L slightly ≥ sepals	PET L 2× sepals
COMMON CHICKWEED	BOG STITCHWORT	LESSER STITCHWORT	GREATER STITCHWORT

Ⓐ Common Chickweed　*Stellaria media*

H to 20 cm. **Form** sprawling (SPREAD to 50 cm). **Fls** D 6–9 mm; **Lvs** L to 20 mm; broadly oval; stalked. **Hab** arable, disturbed ground. **SS** Three-nerved Sandwort (*p. 150*) [LVS with 3 distinct veins; PET not split].

J F M A M J J A S O N D

STEM with **line of hairs** running down

LVS oval

Ⓟ Bog Stitchwort　*Stellaria alsine*

H to 40 cm. **Form** weakly erect. **Fls** D 6 mm. **Lvs** L to 15 mm; narrowly oval; stalkless on flowering shoots. **Hab** marshes, streamsides, tracks and depressions in acid grassland.

J F M A M J J A S O N D

STEM angled or round

LVS narrowly oval

Ⓟ Lesser Stitchwort　*Stellaria graminea*

H to 80 cm. **Form** weakly erect. **Fls** D 5–15 mm. **Lvs** L to 40 mm; narrowly oval; MARGINS hairy. **Hab** dry grassland, heaths. **SS** Marsh Stitchwort *S. palustris* (N/I) [PET L 2× SEP; BRACTS pale; green central stripe].

J F M A M J J A S O N D

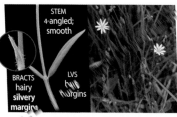

STEM 4-angled; smooth

BRACTS **hairy silvery margins**

LVS **margins**

Ⓟ Greater Stitchwort　*Stellaria holostea*

H to 60 cm. **Form** weakly erect. **Fls** D 20–30 mm. **Lvs** L to 50 mm; narrow; MARGINS rough. **Hab** woods, hedgerows. **SS** Wood Stitchwort *S. nemorum* (N/I) [LVS oval].

J F M A M J J A S

STEM 4-angled; rough

BRACTS **entirely green**

LVS rough margins

PETALS **white, deeply split;** STYLES **4 or 5;** SEPALS **separate, not forming a tube**

Ⓐ Sea Mouse-ear *Cerastium diffusum*

H to 10 cm. **Form** largely prostrate (SPREAD to 30 cm); stickily hairy. **Fls** D 3–6 mm; L PET < SEP; BRACTS entirely green. **Lvs** ± broadly oval. **Hab** sand dunes, open grassland, gravelly areas, predominantly coastal. **SS** Little Mouse-ear *C. semidecandrum* (N/i) [PET + SEP 5; BRACTS silvery with green central stripe].

J F M A M J J A S O N D

PETALS; SEPALS; STAMENS; STYLES **usually 4**

Ⓟ Snow-in-summer *Cerastium tomentosum*

H to 20 cm. **Form** densely **mat-forming** (SPREAD to 30 cm); densely hairy. **Fls** D 12–18 mm; L PET > SEP. **Lvs** narrowly oval. **Hab** grassland, dunes, brownfield, road verges. **SS** Field Mouse-ear *C. arvense* (N/i) [FLS fewer; FORM pubescent; not so mat-forming]

J F M A M J J A S O N D

PET L at least 2× sepals

LVS densely covered in soft white hairs

SEPARATE BY: ► sepal hairs ► bract colour ► inflorescence ► leaf colour + shape

Ⓟ Common Mouse-ear *Cerastium fontanum*

H to 30 cm. **Form** mat-forming with erect fl shoots; hairy. **Fls** D 7–10 mm on relatively long stalks. **Lvs** ± broadly oval; green; with long white, mostly non-sticky, hairs. **Hab** grassland, arable margins.

J F M A M J J A S O N D

Ⓐ Sticky Mouse-ear *Cerastium glomeratum*

H to 40 cm. **Form** erect; stickily hairy. **Fls** D 8–10 mm on short stalks. **Lvs** broadly oval; yellow-green; hairy. **Hab** bare patches (*e.g.* molehills, anthills) in grassland; arable margins; walls and sand dunes.

J F M A M J J A S O N D

L ≥ sepals

SEPALS hairs mostly non-sticky

INFL relatively unclustered

BRACT margins silvery

LVS oval; green

PET L ≥ sepals

INFL clustered; compact

SEPALS hairs long; white; sticky (often with adhering particles)

BRACTS entirely green

LVS broad oval; yellow-green

LEAVES without stipules

SEPALS **separate, not forming a tube;** PETALS **entire or slightly notched**

Ⓑ Thyme-leaved Sandwort
Arenaria serpyllifolia

H to 25 cm. **Form** sprawling to erect and bushy; widely branched. **Fls** D 5–8 mm; L PET < SEP; STYLES 3. **Lvs** L to 25 mm; narrowly oval; pointed. **Hab** woods, hedgebanks. **SS** Slender Sandwort *A. leptoclados* (INSET) [generally smaller; FR lacks neck]; often grow together.

FR flask-shaped, with short neck

SLENDER

Ⓐ Three-nerved Sandwort
Moehringia trinervia

H to 40 cm. **Form** weakly erect; widely branched. **Fls** D 6 mm; L PET < SEP; STYLES 3. **Lvs** L to 5 mm; broadly oval. **Fr** seed shiny; smooth with attached oil-body. **Hab** woods, dry shady places. **SS** Wood Stitchwort *Stellaria nemorum* (N/I) [PET divided; L > SEP].

FL

LF 3-veined (most obvious from below)

Ⓟ Knotted Pearlwort *Sagina nodosa*

H to 15 cm. **Form** diffusely spreading; erect flowering stem. **Fls** D 5–10 mm; L PET 2× SEP; STYLES 5. **Lvs** linear with a tiny terminal point (L <0·1 mm); LOWER L >3× UPPER. **Hab** short turf, damp areas.

STYLES **5**

FL

LVS with additional short leaves ('knotted') at base

SEPALS **separate, not forming a tube;** PETALS **minute or absent**

● **'Tiny/absent' flowered pearlworts** | **Fls** D 5 mm; SEPALS 4 or 5 with white lobes. **Lvs** linear; tip with short point in most species.

Ⓟ Procumbent Pearlwort *Sagina procumbens*

Spread to 20 cm. **Form** mat-forming; rooting at nodes; many short non-flowering shoots; **central rosette**. **Fls** D 5 mm; SEP blunt, hooded at tip. **Lvs** linear; point L <0·2 mm. **Hab** paths, lawns, short turf.

FORM central rosette diagnostic of species

SEPALS blunt; hooded at tip

LF point L <0·2 mm

Annual pearlwort ID	Ⓐ **Slender Pearlwort** *Sagina filicaulis*	Ⓐ **Annual Pearlwort** *Sagina apetala*
Form	erect, diffuse; H to 15 cm	
Sepals	INNER + OUTER both hooded; may have red margins; spread in fruit	INNER hooded; OUTER pointed; never with red margins; erect in fruit
Leaf point	L <0·4 mm	
Fruit	L 1·8–2·4 mm	L 1·6–2·2 mm
Habitat	paths, bare ground, cliffs	

SLENDER   ANNUAL

J F M A M J J A S O N D   J F M A M J J A S O N D

SS Sea Pearlwort *S. maritima* (N/I) [LVS blunter; tip L <0·1 mm].

ALL SEPALS hooded; margins can be red

INNER SEPALS hooded; margins never red

SEPALS spread in fruit

SEPALS erect in fruit

LF point L <0·4 mm in both species

---

45 Amaranthaceae | **Goosefoot + Orache** family   10 spp. | 52 spp. B&I

Form herbaceous, can have woody base; typically greenish. **Fls** single whorl of green or straw-coloured segments. **Lvs** with fleshy tissues or salt-excretion cells in those species that are adapted to salty habitats. NOTE: recent molecular research has led to many changes of scientific names.

Goosefoot groups and species identification

LEAVES **strongly flattened; fleshy in some**

FLOWERS bisexual; FRUIT not enclosed

FLOWERS unisexual; FRUIT between two distinctively shaped bracts

FORM not woody

FORM woody shrub

FL

BRACT

GOOSEFOOTS + **Beet** *pp. 152–153*

ORACHES *pp. 152–153*

Sea-purslane *p. 154*

LEAVES **fleshy; swollen; at least half-cylindrical**

X-SECTION

Annual Sea-blite *p. 154*

● **Glassworts (Samphires)** *Salicornia* and *Sarcocornia* spp. | Spiky, saltmarsh mud specialists that look like no other plant. The stems are entirely enclosed in 'leaves' fused into a succulent sheath, forming apparent segments; green in spring, turning yellow, orange or red in autumn. Flowers, hidden behind the fleshy leaves, are a single anther which emerges through a tiny pore, and an ovary with a forked stigma. There are broadly three Glasswort types – **one perennial species** and two groups of annuals (**one with single flowers**, the other with **flowers in groups of 3**). Some are very tricky to identify to species and best tackled in Sep–Oct when the colours can be helpful.

151

FLOWERS **bisexual;** FRUIT **not enclosed** **SEPARATE BY:** ▶ infl structure + fruit details

🅐 **Red Goosefoot** *Oxybasis rubra*

J F M A M J J A S O N D

H to 80 cm. **Form** prostrate to erect. **Infl** greenish-red; not mealy. **Lvs** L to 50 mm; broadly diamond-shaped; usually deeply toothed; **glossy**. **Hab** manure heaps, farmyards, fallow fields, drying mud, coastal marshes, dune slacks.

🅐 **Fat-hen** *Chenopodium album*

J F M A M J J A S O N D

H to 150 cm. **Form** erect; STEM can be suffused with purple. **Infl** greenish; **mealy**. **Lvs** L to 50 mm; triangular to diamond-shaped; usually toothed; **mealy** (especially on underside). **Hab** arable margins, other cultivated ground, brownfield.

LF deeply and irregularly toothed; often with strong red tinge

LF variable; broadly triangular; toothed but not 3-lobed

FLOWERS **bisexual;** FRUIT **between two distinctively shaped bracts**

🅐 **Common Orache** *Atriplex patula*

J F M A M J J A S O N D

H to 80 cm. **Form** sprawling to erect. **Infl** greenish; not mealy. **Lvs** L to 60 mm; diamond-shaped, STALK L <10 mm. **Hab** disturbed areas, both inland and coastal.

🅐 **Spear-leaved Orache** *Atriplex prostrata*

J F M A M J J A S O N D

H to 80 cm. **Form** sprawling to erect; often reddish. **Infl** greenish, slightly mealy. **Lvs** L to 60 mm; triangular; STALK L >10 mm. **Hab** saltmarsh, upper beaches, arable margins, brownfield.

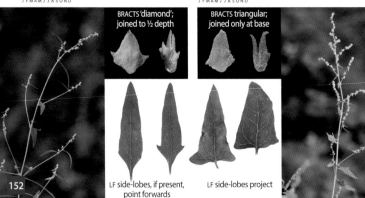

BRACTS 'diamond'; joined to ½ depth

BRACTS triangular; joined only at base

LF side-lobes, if present, point forwards

LF side-lobes project

Ⓐ Fig-leaved Goosefoot
Chenopodium ficifolium

H to 100 cm. **Form** erect. **Infl** greenish; mealy; TEPALS slightly keeled. **Lvs** L to 80 mm; **deeply 3-lobed**; mealy. **Hab** manure heaps, arable margins, other cultivated ground, brownfield.

J F M A M J J A S O N D

Ⓐ Many-seeded Goosefoot
Lipandra polysperma

H to 100 cm. **Form** sprawling to erect; well-branched; many red-tinged; STEM square. **Infl** greenish; not mealy. **Lvs** L to 80 mm; **broadly oval, almost untoothed**. **Hab** arable margins, other cultivated ground, brownfield.

J F M A M J J A S O N D

LF variable; central lobe narrow, almost parallel-sided; L 2–3× lateral lobes

LF broadly oval; ± untoothed

Ⓐ Grass-leaved Orache   *Atriplex littoralis*

H to 100 cm. **Form** erect. **Infl** greenish; slightly mealy. **Lvs** L to 60 mm; **very narrowly diamond-shaped or linear**, STALK very short or absent. **Hab** saltmarsh, sea walls and other coastal habitats, salted road verges.

J F M A M J J A S O N D

ⓐⓟ Beet   *Beta vulgaris*

L to 100 cm. **Form** sprawling; woody at base. **Fls** small; greenish; 1–3. **Lvs** L to 20 cm; broadly oval; glossy green (fleshy in coastal ssp.); edges can be wavy. **Hab** cultivated / disturbed ground, sea walls, saltmarshes, coastal shingle.

J F M A M J J A S O N D

BRACTS triangular; joined at base; rough

LF narrow; side-lobes absent

FL with 5 equal green or purple tepals

LF ± fleshy; can be red or purplish

FR

153

LEAVES **fleshy; strongly flattened**

FORM **mounded shrub**

ⓟ Sea-purslane *Atriplex portulacoides* **WS**

FL

H to 80 cm. **Infl** greenish-yellow, dense, not mealy. **Lvs** L to 60 mm; narrowly oval; entire; fleshy; can be **covered in crystalline salt**. **Hab** saltmarsh (especially creek-sides), splash-zone marshes on cliffs.

J F M A M J J A S O N D

LF narrowly oval

LEAVES **fleshy; swollen; at least half-cylindrical**

ⓐ Annual Sea-blite *Suaeda maritima*

FL 5 tepals; 5 stamens; 2 stigmas

H to 50 cm. **Form** succulent; hairless; very variable – from erect to heavily branched to spreading; base can be woody. **Infl** small; greenish; in groups of 1–3. **Lvs** L to 25 mm; linear; fleshy; autumn colour ranges from **yellow-green to russet or purple**. **Hab** muddy saltmarshes, salted road verges.

J F M A M J J A S O N D

LF flat above, rounded below; tip sharply pointed

---

46 Montiaceae | **Blinks** family

2 spp. | 3 spp. B&I

Form most species fleshy. **Fls** PETALS 5; SEPALS 2; OVARY superior – the two sepals differentiate this family from other families with 5-petalled flowers and a superior ovary.

ⓐ Pink Purslane *Claytonia sibirica*

H to 40 cm. **Form** erect. **Fls** D 15–20 mm; **pink**; PET deeply notched; STAMENS 5. **Lvs** BASAL LVS L to 30 mm, oval, long-stalked; STEM-LVS opposite; **just one stalkless pair**. **Hab** damp woodland, shaded stream banks.

J F M A M J J A S O N D

ⓐⓟ Blinks *Montia fontana*

H to 40 cm. **Form** tiny and erect, or trailing, or aquatic f. **Fls** D 3 mm; white. **Lvs** STEM-LVS L to 20 mm; in several opposite pairs. **Fr** SEEDS blackish; smooth (ssp. *amporitana*) or warty (ssp. *chondrosperma*). **Hab** damp to wet places (*e.g.* pond margins; damp depressions in heathland). **SS** water-starworts (*p.228*) [PET + SEP absent]; **Allseed** *Radiola linoides* (N/I) [FL-parts in 4s].

J F M A M J J A S O N D

FL petals deeply notched

FL tiny

FORM repeatedly forked branching structure

154

47 Cornaceae | **Dogwood** family | 2 spp. | 5 spp. B&I

Form shrubs. **Infl** branched, ± flat-topped. **Fls** PETALS 4; SEPALS 4; STAMENS 4; STYLES 1; OVARY inferior. **Lvs** opposite; **prominent veins** contain threads that persist when leaf torn. **Fr** a 1-seeded berry.

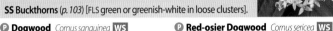

● **Dogwoods** | **Form** deciduous shrubs; patch forming by suckers.

SS Buckthorns (*p.103*) [FLS green or greenish-white in loose clusters].

Ⓟ Dogwood *Cornus sanguinea* WS

H to 4 m. **Fls** D 5–10 mm; creamy-white; in flat-topped clusters. **Lvs** L to 80 mm; broadly oval with abrupt, sharp point. **Fr** shiny, purple-black berry; D 5–8 mm. **Hab** woods, scrub, hedges especially on calcareous soils. **SS** ssp. *australis* (N/I) [LF-HAIRS 2-branched; appressed to lf-surface] is often planted in urban areas/road verges.

J F M A M J J A S O N D

Ⓟ Red-osier Dogwood *Cornus sericea* WS

H to 3 m. **Fls** D 4–6 mm; white; in usually flat-topped clusters. **Lvs** L to 100 mm; broadly oval with tapering point. **Fr** whitish berry (if it ripens); D 4–7 mm; SEED L = W. **Hab** parks, gardens, roadsides, riverbanks, damp woodland. **SS White Dogwood** *C. alba* (N/I) [seed L>W; 1ST-YEAR TWIGS bright red in winter] is planted in similar areas.

J F M A M J J A S O N D

1ST-Y TWIG purplish red

FR **purple-black**

LF 2–5 pairs of lateral veins

1ST-Y TWIG dark red or greenish-yellow

FR **whitish**

LF 6–7 pairs of lateral veins

48 Primulaceae | **Primrose** family

Fls PETALS (where present) 5; SEPALS fused into tubes, at least at the base – a combination of features that is characteristic of this family.

Primrose groups and species identification

LEAVES **all basal**; SEPAL-TUBE **long**; largely fused	LEAVES **at least some on the stem**; SEPAL-TUBE **divided ≥ half way**		
PETAL-TUBE **showy**; long	PETAL-TUBE **very short**		PETAL-TUBE **absent**
	FORM erect, tall; oval to oblong (L ≥ 2×W)	FORM low-growing or prostrate; LVS rounded to oval (L ≤ 2×W)	LVS **fleshy**
			SEPALS petal-like
Cowslip + Primrose *p.157*	**LOOSESTRIFES** *p.157*	**PIMPERNELS** *p.156*	**Sea Milkwort** *p.156*

℗ **Sea Milkwort** *Lysimachia maritima*

L to 30 cm. **Form** rooting at some nodes. **Fls** D 5 mm; PET absent; SEP **petal-like, pale pink** or white; solitary. **Lvs** L to 12 mm; slightly fleshy. **Fr** rounded; D 2·5–4·0 mm. **Hab** salty or brackish mud and sand, coastal rock crevices.

LF **slightly fleshy**; in opposite pairs; each pair at right-angles to the adjacent pair

● **Pimpernels** | **Form** low-growing or prostrate. **Fls** PETAL-TUBE short; **Lvs** on stem; L ≤ 2× W.

℗ **Bog Pimpernel** *Lysimachia tenella*

L to 20 cm. **Form** not rooting at nodes. **Fls** D 6–10 mm; solitary; PET **pale pink**; bell-shaped. **Lvs** L to 10 mm; ± round; STALK short, slender. **Fr** rounded; D 2·5–4·0 mm. **Hab** damp ground on acid soils.

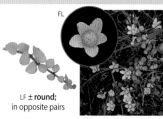

LF ± **round**; in opposite pairs

℗ **Scarlet Pimpernel** *Lysimachia arvensis*

L to 40 cm; **Fls** D 4–7 mm; solitary; PET **scarlet** (rarely pink or blue), with purple eye. **Lvs** L to 28 mm; broadly oval; STALK short, slender. **Fr** rounded; D 4–6 mm. **Hab** free-draining, sandy soils, disturbed and cultivated ground. **SS Blue Pimpernel** *L. foemina* (N/I) [PET hairs fewer, shorter + with more globular tip].

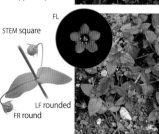

STEM square

LF rounded
FR round

℗ **Yellow Pimpernel** *Lysimachia nemorum*

L to 40 cm; **Form** sprawling. **Fls** D 5–12 mm; STALK **slender**, L ≥ leaf; PET yellow. **Lvs** L to 30 mm; broadly oval; pointed; STALK short, slender. **Fr** rare; D to 3 mm. **Hab** damp, shady places: woodland rides and flushes, hedges, ditches, marshes.

SEPALS very narrow

LF pointed oval

℗ **Creeping-Jenny** *Lysimachia nummularia*

L to 60 cm. **Form** rooting at nodes. **Fls** D 15–30 mm; STALK **stout**, L < leaf; PET yellow. **Lvs** L to 30 mm; round to broadly oval; black glands; STALK short, stout. **Fr** D to 3 mm (rarely produced). **Hab** shady and damp hedges and ditches, river and stream banks, damp grassland.

SEPALS broad

LF round

● **Loosestrifes** | **Form** erect. **Lvs** some on stem; L ≥ 2×W. **Fls** PETAL-TUBE very short; SEPAL-TUBE divided to halfway or more. **Lvs** dotted with glands.

℗ **Yellow Loosestrife** *Lysimachia vulgaris*

H to 150 cm. **Fls** D 8–18 mm; yellow (can have orange 'eye'); in loose, leafy infl; PETAL-LOBES L to 10 mm, hairless, tip rounded. **Lvs** L to 10 cm; narrowly oval; quite hairy. **Fr** D to 5 mm; ± **globular**; L ≥ sepals. **Hab** fens, marshes, banks of waterbodies. **SS** Tufted Loosestrife *L. thyrsiflora* (N/i) [FLS small in dense, long-stalked clusters].

FL

PET-LOBES hairless

SEPALS **orange margin**

LF ± stalkless

℗ **Dotted Loosestrife** *Lysimachia punctata*

H to 120 cm. **Fls** D 10–16 mm; yellow, with dark orange 'eye'; clustered in ± dense leafy infl; PETAL-LOBES L 9–15 mm, margins with glandular hairs, tip ragged. **Lvs** L to 12 cm; narrowly oval; very hairy. **Fr** L to 5 mm; **egg-shaped**; L < sepals. **Hab** damp places, rough grassland, roadsides. streamsides in woodland.

FL

PET-LOBES glandular

SEPALS all green

LF short-stalked

● **Primrose + Cowslip** | **Lvs** all basal (rosette). **Fls** PETAL-TUBE showy, long; SEPAL-TUBE long, fused for much of its length.

SS Oxlip *P. elatior* (N/i) [LVS abruptly contracted to stalk; FLS size as Primrose, several on common stalk like Cowslip; AROMA peachy notes]; also beware complex hybrids.

℗ **Primrose** *Primula vulgaris*

H to 15 cm. **Fls** D 20–40 mm; usually solitary; PET **pale yellow** with darker centre and orange honey-guides; AROMA (when warm) spicy, with lemon, honey and violet. **Lvs** L to 15 cm; UPPERSIDE smooth; UNDERSIDE downy. **Fr** L to 20 mm; SEEDS sticky when fresh. **Hab** woods, hedgerows, damp grassland, often on heavy soils.

FL

LF **tapering gradually to base**; stalk very short or absent

℗ **Cowslip** *Primula veris*

H to 30 cm. **Fls** D 8–15 mm; up to 30 in an umbel nodding to one side; PET **deep yellow** with orange mark on each; AROMA (when warm) sweet and spicy, with apricot. **Lvs** L to 15 cm; UPPERSIDE + UNDERSIDE downy. **Fr** L to 10 mm; enclosed by inflated sepal-tube; SEEDS dry when fresh. **Hab** grassy places usually on light, calcareous soils.

FL

LF ± **abruptly contracted** to stalk

49 Ericaceae | **Heather** family

Form shrubs, often low-growing and mostly avoiding limy soils. **Fls** typically pale pink to mauve; PETALS typically fused into a tube or bell. **Lvs** dark green; tiny and scale-like in some.

Heather groups and species identification

FORM small shrubs (H <2 m); LEAVES small (L<30 mm)			FORM **large shrubs (H >2 m)**; LVS **large, (L<30 mm); leathery**
FLOWER PARTS **in 4s or 5s**; PETALS **fused**		FLOWER PARTS **tiny; in 3s**; PETALS **free**	
LVS **oval**	LVS **'needles' or 'scales'**		

ERICACEOUS 'BERRIES' *below*

HEATHERS *opposite*

Crowberry *opposite*

RHODODENDRONS *opposite*

FORM **small shrubs (H <2 m)**; FL PARTS **in 4s or 5s**; PETALS **fused**; LEAVES **oval**; FRUIT **globular berry**

ⓟ **Bilberry** *Vaccinium myrtillus* WS

H to 60 cm. **Form** spreading with many erect stem; TWIGS 3-angled; green. **Fls** L 4–6 mm; **deep pink**; 1–2 in leaf-axils. **Lvs** L to 30 mm; bright green; thin. **Fr** bluish-black; D 6–10 mm. **Hab** upland moorland, woodland on acid soils.

J F M A M J J A S O N D

FL **lantern-shaped**

LF-MARGIN **toothed**

FR bluish-black with whitish coating

ⓟ **Cowberry** *Vaccinium vitis-idaea* WS

H to 30 cm; SPREAD to 150 cm. **Form** ± prostrate, creeping dwarf shrub. **Fls** L 5–8 mm; **pale pink**; 6–10 in terminal cluster. **Lvs** L to 30 mm; dark green; leathery. **Fr** red; D 5–10 mm. **Hab** upland moorland, woods, on acid soils.
SS Bog Bilberry *V. uliginosum* (N/I) [FORM deciduous; FLS fewer; constricted at mouth].

J F M A M J J A S O N D

FL **bell-shaped; mouth splayed**

LF margin **curled under**

LF UND **gland dots**

FR red berry

ⓟ **Bearberry** *Arctostaphylos uva-ursi* WS

H to 30 cm; SPREAD to 150 cm. **Form** prostrate, with long, rooting branches. **Fls** L 5–6 mm; **pale pink**; 4–12 in terminal cluster. **Lvs** L to 30 mm; **broadest near tip**; UPPERSIDE dark green; UNDERSIDE paler. **Fr** bright red; D 6–10 mm. **Hab** upland moorland, rocks, in exposed places on thin peaty soils.

J F M A M J J A S O N D

FL **bell-shaped; mouth not splayed**

LF margin **flat**

LF UND **net-veined**

FR red berry

FORM **small shrubs (H <2 m);** FL PARTS **in 4s or 5s;** PETALS **fused;** LEAVES **needle- or scale-like;** FRUIT **dry capsule**

℗ Heather
Calluna vulgaris **WS**

H to 80 cm. **Form** erect, multi-stemmed, becoming straggly. **Fls** L 3–4 mm; numerous on short stalks in leaf-axils. **Lvs** L to 3·5 mm; **scale-like,** ± overlapping. **Hab** moors, heaths, on acid, well-drained soils.

℗ Bell Heather
Erica cinerea **WS**

H to 75 cm. **Form** erect, becoming straggly. **Fls** L 4–7 mm; numerous in rounded clusters both terminal and in leaf-axils. **Lvs** L to 7 mm; needle-like, dark green. **Hab** dry heaths and moors.

℗ Cross-leaved Heath
Erica tetralix **WS**

H to 70 cm. **Form** erect; sparsely branched; greyish-hairy. **Fls** L 6–9 mm; up to 12 in rounded terminal clusters. **Lvs** L to 5 mm; needle-like, **hairy.** **Hab** heaths, moors, bogs, in damp areas.

HEATHER

FL **pale pink;** elongated-globular

LVS in **whorls of 4**

FL **magenta;** bell-shaped

LVS in **whorls of 3**

FL **pinkish-lilac;** bell-shaped

LVS in **opposite pairs**

BELL HEATHER

CROSS-LEAVED

FORM **small shrubs (H <2 m);** FL PARTS **in 3s;** PETALS **free;** ♂ + ♀ on separate plants

℗ Crowberry *Empetrum nigrum* **WS**

H to 15 cm; SPREAD to 120 cm. **Form** mat-forming; TWIGS reddish when young. **Fls** tiny (L 1–2 mm); stalkless; 1–3 in leaf-axils. **Lvs** L to 7 mm; narrow; clustered. **Fr** berry; D 5–6 mm; **black when ripe. Hab** moors, bogs, pine and birch woods, on dry, peaty soils.

♂ FLS long filaments

FR

♀ FLS

LVS shiny; leathery

FORM **large shrubs (H >2 m);** LEAVES **large, leathery**

℗ Rhododendron *Rhododendron ponticum* **WS**

H to 4 m. **Form** patch-forming; evergreen. **Fls** D 40–60 mm; **purple; yellow-spotted inside. Lvs** L to 20 cm; UPPERSIDE dark green; UNDERSIDE paler. **Fr** rounded woody capsule; D 2·5–4·0 mm; slightly furry. **Hab** woods and heaths on acid or peaty soils. **SS** leaves – **Cherry Laurel** (*p. 93*) [CRUSHED LVS scentless].

unmistakeable in flower

PET-LOBES spreading, SYM slight bilateral (3 upper petals, 2 lower)

159

50 Rubiaceae | Bedstraw family `10 spp.` | 20 spp. B&I

Easily recognizable family. **Form** trailing, climbing or upright.
Fls small (D ≤6mm); in loose clusters or in leaf-axils; PETALS 4–5;
fused at base into a short tube, SEPALS tiny or absent.
Lvs 2, together with 2 or more leaf-like stipules in a whorl.
Fr SEEDS paired and partially fused.

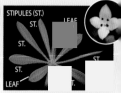

STIPULES (ST.)

Key to Bedstraws

Fls	Infl	SPECIES	
yellow	dense clusters in leaf-axils	**Crosswort**	*below*
	terminal cluster	**Lady's Bedstraw**	*below*
pink	terminal cluster	**Field Madder**	*below*
white	flat-topped clusters	**Woodruff**	*p. 162*
	loose terminal + leaf-axil clusters	**Other *Galium* bedstraws** [6 species]	*opposite*

IDENTIFY BY: ► flower colour ► form ► infl shape ► leaf, stem + seed details

---

FLOWERS **yellow**; SEPALS **absent**

Ⓟ Lady's Bedstraw *Galium verum*

H to 60cm. **Form** creeping
to erect; patch forming. **Fls**
D 2–3mm. **Lvs** L to 25mm;
linear; short projecting point;
AROMA of hay when crushed
or dried. **Fr** L to 1·5mm;
smooth; black when ripe.
Hab dry grassland, especially
calcareous soils and coastal.

J F M A M J J A S O N D

FLS in
terminal
clusters

LVS in whorls of 8–12;
margins inrolled

Ⓟ Crosswort *Cruciata laevipes*

H to 60cm. **Form** erect, hairy,
patch-forming. **Fls** D 2–3mm.
Lvs L to 25mm; **narrowly oval**;
3-veined; AROMA strongly
of honey. **Fr** L to 1·5mm;
smooth; black when ripe.
Hab grassland and scrub,
especially on calcareous soils.

J F M A M J J A S O N D

FLS in
axillary
clusters

FLS

LVS in whorls of 4;
3-veined

FLOWERS **pink**; INFLORESCENCE **terminal clusters**; SEPALS **present**

Ⓐ Field Madder *Sherardia arvensis*

L to 40cm. **Form** trailing.
Fls D 2–3mm. **Lvs** L to 20mm;
narrowly oval; pointed; LOWER
STEM whorls of 4; HIGHER
STEM 5–6. **Fr** L to 4mm paired;
bristly. **Hab** arable margins,
open grassland, especially on
calcareous or free-draining
soils.

J F M A M J J A S O N D

FLS

LVS with forward-pointing prickles

FLOWERS **white**; INFLORESCENCE **loose clusters (terminal and/or in leaf-axils)**; SEPALS **absent**

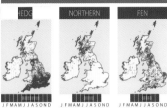

HEDGE | NORTHERN | FEN | MARSH | HEATH | CLEAVER

J F M A M J J A S O N D   J F M A M J J A S O N D   J F M A M J J A S O N D   J F M A M J J A S O N D   J F M A M J J A S O N D   J F M A M J J A S O N D

🅟 Hedge Bedstraw *Galium album*

H to 100 cm. **Form** sprawling to erect.
Infl loose. **Fls** D 3–4 mm with pointed lobes.
Lvs L to 25 mm; narrowly oval. **Fr** L 1–2 mm.
Hab rough grassland, hedgebanks.

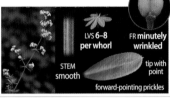

LVS **6–8** per whorl | FR **minutely wrinkled**
STEM smooth | tip with point
forward-pointing prickles

🅟 Northern Bedstraw *Galium boreale*

H to 45 cm. **Form** erect, rigid. **Infl** leafy,
pyramidal clusters. Fls D 4 mm. **Lvs** L to 40 mm;
narrowly oval, widest below middle; **3-veined**.
Fr L 2·5 mm. **Hab** grassy, rocky and gravelly
places in the uplands; sand dunes.

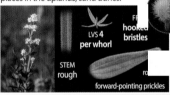

LVS **4** per whorl | FR hooked bristles
STEM rough | r...
forward-pointing prickles

🅟 Fen Bedstraw *Galium uliginosum*

H to 60 cm. **Form** slender; sprawling to
scrambling. **Infl** narrow. **Fls** D 2·5–3·0 mm with
pointed lobes. **Lvs** L to 10 mm; narrowly oval.
Fr L 1–2 mm. **Hab** fens, marshes.

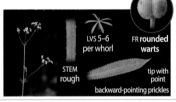

LVS **5–6** per whorl | FR **rounded warts**
STEM rough | tip with point
backward-pointing prickles

🅟 Common Marsh-bedstraw *Galium palustre*

H to 100 cm. **Form** sprawling to weakly erect.
Infl loose, spreading. **Fls** D 3·0–4·5 mm with
pointed lobes. **Lvs** L to 35 mm; narrowly oval.
Fr L 1–2 mm. **Hab** marshes, fens, edge of
waterbodies, wet woodland.

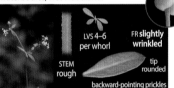

LVS **4–6** per whorl | FR **slightly wrinkled**
STEM rough | tip rounded
backward-pointing prickles

🅟 Heath Bedstraw *Galium saxatile*

H to 25 cm. **Form** erect, mat-forming. **Infl**
leafy clusters. **Fls** D 3 mm. **Lvs** L to 10 mm. **Fr**
L 1·5–2·0 mm. **Hab** acid grassland, heathland,
moorland. **SS** Limestone Bedstraw *G. sterneri*
(N/I) [LF-MARGIN with some backward-pointing
prickles; HAB calcareous soils only].

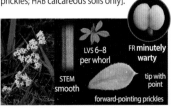

LVS **6–8** per whorl | FR **minutely warty**
STEM smooth | tip with point
forward-pointing prickles

🅐 Cleavers *Galium aparine*

L to 300 cm. **Form** scrambling. **Infl** few-
flowered. **Fls** D 1·5–2·0 mm; 2–5 in clusters
in leaf-axils. **Lvs** narrowly oval; L to 50 mm.
Fr L 1·5–2·0 mm. **Hab** hedgebanks, arable
margins, scrub, brownfield.

LVS **6–8** per whorl | FR **hooked bristles**
STEM rough | tip with point
backward-pointing prickles

161

FLOWERS white; INFLORESCENCE flat-topped clusters; SEPALS absent

Ⓟ Woodruff *Galium odoratum*

H to 40 cm. **Form** erect shoots; patch-forming. **Infl** terminal and lateral flat-topped heads. **Fls** D 4–6 mm. **Lvs** L to 20 mm; 1 main vein; narrowly oval; AROMA hay-scented when bruised or dried. **Fr** L 2–3 mm; with hooked black-tipped bristles. **Hab** woods and hedges, especially on calcareous soils.

LVS in whorls of 6–8; margin with forward-pointing prickles

J F M A M J J A S O N D

---

51 Convolvulaceae | Bindweed + Dodder family   **4 spp. | 8 spp. B&I**

Two very divergent subfamilies that have little in common but their anti-clockwise twining habit. The parasitic dodders lack green pigmentation, showy flowers and roots.

● **Bindweeds | Form** green twining stem and large leaves. **Fls** funnel-shaped; PETALS generally fused; SEPALS small and not joined; BRACTS below flowers sometimes large and encompassing flower-base. **Fr** rounded capsule.

HEDGE   LARGE

J F M A M J J A S O N D   J F M A M J J A S O N D

SEPARATE BY:
► flower size + colour
► sepals + bracts

Larger bindweeds	Ⓟ **Hedge Bindweed** Ⓒ *Calystegia sepium*	Ⓟ **Large Bindweed** Ⓒ *Calystegia silvatica*
Form	vigorous twining climber or trailer; sometimes woody at base; L to 3 m	
Flower	D 10–30 mm	D 50–80 mm
Sepals	visible between bracts	± concealed by bracts
Leaf	broadly triangular; deep angle where stalk joins blade	
	L 5–10 cm	L 15–15 cm
Fruit	D 7–12 mm	D 10–15 mm
Habitat	hedgerows, gardens	
	riverbanks, wet woodland	brownfield

LARGE BINDWEED

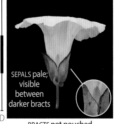

SEPALS pale; visible between darker bracts

BRACTS not pouched

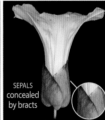

SEPALS concealed by bracts

BRACTS pouched

LVS are a similar shape in both species, with some size overlap; although Large Bindweed leaves are usually much larger

P Field Bindweed *Convolvulus arvensis* **C**

L usually to 75 cm (up to 2 m).
Form twining climber or sprawler. **Fls** D 10–30 mm; white or pink-and-white; 1–3 together; FL-STALK longer than lvs. **Lvs** L to 50 mm; triangular with basal lobes. **Fr** D 3 mm. **Hab** arable margins, road verges, rough grassland.

FL usually pink-and-white

FL-STALK pair of small bracts halfway

LF usually ± triangular

● **Dodder** | **Form lacking chlorophyll**; twining parasites; tangle of **fine reddish stems** on host plants. **Lvs** reduced; scale-like. **Fls** tiny, in **dense globular clusters**.

A Dodder *Cuscuta epithymum* **C**

L to 1 m. **Fls** D 3–4 mm; in dense clusters D 5–10 mm. **Fr** rounded capsule; D 2·0–2·5 mm. **Hab** heathland, downland (main hosts are Gorse (*p.81*) and Heather). (*p.159*). **SS Greater Dodder** *C. europaea* (N/I) [FLS paler, in larger clusters; parasitic mainly on Common Nettle (*p.105*)].

FL bell-shaped, with spreading lobes; STAMENS L > petals

FL-STEM reddish and scrambling

52 Apocynaceae | Periwinkle family

2 spp. | 3 spp. B&I

Form trailing, woody-based, evergreen perennials. **Fls** PETALS 5; fused into a tube; distinctively twisted in one direction in bud. **Lvs** opposite.

SEPARATE BY: ► leaf-stalk + leaf-margins ► sepal hairiness

P Greater Periwinkle *Vinca major*

H to 30 cm; **L** to 200 cm. **Form** spreading (some arching), many semi-ascending; rooting at tip. **Fls** purplish-blue; D 40–50 mm. **Lvs** L to 80 mm; broadly oval; LF-STALK L to 10 mm. **Hab** hedgebanks, scrub, woodland.

LF-STALK L to 10 mm; fringed with minute hairs

SEPAL MARGIN hairy

P Lesser Periwinkle *Vinca minor*

H to 20 cm; **L** to 100 cm. **Form** spreading (some arching); rooting at tip and nodes. **Fls** blue; D 25–30 mm. **Lvs** L to 45 mm; oval; LF-STALK minute. **Hab** woods, hedgebanks, shady places; frequently planted for Pheasant cover.

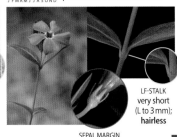

LF-STALK very short (L to 3 mm); hairless

SEPAL MARGIN hairless

53 Veronicaceae | **Speedwell + Toadflax** family | 21 spp. | 43 spp. B&I

A disparate family defined mainly on molecular evidence – the genera are each readily recognizable with common features as follows: **Fls** SYMMETRY bilateral (some almost radial); PETALS + SEPALS each fused into a tube; some with a distinct spur at the base of the petal-tube.

IDENTIFY TO TYPE BY: ► flower shape

Speedwell groups and species identification

FLS **broadly tubular, without spur**; STAMENS 4	FLS **with basal pouch or spur**; STAMENS 4

Foxglove p.169	**FLUELLENS + TOADFLAXES** below and opposite

FL PETALS **4**; fused at the base into a very short tube; STAMENS 2

FL arising singly from leaf-axils	FL in clusters or spikes arising from leaf-axils	FL in terminal leafy spike

SPEEDWELLS p.166

FLS **with basal pouch or spur**; STAMENS 4

● **Fluellens** | **Form** creeping to scrambling. **L** to 50 cm. **Fls** L 7–12 mm (incl. long, conical spur) strongly 2-lipped (UPPER LIP purple, LOWER yellow); PETAL-TUBE closed by swellings on the lower lip. **Hab** mainly along arable margins, especially on light and calcareous soils – often growing together.

SHARP-LEAVED ROUND-LEAVED

Ⓐ Sharp-leaved Fluellen **Ⓐ Round-leaved Fluellen**
Kickxia elatine *Kickxia spuria*

Flower	yellow; UPPER LIP **pale purple**; SPUR **straight**; STALK **hairless**	yellow; UPPER LIP **dark purple**; SPUR **curved**; STALK **hairy**
Leaves	triangular with basal lobes	broadly oval

J F M A M J J A S O N D   J F M A M J J A S O N D

FL-STALK **hairless**
FL
LF **basal lobes**

FL-STALK **hairy**
FL
LF **unlobed**

FLS with basal pouch or spur; STAMENS 4

🔵 **Toadflaxes** | **Form** erect or trailing over rocks. **Fls** size, colour and spur-shape are distinctive for each species, as are their arrangement – spike-like or arising singly from leaf-axils. **Lvs** linear to narrowly oblong or palmately lobed. **Hab** usually disturbed or naturally bare.

🅿 **Snapdragon** *Antirrhinum majus* is a familar garden plant (cultivated since the late 16th century) which can become naturalized (*e.g.* on old walls and buildings, rock faces, pavements and brownfield). Flower structure is as other toadflaxes but it has much larger flowers (L 30–50mm including the spur). Flowers are typically pink-purple, with other colours indicative of a garden cultivar origin.

🅿 **Common Toadflax** *Linaria vulgaris*

H to 80cm. **Form** erect. **Fls** L 18–35mm (incl. spur); **yellow with orange throat swelling**; in spikes which can be dense. **Lvs** grey-green; hairless. **Fr** capsule egg-shaped; L 2× sepals. **Hab** open dry grassland, railway embankments, brownfield.

J F M A M J J A S O N D

SPUR **almost straight** (L to 13mm)

LF ± linear; L to 80mm

🅿 **Purple Toadflax** *Linaria purpurea*

H to 100cm. **Form** erect. **Fls** L 7–15mm (incl. spur); **purple**, mauve or pink (can show darker veins); throat swelling can be white. **Lvs** grey-green; hairless. **Fr** capsule globular; L ± sepals. **Hab** brownfield, rough areas, walls, rocks.

J F M A M J J A S O N D

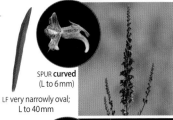

SPUR **curved** (L to 6mm)

LF very narrowly oval; L to 40mm

🅰 **Small Toadflax** *Chaenorhinum minus*

H to 25cm. **Form** erect. **Fls** L 6–9mm (incl. spur); **pink-purple with yellow throat**. **Lvs** often stickily hairy. **Hab** arable margins, brownfield, railway sidings.

J F M A M J J A S O N D

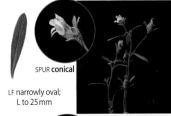

SPUR **conical**

LF narrowly oval; L to 25mm

🅿 **Ivy-leaved Toadflax** *Cymbalaria muralis*

L to 60cm. **Form** trailing. **Fls** L 9–12mm (incl. spur); mauve to purple (a few white); **yellow or white in closed throat**. **Lvs** L to 25mm; palmately lobed (ivy-like). **Hab** walls, rocks, shingle.

J F M A M J J A S O N D

LF **3–5-lobed**, often purple beneath

165

● **Speedwells** | A readily recognizable genus. **Fls** PETALS 4 of unequal size (SYMMETRY bilateral); fused at the base into a very short tube (around the plant you may find whole fallen flowers rather than only fallen petals); STAMENS 2.

FLOWERS **in clusters or spikes arising from base of leaves** | Wetland plants

℗ Brooklime *Veronica beccabunga*

H to 60 cm. **Form** spreading to erect. **Fls** D 6–10 mm; mostly 2 spikes per node. **Lvs** L to 60 mm; broadly oval, **fleshy**, hairless. **Hab** ponds, ditches, riverbanks.

J F M A M J J A S O N D

FLS mostly **2 spikes per node**; PET **bright blue**

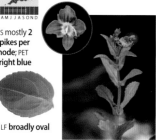

LF broadly oval

℗ Marsh Speedwell *Veronica scutellata*

H to 60 cm. **Form** slender; spreading to scrambling. **Fls** D 5–8 mm. **Lvs** L to 40 mm; linear to narrowly oval, pointed, stalkless. **Hab** marshes, water margins, bogs.

J F M A M J J A S O N D

FLS **1 spike per node**; PET **lilac to whitish**

LF **narrow; stalkless**

℗ Blue Water-speedwell
Veronica anagallis-aquatica

H to 50 cm. **Form** erect; STEM green. **Fls** D 5–6 mm. **Lvs** L to 120 mm; AT STEM-BASE narrowly oval; short-stalked; UPPER STEM longer, pointed; stalkless. **Hab** ponds, slow-moving streams, marshes.

J F M A M J J A S O N D

FLS mostly 2 spikes per node; PET **pale blue with darker veins**

FL-STALK L ≥ **bract**

LF green

℗ Pink Water-speedwell
Veronica catenata

H to 50 cm. **Form** erect; STEM commonly with reddish tinge. **Fls** D 5–6 mm. **Lvs** L to 120 mm; narrowly oval and stalked. **Hab** ponds, ditches, marshes, often on bare mud.

J F M A M J J A S O N D

FLS mostly 2 spikes per node; PET **pinkish**

FL-STALK L < **bract**

LF can have reddish tinge

STEM can be reddish

SS Blue and **Pink Water-speedwell** often grow together and hybridize freely. Hybrids are often more robust plants than either parent and have a less-inflated capsule without, or with only small and deformed, seeds.

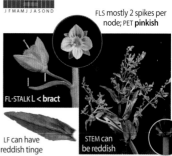

HYBRID | BLUE

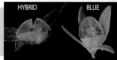

FLOWERS **in clusters or spikes arising from base of leaves**　　Plants of drier habitats

ⓟ Germander Sp'well
Veronica chamaedrys

H to 50 cm. **Form** spreading to erect. **Fls** D 8–12 mm. **Lvs** L to 25 mm; rounded triangle; coarsely toothed; LF-STALK L to 5 mm. **Hab** woodlands, hedgebanks, dampish grassland.

STEM **hairs in two opposite lines**

SPIKE **with up to 20 flowers**; PETALS **blue**

ⓟ Wood Speedwell
Veronica montana

H to 40 cm. **Form** spreading to erect. **Fls** D 7–10 mm. **Lvs** L to 25 mm; rounded triangle; coarsely toothed; LF-STALK L 5–15 mm. **Hab** woodlands, especially dampish.

STEM **hairy all round**

SPIKE **with up to 5 flowers**; PETALS **lilac-blue**

ⓟ Heath Speedwell
Veronica officinalis

H to 40 cm. **Form** creeping with erect fl shoots. **Fls** D 5–9 mm; in dense spikes. **Lvs** L to 30 mm; **broadly oval**; rounded teeth, stalkless; hairy. **Hab** dry grassland, clearings, heathland.

STEM **hairy all round**

SPIKE with up to 25 flowers; PETALS lilac to greyish-blue

GERMANDER
J F M A M J J A S O N D

WOOD
J F M A M J J A S O N D

HEATH
J F M A M J J A S O N D

FLOWERS **in terminal leafy spike**

Ⓐ Wall Speedwell *Veronica arvensis*

J F M A M J J A S O N D

H to 25 cm. **Form** erect, downy. **Fls** D 2–4 mm. **Lvs** L to 10 mm; rounded triangle; coarsely toothed. **Fr** 2-lobed; glandular-hairy; **W** = **L**. **Hab** arable margins, dry open grassland, walls.

LF triangular-oval; coarsely toothed

FR
FL **bright blue**; almost stalkless

ⓟ Thyme-leaved Speedwell
Veronica serpyllifolia ssp. *serpyllifolia*

J F M A M J J A S O N D

H to 30 cm. **Form** spreading, rooting at nodes, with at least half of flowering stem erect, hairless. **Fls** D 5–10 mm. **Lvs** L to 20 mm; broadly oval; hairless; entire. **Fr** 2-lobed; **W** > **L**. **Hab** arable margins, open grassland, brownfield, mountain ledges and gravels.

LF oval; entire

FR
FL **pale blue with darker veins**

NOTE: flowers bright blue in more creeping montane form (ssp. *humifusa*)

167

FLOWERS borne singly in the leaf axils

SLENDER

ⓟ Slender Speedwell *Veronica filiformis*

H to 5 cm. **Form** far-creeping; stem minutely hairy. **Fls** D8–12mm; blue; LOWER LIP can be paler; on wiry stalks much longer than leaves.
Lvs L 5–10mm **rounded to kidney-shaped**; shallowly lobed. **Fr** 2-lobed (very rarely produced). **Hab** lawns, parks and other grassy sites.

LF **rounded to kidney-shaped**

ⓐ Ivy-leaved Speedwell *Veronica hederifolia*

IVY-LEAVED

L to 60 cm. **Form** scrambling. **Fls** D4–9mm; on stalks shorter than leaves. **Lvs** 5–10mm **palmately lobed**; hairy.
Fr slightly 2-lobed; W>L; hairless. **Hab** ssp. *hederifolia* disturbed habitats, gardens; ssp. *lucorum* more shaded habitats *e.g.* woodland edges. **SS** the two sspp. differ slightly (*see table*) but are often not easy to confirm.

subspecies	FLOWER	LF: APICAL LOBE
ssp. *hederifolia*	D ≥6mm; very pale blue; ANTHERS **blue**	W > L
ssp. *lucorum*	D ≤6mm; pale lilac; ANTHERS white/pale blue	W < L

ssp. *hederifolia*　　　　ssp. *lucorum*

SS The **field-speedwells**, especially **Green** and **Grey**, can be confusing.

ⓐ Common Field-speedwell
Veronica persica

COMMON

L to 50 cm. **Form** scrambling to creeping. **Fls** D8–12mm. **Lvs** L to 25mm; rounded triangle, toothed.
Fr W 2× L; 2-lobed; lobes diverging; **with glandular hairs and shorter non-glandular ones**. **Hab** arable margins, gardens, brownfield.

FL bright blue; LOWER LIP **usually white**

ⓐ Grey Field-speedwell
Veronica polita

GREY

L to 30 cm. **Form** scrambling.
Fls D3–8mm. **Lvs** L to 15mm W>L; ± triangular; toothed; greyish-green.
Fr W <2× L; 2-lobed; **with some long glandular hairs and many shorter, curved non-glandular ones**. **Hab** arable margins, gardens.

FL bright blue; LOWER LIP **bright blue**

ⓐ Green Field-speedwell
Veronica agrestis

GREEN

L to 30 cm. **Form** scrambling to erect. **Fls** D3–8mm. **Lvs** L to 15mm; rounded triangle; toothed; light green.
Fr W <2× L; 2-lobed; **with long glandular hairs only**. **Hab** arable margins, gardens.

FL **pale blue or lilac**; LOWER LIP **white**

FLS **broadly tubular, without spur** STAMENS 4

B Foxglove *Digitalis purpurea* ✕

H to 50 cm. **Form** erect; hairy; rosette-forming. **Infl** tall, unbranched spike.
Fls PETAL-TUBE L 40–55 mm; weakly 2-lipped; pink to purple with variable dark spots and blotches on paler pink inside. **Lvs** broadly oval; hairy.
Hab woodland clearings and edges; heathland, uplands. **SS** rosette leaves can resemble other species *e.g.* **mulleins** (*p. 179*) [ROSETTES usually mealy-white] and **Ploughman's-spikenard** *Inula conyzae* (N/I) [LVS sweetly aromatic when rubbed].

PETAL-TUBE weakly 2-lipped, lobes much shorter than the pouch-shaped tube; FL with variable dark spots and blotches on paler pink inside

---

54 Aquifoliaceae | **Holly** family   `1 sp. B&I`

Form evergreen tree. **Fls** each tree has predominantly ♂ or ♀ flowers. **Lvs** leathery; most with very spiny margins.

P Holly *Ilex aquifolium* WT

H to 20 m. **Form** TWIGS green. **Fls** D 6 mm; PET 4; white, fused at extreme base; ♂ FLS 4 stamens; ♀ FLS 4 abortive stamens, STIGMA 4-lobed. **Lvs** oval; L to 100 mm; glossy, dark green; LF-MARGIN usually wavy with strong spines. **Fr** ± globular **scarlet berry**; D 6–10 mm. **Hab** woods, hedges, scrub, gardens. **SS** cultivars with variegation and/or differences in spininess; leaves – **Oregon-grape** (*p. 71*) [LF spines fewer; weaker; all in one plane]; **Evergreen Oak** (*p. 106*) [LF spines fewer; weaker; BARK dark brown, fissured].

♀ FL   ♂ FL

LF **strong spines distinctive**

FR + LVS **unmistakable together**

BARK smooth; pale

---

55 Boraginaceae | **Borage + Forget-me-not** family   `14 spp. | 43 spp. B&I`

Form usually rough or bristly. **Fls** PETALS 5; fused into a tube, at least at the base. **Fr** SEEDS 4 per flower; seeds (often termed 'nutlets') hard and typically coloured.

Borage groups and species identification

FL **symmetry clearly bilateral**	FL **symmetry ± radial**		
	FL **L tubular part ≥ petal-lobes**	FL **L tubular part < petal-lobes**	
		FL throat constricted by white scales	FL throat constricted by yellow scales

Viper's-bugloss *p. 170*	**COMFREYS + LUNGWORTS** *pp. 170–171*	**BORAGES** *p. 171*	**FORGET-ME-NOTS** *p. 173*

FL symmetry clearly bilateral

B **Viper's-bugloss** *Echium vulgare*

H to 80 cm. **Form** erect, branched. **Fls** funnel-shaped; D 12–18 mm; blue; in dense infl (coiled in bud (**buds pink**)). **Lvs** L to 15 cm; narrowly oval with prominent midrib and no apparent lateral veins. **Fr** seed; blackish with irregular ridges. **Hab** sandy or calcareous grassland, cliff slopes, sand dunes, shingle banks.

STAMENS
pink;
unequal;
protruding
from petal-
tube

LVS bristly; hairs
with red bases

FL symmetry ± radial | **FL L tubular part ≥ petal-lobes**   1/2

FORM erect; INFL **nodding heads**

SEPARATE BY: ▶ **leaf-base and stem details** ▶ **seed colour and sheen**

P **Russian Comfrey**
Symphytum asperum × officinale
= *S. × uplandicum*

FLS

H to 150 cm. **Form** very bristly. **Fls** D 15–18 mm; blue to purple. **Lvs** L to 25 cm; ± broadly oval; BASE **not, or barely, running down stem.** **Fr** seed; blackish; dull; rough. **Hab** banks, rough grassland, hedgebanks, road verges, brownfield.

STEM
± unwinged;
very bristly

SEEDS blackish; dull

P **Common Comfrey**
Symphytum officinale

FLS

H to 150 cm. **Fls** D 15–18 mm; purplish-blue, or **cream** (especially in wetland habitats). **Lvs** L to 25 cm; ± broadly oval; BASE **runs down stem as broad wings, at least to the next node.** **Fr** seed; black; shiny. **Hab** rough grassland, hedgebanks, river banks, drier reedbeds.

STEM
winged;
long hairs

SEEDS black; shiny

FL symmetry ± radial | FL L tubular part ≥ petal-lobes 2/2

FORM erect, spreading, evergreen; INFL dense cluster

ⓟ Lungwort *Pulmonaria officinalis*

H to 30 cm. **Fls** D 10–12 mm; reddish-purple. **Lvs** L to 15 cm; broadly oval; BASE heart-shaped or abruptly contracted. **Hab** hedgebanks, woodland edges, scrub, gardens. **SS** several similar spp. and cultivars [LVS shape + pattern].

J F M A M J J A S O N D

LVS with variable white spots and patches

FL symmetry ± radial | FL L tubular part < petal-lobes 1/2

FL throat constricted by white scales

ⓟ Green Alkanet *Pentaglottis sempervirens*

H to 100 cm. **Form** erect; patch-forming. **Fls** D 8–10 mm; bright blue in a clustered, leafy infl. **Lvs** L to 40 cm; ± broadly oval; ± pointed; remain through winter. **Fr** blackish, ridged seed. **Hab** brownfield, rough ground, woodland edges, riverbanks. **SS** leaves – **Bristly Oxtongue** (*p. 205*) [LF bristle pustules larger and more whitish].

J F M A M J J A S O N D

LF bristles often with pustule at base

ⓐ Bugloss *Lycopsis arvensis*

H to 50 cm. **Form** spreading to erect. **Fls** D 5–6 mm; bright blue; PETAL-TUBE curved. **Lvs** L to 15 cm; linear to narrowly oval; margin strongly wavy. **Fr** blackish, ridged, warty seed. **Hab** arable margins, disturbed ground, especially on calcareous or freely draining soils.

J F M A M J J A S O N D

FL petal-lobes slightly unequal

LF-MARGIN strongly wavy

ⓐ Borage *Borago officinalis*

H to 60 cm. **Form** erect. **Fls** D 15–20 mm; reddish- to bright blue (can be white). **Lvs** L to 25 cm; broadly oval; LOWER LVS distinctly stalked; roughly hairy. **Fr** L 7–10 mm; blackish; ridged; with a collar-like ring. **Hab** brownfield, arable margins, disturbed areas.

J F M A M J J A S O N D

LF

ANTHERS **fully exposed**

FL petal-lobes pointed; flat or bent backwards

FL **symmetry ± radial** | FL **L tubular part < petal-lobes**

● **Forget-me-nots** | **Fls** blue; **PETALS** throat constricted by yellow scales.

SS Other rare and localized forget-me-nots of specialized habitats.

SEPALS with hairs (at least some curved/hooked) sticking out; HABITAT dry ground

Form		Flowers		Fruit stalk	Other	SPECIES
		DIAMETER	PETAL-TUBE	L vs SEPAL-TUBE		
erect	tufted	3–5 mm	**L < SEPALS**	L > SEPAL-TUBE	FL gently saucer-shaped	**Field**
	well-branched	6–8 mm	**L ≥ SEPALS**		FL flat	**Wood**
	rosette-forming	1·5–3·0 mm		L < SEPAL-TUBE (can be >)	FL usually more intensely blue than other forget-me-nots	**Early**
spreading		2–3 mm		L < SEPAL-TUBE	FL **colour changes**: cream/yellowish at emergence, gradually becoming yellow, (some can be pink), finally sky-blue	**Changing**

⑳ Field Forget-me-not
Myosotis arvensis

H to 40 cm. **Lvs** L to 60 mm.
Hab open well-drained grassland, fields, gardens.

FR-STALK > SEP-TUBE

FL D <5 mm; saucer-shaped

⑳ Wood Forget-me-not
Myosotis sylvatica

H to 50 cm. **Lvs** L to 80 mm.
Hab gardens; disturbed areas in towns, woods, mountains.

FR-STALK > SEP-TUBE

FL D >5 mm; flat

Ⓐ Early Forget-me-not
Myosotis ramosissima

H to 15 cm. **Lvs** L to 40 mm.
Hab dry, open grassland, heaths, sand dunes.

FR-STALK < SEP-TUBE

FL D <3 mm; intense blue

Ⓐ Changing Forget-me-not
Myosotis discolor

H to 25 cm. **Lvs** L to 40 mm.
Hab dry, open grassland, arable margins.

FR-STALK < SEP-TUBE

FL colour changes

Form		Flowers			Form	SPECIES
		DIAMETER	STYLE	SEPAL-TEETH		
erect	branched; surface or underground runners	6–8mm	L > SEPAL-TUBE	**broadly triangular (equilateral)**		**Water**
	runners absent	3–4mm	L < SEPAL-TUBE	narrowly triangular (isosceles)	STEM lower part with **appressed hairs**	**Tufted**
	surface runners	5–6mm			STEM lower part with **spreading hairs**	**Creeping**

SEPALS **all hairs straight and closely appressed**; HABITAT **damp/wet habitats**

Hab margins of watercourses and bodies, wet marshes.

Ⓟ Water Forget-me-not
Myosotis scorpioides

H to 50 cm. **Lvs** L to 100 mm.

STYLE long; beyond SEP-TUBE

J F M A M J J A S O N D

FL D 6–8mm

SEPAL-TEETH broadly **triangular** (equilateral)

Ⓣ Tufted Forget-me-not
Myosotis laxa

H to 50 cm. **Lvs** L to 40 mm.

STYLE short; not beyond SEP-TUBE

J F M A M J J A S O N D

FL D 3–4mm

LWR STEM HAIRS **appressed**

FR-STALK L <2× SEP-TUBE L

SEPAL-TEETH narrowly triangular (isosceles)

Hab margins of acidic waterbodies, bogs.

Ⓒ Creeping Forget-me-not
Myosotis secunda

H to 50 cm. **Lvs** L to 40 mm.

STYLE short; not beyond SEP-TUBE

J F M A M J J A S O N D

FL D 5–6mm

LWR STEM HAIRS **spreading**

FR-STALK L 3–5× SEP-TUBE L (longest of the forget-me-nots)

SEPAL-TEETH narrowly triangular (isosceles)

56 Plantaginaceae | **Plantain** family

A family with two divergent groups. **Plantains** have a basal rosette and a narrow, unbranched spike-like inflorescence of tiny, bisexual flowers with long stamens. The anomalous **Shoreweed** is highly distinctive in flower by virtue of its very long-stalked male flowers.

FL

FR

IDENTIFY BY: ► leaf shape ► stamen (filament + anther) colour

SS Arrowgrasses (*p. 233*) [FLS on short stalks].

● **Shoreweed** | **Fls** tiny; only flowers when growing on exposed mud.

ⓟ **Shoreweed** *Littorella uniflora*

♂ FLS

H to 10 cm. **Form** rosette (D to 20 cm), patch-forming. **Fls** unisexual; tiny, inconspicuous; ♂ **solitary on stalks as long as leaves;** ♀ stalkless, at base of ♂ stalk. **Lvs** L to 10 cm; linear; half-cylindrical, **solid. Hab** shallow water at lake edges (to 4 m depth); adjacent exposed shores. **SS** non-flowering rosettes in water could be mistaken for a number of rarer aquatic plants such as **quillworts** *Isoëtes* spp. (N/i) [LEAF CROSS-SECTION 4-channelled] and **Water Lobelia** *Lobelia dortmanna* (N/i) [LEAF CROSS-SECTION 2-channelled].

J F M A M J J A S O N D

♂ FLS

♀ FL

LF half-cylindrical and solid in section

● **Plantains** | **Form** rosette-forming. **Fls** tiny; SEPALS 4, papery; PETALS 4, green or brown.

ⓟ **Sea Plantain** *Plantago maritima*

H to 40 cm. **Form** erect, rosette-forming **Fls** SPIKE L to 10 cm on stalk (L to 30 cm); STAMENS FILAMENTS + ANTHERS pale yellow. **Lvs** linear; fleshy; can be toothed. **Hab** saltmarshes, coastal rocks, salt-sprayed grassland, wet mountain rocks. **SS Sea Arrowgrass** (*p. 233*) – in saltmarsh; [FLS 6 tepals and fruits; LVS spicy-sweet coriander-like scent when crushed].

J F M A M J J A S O N D

FILAMENTS + ANTHERS pale yellow

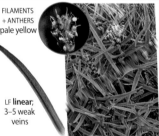

LF **linear;** 3–5 weak veins

ⓐⓟ **Buck's-horn Plantain** *Plantago coronopus*

H to 20 cm. **Form** prostrate to erect. **Fls** SPIKE L to 4 cm on stalk (L to 20 cm); STAMENS FILAMENTS + ANTHERS yellow. Lvs deeply pinnately lobed to narrowly oval and toothed; usually hairy. **Hab** bare sandy locations, cracks in walls or pavements; especially in coastal areas.

J F M A M J J A S O N D

FILAMENTS + ANTHERS yellow

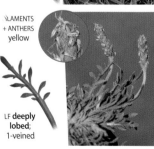

LF **deeply lobed;** 1-veined

ⓟ Greater Plantain *Plantago major*

H to 60 cm. **Form** erect. **Fls** SPIKE L to 20 cm on stalk (L to 40 cm); STAMENS FILAMENTS yellow; ANTHERS typically purplish. **Lvs** L to 15 cm; broadly-oval; 5–9 veins (can be fewer). **Hab** open, trampled grassland, arable margins, brownfield, depressions in coastal marshes.

JFMAMJJASOND

FILAMENTS yellow; ANTHERS usually purplish

LF broad; long-stalked

ⓟ Ribwort Plantain *Plantago lanceolata*

H to 50 cm. **Form** erect; hairy; rosette-forming; FL-STALK furrowed. **Fls** SPIKE egg-shaped; L to 4 cm on **furrowed stalk** (L to 50 cm); STAMENS FILAMENTS + ANTHERS pale yellow. **Lvs** L to 15 cm; linear to narrowly oval; ± pointed; can have shallow teeth; 3–5 obvious veins. **Hab** grassland.

JFMAMJJASOND

FILAMENTS + ANTHERS pale yellow

LF gradually narrowed to short stalk

ⓟ Hoary Plantain *Plantago media*

H to 50 cm. **Form** erect; hairy; rosette-forming. **Fls** SPIKE L to 6 cm on stalk (L to 40 cm); STAMENS FILAMENTS pink-purple, prominent; ANTHERS pale yellow; FLS scented. **Lvs** L to 8 cm; greyish-hairy; narrowly to broadly oval (can have shallow teeth); 5–9 veins; distinct short stalk. **Hab** calcareous to neutral dry grassland.

JFMAMJJASOND

FILAMENTS pink-purple; ANTHERS pale yellow

LF broad; **greyish-hairy**

57 Solanaceae | **Nightshade** family

8 spp. | 17 spp. B&I

Fls flower-parts in 5s; large and funnel-shaped or smaller and star-shaped; PETALS and SEPALS both fused at their bases. **Fr** a berry in most. NOTE: **many have parts that are toxic or, conversely, edible** – whole plants or toxic parts of plants [as indicated] are marked with an ✕.

FORM **woody shrub**; FL **small** (D < 20 mm); **anthers not bunched** ANTHERS not bunched	**Duke of Argyll's Teaplant** p. 176
FL **large**; D > 20 mm; showy; bell- or funnel-shaped; petal-lobes broad; at most only slightly splayed; solitary ANTHERS separated; not bunched around the style	**GROUP 1** p. 176
FL **typically** star-shaped; petal-lobes widely splayed (can be bent back); usually in clusters of 2 or more ANTHERS bunched into a cone-shaped bundle around the style or slightly spreading	**GROUP 2** p. 177

175

Ⓟ Duke of Argyll's Teaplant WS
Lycium barbarum

H to 250cm. **Form** deciduous arching, suckering spiny shrub. **Fls** D <17mm; **purplish**; 1–4 in leaf-axils. **Lvs** L to 10cm; greyish; narrowly oval; widest near middle. **Fr** egg-shaped **red berry**; L 10–20mm. **Hab** hedgerows, scrub. **SS** Chinese Teaplant *L. chinense* (N/i) [FLS D >17mm; lobed >½ way, veins in throat branched; LVS widest below middle].

J F M A M J J A S O N D

FL funnel-shaped with deep lobes ≤ ½ way; throat with fine, dark, ± straight veins

LF widest near middle

FR red berry

GROUP 1

Ⓟ Deadly Nightshade *Atropa belladonna* ✕

H to 150cm. **Form** erect; stout; branched; moderately bushy. **Fls** D 24–30mm; **greenish, purple and brown**; solitary, drooping, bell-shaped. **Lvs** L to 20cm; broadly oval, can be lobed. **Fr** berry; L 15–20mm; **shiny black** when ripe. **Hab** damp shady places, woodland clearings, scrub.

J F M A M J J A S O N D

PET-LOBES short, splayed or recurved

LF broadly oval

FR shiny black berry

ⓐⒷ Henbane *Hyoscyamus niger* ✕

H to 100cm. **Form** erect; branched; stickily hairy; foul-smelling. **Fls** D 20–30mm; yellowish with **network of strong purple veins**; solitary in 2 rows on 1 side of stem. **Lvs** L to 10cm; broadly oval; can be lobed; base clasping stem. **Fr** dry capsule; L to 10mm. **Hab** bare, disturbed ground, near farm buildings, often coastal.

J F M A M J J A S O N D

PET-LOBES deeply lobed

LF oval to lobed; clasping stem

Ⓐ Thorn-apple *Datura stramonium* ✕

H to 200cm. **Form** erect, hairless; fast-growing; can form a substantial bush. **Fls** white; L 50–100mm; solitary in branch-forks; drooping or upright; lobes can be pointed. **Lvs** L to 20cm; broadly oval, with jagged lobes and teeth. **Fr** spiny egg-shaped capsule; D to 70mm (incl. spines). **Hab** bare, disturbed ground, field margins.

J F M A M J J A S O N D

FL trumpet-shaped

LF edges jagged

FR large; spiny

GROUP 2

A Black Nightshade *Solanum nigrum* ✗

H to 70 cm. **Form** weakly erect to spreading. **Fls** D 10–14 mm; white; oval (can be slightly lobed); in branched clusters of 2–10. **Fr** glossy black berry; D 6–10 mm. **Hab** disturbed ground, arable margins, gardens, brownfield.

J F M A M J J A S O N D

FR glossy black when ripe

P Bittersweet *Solanum dulcamara* ✗

L to 200 cm. **Form** scrambler; can be woody at base. **Fls** D 10–15 mm; purple; in branched clusters of 2–25. **Lvs** triangular; can have up to 4 small basal lobes. **Fr** egg-shaped red berry; L 8–12 mm. **Hab** damp woods, fens, stream banks.

J F M A M J J A S O N D

FR red when ripe

aP Potato *Solanum tuberosum* ✗ [FR, STEM, LVS]

L to 200 cm. **Form** erect; tuberous; STEM winged. **Fls** D 20–35 mm; white or pinkish; in branched clusters of 2–20. **Lvs** L to 40 cm; irregularly 1–2-pinnate; LFLTS broadly oval. **Fr** flattened globular berry; D 20–40 mm. **Hab** cultivated ground, brownfield, rubbish tips.

J F M A M J J A S O N D

FR ✗ green or purplish when ripe

aP Tomato *Solanum lycopersicum* ✗ [STEM, LVS]

H to 200 cm. **Form** bushy; sprawling; hairy; spicily aromatic. **Fls** D 18–25 mm; yellow; in branched clusters of 2–20. **Lvs** L to 40 cm; pinnate; LFLTS toothed or lobed. **Fr** globular berry; D 20–100 mm. **Hab** cultivated ground, brownfield, rubbish tips, sewage farms and outfalls.

J F M A M J J A S O N D

FR red when ripe

LF ✗

177

58 Oleaceae | **Ash** family

4 spp. | 8 spp. B&I

Form trees and shrubs. **Fls** a wide range, from inconspicuous with PETALS + SEPALS absent to conspicuous with PETALS 4, fused + SEPALS 4, fused; all have 2 stamens per flower. **Lvs** opposite. **Fr** varied – berries, capsules and winged seeds.

● **Privets** | **Form** semi- to almost fully evergreen shrubs. **Fls** white; scented (particularly in the evening); in spikes. **Lvs** oval (L to 6 cm). **Fr** black berry D 6–8 mm.

IDENTIFY BY: ▶ leaf, flower, stem and twig details

Privet identification		Wild Privet	Garden Privet
Flower	PETAL-TUBE	L 3 mm; **L ± petal-lobes**	L 5 mm; **L > petal-lobes**
	WINTER BUD SCALES	fringed with hairs	hairless
Leaf	COLOUR	green; dull	green [can be variegated]; glossy, leathery
	SHAPE	narrowly oval	broadly oval
Twig	FIRST-YEAR	minutely hairy	hairless

WILD   GARDEN

J F M A M J J A S O N D   J F M A M J J A S O N D

ⓟ **Wild Privet** WS
Ligustrum vulgare

H to 3 m. **Form** semi-evergreen. **Hab** hedges, scrub, especially on calcareous soils.

ⓟ **Garden Privet** WS
Ligustrum ovalifolium

H to 5 m. **Hab** gardens, hedges, brownfield.

WILD PRIVET

FR black berries in clusters

GARDEN PRIVET

WINTER BUD-SCALES with fringe of hairs; 1ST-YR TWIGS minutely hairy

LOBE

TUBE

LF narrowly oval

WINTER BUD-SCALES hairless; 1ST-YR TWIGS hairless

LOBE

TUBE

LF broadly oval

ⓟ **Lilac** *Syringa vulgaris*

J F M A M J J A S O N D

H to 7 m. **Form** deciduous, suckering tree. **Fls** L 8–12 mm; purple to white; sweetly scented; in terminal, conical to cylindrical infl. **Lvs** L to 12 cm. **Fr** capsule; L 8–12 mm. **Hab** gardens, hedges, railway embankments and brownfield. **SS** Butterfly-bush superficially similar in flower [FORM arching; LVS longer; FLS Jul–Sep].

FL

LF ± heart-shaped

P Ash *Fraxinus excelsior* **WT**

H to 35 m. **Form** deciduous tree. **Fls** in **compact heads; before leaves**; PET + SEP absent; ♂ 2 purple stamens in spiky purple globe; ♀ in looser (can be separate) green clusters. **Lvs pinnate;** LFLTS L to 7 cm. **Fr** SEED 1-winged 'key'; in clusters. **Hab** woods, scrub, hedges, parks, especially on more calcareous soils.

J F M A M J J A S O N D

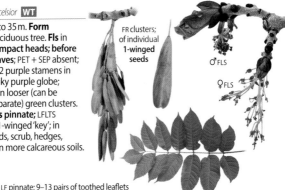

FR clusters; of individual **1-winged seeds**

♂ FLS

♀ FLS

LF pinnate; 9–13 pairs of toothed leaflets

---

59 Scrophulariaceae | Figwort + Mullein family   **4 spp. | 23 spp. B&I**

A disparate family defined mainly on molecular evidence. **Fls** SYMMETRY bilateral. **Fr** 2-celled capsule – the genera included in this family are each readily recognizable even though their shared defining features are found in other families.

● **Buddleja | Form** woody shrub. **Infl** pyramidal; distinctive in late summer.

P Butterfly-bush *Buddleja davidii* **WS**

H to 5 m. **Form** arching, semi-evergreen shrub. **Fls** L 9–11 mm; **lilac** (darker or paler/white, in some cultivars) in dense, pyramidal clusters; AROMA musty, musky. **Lvs** L to 25 cm; narrowly to broadly oval. **Hab** walls, gardens, brownfield. **SS** Lilac superficially similar in flower [FORM not arching; LVS broader; FLS May–Jun].

J F M A M J J A S O N D

LF with long, drawn-out point

PET-TUBE 4-lobed; throat of flower **orange**

● **Mulleins | distinctive form and flowers. Form** tall. **Fls** PETALS 5.

B Great Mullein *Verbascum thapsus*

H to 200 cm. **Form** rosette in 1st year; erect in 2nd; **white-woolly. Fls** D 20–30 mm; yellow; 2–4 at each bract base; in tall, usually **unbranched** spike. **Lvs** BASAL L to 50 cm; broadly oval; STEM bases run down stem. **Hab** rough grassland, brownfield. **SS** many scarcer (or garden) species and hybrids [STAMEN details; especially hairs on filaments].

J F M A M J J A S O N D

FL lowest lobe slightly larger than others; 3 upper filaments covered in whitish hairs, 3 lower ones almost hairless

179

● **Figworts** | **Form** tall; erect. **Fls** in loose clusters forming a branched, leafy spike; **distinctive**; L 7–10mm; greenish to **purple-brown**; CENTRAL STAMEN sterile; enlarged; scale-like (staminode).

SEPARATE BY: ► sepal + staminode details ► stem + leaf details

STAMINODE

ⓟ **Common Figwort** *Scrophularia nodosa*

J F M A M J J A S O N D

H to 100cm. **Lvs** L to 15cm; broadly oval; irregularly toothed; blade can run a little way down lf-stalk. **Fr** egg-shaped capsule; L 6–10mm. **Hab** woods, hedges, scrub, usually damp.

LF-TEETH irregular

STAMINODE tip usually squarish

SEP **narrow** silvery border W < 0·3 mm

LF-STALK short weak wings at most

STEM square; unwinged

ⓟ **Water Figwort** *Scrophularia auriculata*

J F M A M J J A S O N D

H to 120cm. **Lvs** L to 15cm; broadly oval; teeth rounded; blade running as wings down lf-stalk; can be lobed at base. **Fr** globular capsule; L 4–6mm. **Hab** water margins, marshes, wet woodland.

LF-TEETH rounded

STAMINODE tip usually rounded

SEP **broad** silvery border W > 0·3 mm

LF-STALK usually **obviously winged**

STEM square; **winged**

60 Hippuridaceae | **Mare's-tail** family

1 sp. B&I

Form patch-forming aquatic/marginal. **Lvs** linear, in whorls. **Fls** tiny, on emergent shoots.

Horsetails (p. 298) have a superficial similarity and are often erroneously called Mare's-tail; however their jointed, whorled branches and sheathed (often angled) stem are distinctive.

ⓟ **Mare's-tail** *Hippuris vulgaris*

J F M A M J J A S O N D

L IN WATER, to 100cm; aerial portion H to 20cm; ON MUD H to 20cm. **Fls** very small; 1 per leaf-axil on emergent portion of stem; mainly ♀ flowers towards tip; mainly ♂ and bisexual (with a single stamen bearing a reddish anther) lower down. **Form** erect when emergent and marginal. **Lvs** AERIAL L to 80mm; (ON MUD much shorter), linear, in whorls of 6–12; UNDERWATER flaccid. **Fr** 1 seed in each leaf-axil. **Hab** ponds and slow-moving rivers/streams; especially in calcareous areas.

FR

HORSETAIL (EQUISETACEAE) BRANCHES jointed and sheathed

MARE'S-TAIL LEAVES not jointed or sheathed

61 Lamiaceae | Dead-nettle + Mint family

A recognizable family, often strongly aromatic, including many culinary herbs. **Form** STEM square in cross-section. **Fls** PETALS fused; SEPALS fused; PETAL-TUBE typically 2-lipped, the upper lip forming a hood under which are located the STAMENS (4 (2 long; 2 shorter)) and STIGMA (usually forked). **Lvs** in opposite pairs, each pair at right-angles to the ones above and below.

SS Leaf shape similar to **nettles** (*p. 105*) [tiny flowers and usually stinging hairs].

Dead-nettle groups and species identification

PETAL-TUBE with 4 more-or-less equal-sized lobes		PETAL-TUBE with distinct lower lip only – UPPER LIP reduced to a rim round the top of the tube
FERTILE STAMENS 2	FERTILE STAMENS 4	

Gypsywort
p. 182

MINTS
p. 182

Wood Sage, Bugle
p. 183

PETAL-TUBE **upper and lower lips distinct** – UPPER LIP **forming a hood**; STAMENS **4** - (2 long, 2 shorter, under hood); AROMA **pungent when crushed**

1
2
2-lipped   not 2-lipped

SEPAL-TUBE with 5 ± equal lobes; **distinctly 2-lipped**	SEPAL-TUBE with 5 ± equal lobes; **usually not clearly 2-lipped**		
	INFL **erect; held above the vegetative shoots** in ± compact or elongated terminal spikes	INFL **elegant, tapering, terminal spire of ± crowded whorls**	INFL **erect or hidden amongst leaves (in leaf-axils)** in ± well-separated whorls

SKULLCAPS, Selfheal
p. 183

HOREHOUNDS, HEMP-NETTLES
p. 184

WOUNDWORTS, Betony
p. 185

DEAD-NETTLES
pp. 186–187

PETAL-TUBE **upper and lower lips distinct** – UPPER LIP **flat or turned back, not forming a hood**

SEPAL-TUBE with 5 teeth or lobes; **not clearly 2-lipped**	SEPAL-TUBE with 5 teeth; **clearly 2-lipped**	
	FORM herbaceous	FORM woody dwarf shrub

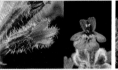
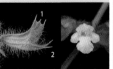

Wild Marjoram, Ground-ivy
p. 188

Balm, Basil
p. 188

Wild Thyme
p. 189

PETAL-TUBE **with 4 more-or-less equal-sized lobes**

● **Gypsywort** | **Form** AROMA faint at most when crushed. **Infl** a terminal spike of separated whorls. **Fls** STAMENS 2 fertile.

Ⓟ Gypsywort *Lycopus europaeus*

FLS

H to 100 cm. **Form** erect, branched, slightly hairy. **Fls** L 4 mm **white with dark red spots;** in clusters spaced up stem at base of large, leaf-like bracts. **Lvs** L to 100 mm; deeply toothed to pinnately lobed. **Fr** L 2 mm. **Hab** fens, wet grassland, wet woodland, beside waterbodies.

J F M A M J J A S O N D

LVS usually **deeply toothed**

● **Mints** | **Form** erect; usually branched; AROMA **mint when crushed. Infl** terminal; crowded; many-flowered whorls forming dense round, cylindrical or tapering heads. **Fls** STAMENS 4 fertile.

SS Another dozen or so mints can be found in B&I comprising native and introduced species, as well as naturally occuring and cultivated hybrids. More comprehensive information is needed for accurate identification. All are mintily aromatic, with creeping surface or underground runners and weakly 2-lipped petal- and sepal-tubes.

Ⓟ Water Mint
Mentha aquatica

H to 90 cm. **Form** erect or sprawling; AROMA pleasantly minty. **Fls** L 4–6 mm; lilac-pink; in **dense oblong head** + 1–2 discrete whorls below; BRACTS tiny; hidden within fls. **Lvs** L to 90 mm; triangular to broadly oval; short-stalked. **Fr** L 2 mm. **Hab** fens, wet grassland, wet woodland, beside waterbodies.

J F M A M J J A S O N D

ⓐⓟ Corn Mint *Mentha arvensis*

H to 60 cm. **Form** erect or sprawling; AROMA unpleasant, sickly sweet. **Fls** L 3–5 mm; lilac; hairy **in discrete whorls**; BRACTS large, leaf-like. **Lvs** L to 65 mm; variable in shape; stalked. **Fr** L <1 mm. **Hab** arable margins, woodland rides, damp grassland, brownfield.

J F M A M J J A S O N D

Ⓟ Spear Mint *Mentha spicata*

H to 100 cm. **Form** erect; AROMA sweet spearmint. **Fls** L 2–4 mm; lilac, pink or white; in **dense, terminal,** slightly tapering to cylindrical spikes; can have slight separation between whorls. **Lvs** L to 90 mm; broadly oval; ± pointed; margin sharply toothed; stalkless. **Fr** L <1 mm. **Hab** gardens, naturalized in rough ground, often damp.

J F M A M J J A S O N D

SEP-TUBE teeth narrow

FLS in oblong head + 1–2 whorls below

LF downy; medium-green to purplish

SEP-TUBE teeth broad

FLS in whorls down stem

LF pale green; hairy

FL bracts longer than fls

FLS in terminal spikes

LF shiny; pale greyish-green; usually hairless

PETAL-TUBE **with distinct lower lip only, upper lip reduced to a rim round the top of the tube**

🅟 Bugle *Ajuga reptans*

H to 30 cm. **Form** spreading; with rosettes; STEM hairy on opposite sides. **Fls** L 14–17 mm; **pale to deep blue with dark veining**; in dense spike; whorls interspersed with unlobed, leafy bracts. **Lvs** L to 70 mm. **Hab** woodland, damp grassland, on neutral to acid soil. **SS** Pyramidal Bugle *A. pyramidalis* (N/I) [BRACTS L > fls; STEM hairy all round].

J F M A M J J A S O N D

LOWER LIP 3-lobed; middle lobe can be notched (as here)

LF broadly oval with slight tip; margins wavy

🅟 Wood Sage *Teucrium scorodonia*

H to 50 cm. **Form** erect; hairy. **Fls** L 8–12 mm; **pale greenish-yellow**; in pairs in usually branched infl. **Lvs** L to 70 mm; deep-set veins on upper surface. **Hab** woods, hedgerows, heaths, on well-drained, sandy soils.

J F M A M J J A S O N D

LOWER LIP 5-lobed; middle lobe large flanked by 2 smaller lobes

LF oval to heart-shaped; stalked

PETAL-TUBE **upper and lower lips distinct** – UPPER LIP **forming a hood;** STAMENS **4 -** (2 long, 2 shorter, under hood); AROMA **pungent when crushed**

1/3

SEPAL-TUBE **distinctly 2-lipped**

🅟 Selfheal *Prunella vulgaris*

H to 30 cm. **Form** creeping, patch-forming. **Infl** cylindrical clustered whorls, interspersed with bracts. **Fls** L 10–15 mm; deep bluish-purple (can be pink or white). **Lvs** L to 90 mm; sometimes toothed. **Hab** open woods, grassland, lawns, road verges, usually dry soils. **SS** Betony (p. 185)

J F M A M J J A S O N D

SEP-TUBE UPPER LIP with 3 tiny teeth; LOWER LIP with ± triangular pointed teeth

LF broadly oval to diamond-shaped

🅟 Skullcap *Scutellaria galericulata*

H to 50 cm. **Form** weakly erect or sprawling. **Infl** leafy; single flower at base of each bract. **Fls** L 10–20 mm; typically pale to bright blue; rarely mauve-pink. **Lvs** L to 50 mm; regular, rounded teeth. **Hab** wet meadows, marshes and fens. **SS** Lesser Skullcap *S. minor* (N/I) [FLS smaller, pinkish-purple; lower lip with dark spots].

J F M A M J J A S O N D

SEP-TUBE UPPER LIP with a **scale on the back**; PET-TUBE LOWER LIP 3-lobed

LF narrowly triangular

PETAL-TUBE **upper and lower lips distinct** – UPPER LIP **forming a hood;**
STAMENS **4 -** (2 long, 2 shorter, under hood); AROMA **pungent when crushed**

SEPAL-TUBE with 5 ± equal lobes; **usually not clearly 2-lipped**

● **Horehounds and Hemp-nettles** | **Form** erect flowering stem; **hemp-nettles** bristly-hairy, with long hairs and long spiny teeth on and around sepal-tube. **Infl** held above the vegetative shoots in ± compact or elongated terminal spikes.

Ⓟ Black Horehound
Ballota nigra

H to 100 cm.
Form erect
or sprawling,
softly downy,
patch-forming;
AROMA
unpleasant
when crushed.
Fls L12–15mm;
dull mauve; in dense whorls
(more crowded at top of
spike). **Lvs** L to 50mm; broadly
oval; regularly toothed
short-stalked. **Fr** 2mm.
Hab woodland margins,
hedgerows, road verges,
brownfield.

Both hemp-nettles: **H** to 100 cm. **Form** erect;
branched, STEM bristly hairy; swollen at lf-nodes. **Fls** lower lip 3-lobed
with 2 rounded humps at the base. **Lvs** L to 100mm;
diamond-shaped to broadly oval; toothed. **Fr** 4mm.

SEPARATE BY: ► flower lower lip details

Ⓐ Common Hemp-nettle
Galeopsis tetrahit

Fls L15–20mm;
pinkish-purple,
with **darker
markings near
centre and
base of lip.**
Hab marshes,
fens, woodland
clearings,
arable margins.

Ⓐ Bifid Hemp-nettle
Galeopsis bifida

Fls L12–15mm;
pinkish-
purple (can
be yellowish),
with darker
markings
**reaching edge
of lip. Hab**
arable margins,
marshes, fens, woodland
clearings.

SEP-TUBE **flared
at the mouth**
with broad, short-
pointed teeth

FL dull mauve   LF short-stalked

COMMON HEMP-NETTLE
LOWER LIP middle lobe **flat
and not notched**

BIFID HEMP-NETTLE
LOWER LIP middle lobe **notched
and with rolled edges**

FL colour variable in
both species

COMMON HEMP-NETTLE

SEPAL-TUBE with 5 ± equal lobes; **usually not clearly 2-lipped**

● **Woundworts** | **Infl** elegant, tapering, terminal spike of ± crowded whorls. **Fls** SEPAL-TEETH pointed, spiny; BRACTS very short or absent.

IDENTIFY BY: ▶ flower colour

ℙ Hedge Woundwort *Stachys sylvatica*

J F M A M J J A S O N D

H to 100 cm. **Form** erect, often branched; AROMA **unpleasant** when crushed. **Fls** L 12–18 mm; **deep red with pale markings**. **Lvs** L to 120 mm; broadly triangular; hairy; teeth coarse, blunt. **Fr** L 2·0–2·5 mm. **Hab** hedgerows, woodland, streamsides, rough ground.

FL deep red with pale markings

STEM-LVS with distinct stalk

ℙ Marsh Woundwort *Stachys palustris*

J F M A M J J A S O N D

H to 100 cm. **Form** erect, with short branches; AROMA absent. **Fls** L 12–15 mm; **dull purple with both dark and pale markings**. **Lvs** L to 120 mm; STEM narrowly triangular; teeth coarse, blunt; BASE heart-shaped. **Fr** L 2 mm. **Hab** marshes, fens, damp places by ponds and watercourses.

FL deep red with pale markings

STEM-LVS unstalked

Ⓐ Field Woundwort *Stachys arvensis*

J F M A M J J A S O N D

H to 40 cm. **Form** slender, erect, hairy stem, branched low down; AROMA absent. **Fls** L 8–10 mm; **very pale mauve with a few darker dots and lines**. **Lvs** L to 40 mm; STEM broadly oval; hairy; BASE can be heart-shaped. **Fr** L 1·5 mm. **Hab** arable margins, gardens, on non-calcareous soils.

FL pale mauve with a few darker markings

STEM-LVS tip rounded

● **Betony** | **Infl** cylindrical (not tapering). **Fls** BRACTS elongate; pointed.

ℙ Betony *Betonica officinalis*

J F M A M J J A S O N D

H to 80 cm. **Form** erect; branched; and short woody underground runners; hairy; sterile leafy rosettes present at flowering. **Fls** L 12–18 mm; reddish-purple or white; INFL ± **flat-topped**. **Lvs** L to 90 mm; hairy and dotted with glands; BASE wedge-shaped; LF-STALKS L to 70 mm; hairy, leafy rosettes present at flowering. **Fr** L 3 mm. **Hab** grassland, heathland, woodland clearings, coastal sites, on sandy soils. **SS** Selfheal (*p. 183*).

FL reddish-purple or white

LVS narrowly triangular; teeth and tip blunt

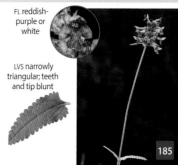

185

PETAL-TUBE **upper and lower lips distinct** – UPPER LIP **forming a hood**; STAMENS **4** - (2 long, 2 shorter, under hood); AROMA **pungent when crushed**

SEPAL-TUBE with 5 ± equal lobes; **usually not clearly 2-lipped**

● **Dead-nettles** | **Form** somewhat sprawling, with soft stem. **Infl** flower-bearing stem erect or hidden amongst leaves. **Fls** in ± well-separated whorls at base of each bract; BRACTS leaf-like. **Lvs** nettle-like but without stinging hairs.

IDENTIFY BY: ▶ flower colour; then by ▶ leaf details

℗ White Dead-nettle
Lamium album

H to 60 cm. **Form** erect or sprawling, softly downy, patch-forming. **Fls** L 25–30 mm; **white**. **Lvs** L to 70 mm; ± triangular to heart-shaped; toothed; hairy. **Fr** L 3 mm. **Hab** hedgerows, waysides, gardens, brownfield.

FLS white; lower lip with **deep central notch**

LVS ± triangular with coarse teeth

℗ Yellow Archangel
Lamiastrum galeobdolon

H to 50 cm. **Form** erect; patch-forming. **Fls** L 20–22 mm; **bright yellow**; in whorls; BRACTS narrow; leaf-like. **Lvs** L to 70 mm ± triangular, coarsely toothed. **Fr** L 4 mm. **Hab** woods, hedges, gardens. **SS** ssp. *argentatum* (N/I) [FLS wider; LVS white blotched].

FLS yellow; lower lip with 3 long lobes, with orange or brown markings

LVS often with whitish marbling in spring

℗ Spotted Dead-nettle
Lamium maculatum

H to 50 cm. **Form** erect or sprawling; softly downy patch-forming. **Fls** L 25–35 mm; pinkish-purple (can be very pale); often shorter and not opening. **Lvs** L to 50 mm; ± triangular to heart-shaped; toothed; hairy. **Fr** L 2·0–2·5 mm. **Hab** cultivated ground, gardens, brownfield (escape from cultivation).

LOWER LIP lateral lobes with only 1 tooth

LVS usually with **large whitish blotch**

SS Red Dead Nettle and **Cut-leaved Dead Nettle** often occur together and some care can be required to differentiate.

Ⓐ Red Dead-nettle
Lamium purpureum

H to 30 cm. **Form** erect; branched; softly downy. **Fls** L 10–18 mm; pinkish-purple. **Lvs** L to 35 mm; triangular to heart-shaped; teeth blunt; hairy; young lvs often reddish-purple.
Fr L 3 mm. **Hab** cultivated ground, gardens, brownfield.

PETAL-TUBE with distinct ring of hairs inside, near base (hard to see)

LVS ± **regularly and shallowly toothed;**

ⒶCut-leaved Dead-nettle
Lamium hybridum

H to 30 cm. **Form** erect; branched; softly downy. **Fls** L 10–18 mm (often shorter and not opening); pinkish-purple. **Lvs** L to 35 mm; triangular to heart-shaped; teeth blunt; hairy; young lvs often reddish-purple. **Fr** L 1·0–2·5 mm. **Hab** cultivated ground, gardens, brownfield.

PETAL-TUBE lacking hairs inside, near base (hard to see)

LVS **deeply and irregularly toothed**

ⒶHenbit Dead-nettle
Lamium amplexicaule

H to 25 cm. **Form** erect; branched; softly downy. **Fls** L 14–20 mm; pinkish-purple (often shorter and not opening); LOWER LIP L ≤ 3 mm. **Lvs** L to 25 mm; round to broadly oval, with rounded teeth; LOWER stalked; UPPER **unstalked and partly clasping stem**. **Fr** L 2–3 mm. **Hab** cultivated ground, gardens, brownfield. **SS Northern Dead-nettle** *L. confertum* (N/I) [FORM larger; LVS all stalked; FLS lower lip L > 3 mm].

FLS look **longer and narrower** than other dead-nettles

UPPER LVS partly clasping stem

LVS round to broadly oval; teeth rounded

LOWER LVS stalked

Form herbaceous plants; which can have a tough base and lower stem.

SEPAL-TUBE with 5 teeth or lobes; **clearly 2-lipped**

Ⓟ Balm *Melissa officinalis*

H to 60 cm. **Form** erect, branched; AROMA **strongly lemon** when crushed. **Fls** L 8–25 mm; **white or pale cream;** in inconspicuous clusters at base of leaf-like bracts. **Lvs** L to 70 mm; broadly triangular; L stalk > ½ L blade. **Hab** frequent naturalized near gardens, brownfield.

J F M A M J J A S O N D

FL cluster packed with long, silky, white hairs

LF long-stalked

Ⓟ Wild Basil *Clinopodium vulgare*

H to 75 cm. **Form** erect, softly hairy; AROMA faintly herby when crushed. **Fls** L 12–22 mm; pink-purple, in **2–3 whorls separated up stem,** each with pair of leaf-like bracts below. **Lvs** L to 45 mm; ± triangular to broadly oval; teeth shallow, rounded; veins deep-set.

J F M A M J J A S O N D

Hab woodland margins, scrub and grassland on light, usually calcareous soils.

FL cluster packed with **long, silky, white hairs**

LF narrowly oval; almost stalkless

SEPAL-TUBE with 5 teeth or lobes; **not clearly 2-lipped**

Ⓟ Wild Marjoram *Origanum vulgare*

H to 50 cm. **Form** tufted; erect; STEM can be purplish. AROMA pleasantly fragrant when crushed. **Fls** L 4–7 mm; **deep purple buds, opening paler or white, in flat-topped clusters. Lvs** L to 40 mm; with glandular hairs. **Hab** dry grassland, banks, road verges. **SS** cultivars

J F M A M J J A S O N D

are found outside the native range.

FL interspersed with small, purple bracts

LF broadly oval; untoothed

Ⓟ Ground-ivy *Glechoma hederacea*

H to 30 cm; **L** to 100 cm. **Form** patch-forming; rooting; AROMA **pungent, like cat urine** when crushed. **Fls** L 15–20 mm; bright **blue-violet with darker spots on lower lip;** in clusters of up to 4 at base of leaf-like bracts. **Lvs** L to 40 mm; with coarse, blunt, irregular teeth. **Hab** woodlands, grassland, hedgerows, disturbed habitats.

J F M A M J J A S O N D

FL interspersed with leaf-like bracts

LF **kidney- to heart-shaped;** long-stalked

Form woody, evergreen tiny dwarf shrub.

ⓟ **Wild Thyme** *Thymus drucei*

H to 70 cm. **Form prostrate or creeping**; AROMA strongly of **thyme.** **Fls** L 6 mm; pink or purple, in ± **crowded infl. Lvs** L to 8 mm; broadly to narrowly oval; hairy. **Hab** short (grazed) grassland on dry acid or calcareous soils, also coastal and upland. **SS** other, rarer thymes [distribution of stem hairs].

STAMENS L > FLS

STEM usually noticeably more **hairy on 2 opposite faces**

J F M A M J J A S O N D

---

62 Gentianaceae | **Gentian** family

2 spp. | 16 spp. B&I

Fls PETALS 4–5 (can be more), fused at base into a tube to which the same number of STAMENS are attached. **Lvs** opposite; hairless; entire. **Fr** a 2-celled capsule.

Ⓐ **Yellow-wort** *Blackstonia perfoliata*

FLS 6–8 narrow petals

H to 50 cm. **Form** erect. **Fls** yellow; D 10–15 mm; PETAL-TUBE L to 10 mm; in terminal **forking infl. Lvs** L to 20 mm; triangular, blue-green. **Hab** calcareous grassland, sand dunes, cliff slopes.

J F M A M J J A S O N D

LVS **fused around the stem**

Ⓑ **Common Centaury** *Centaurium erythraea*

H to 50 cm. **Form** erect; branched, often dwarfed by grazing; basal rosette. **Fls** D 10–12 mm; **soft pink** in terminal forking infl. **Lvs** L to 50 mm; broadly oval. **Hab** dry grassland, heathland, sand dunes, cliff slopes. **SS** Lesser Centaury *C. pulchellum* (N/I) [INFL more open; FL-STALKS >1 mm; FL deeper pink; FORM rosette absent]; several rare species [LF, STEM + FL details].

FLS almost unstalked with 2 basal bracts

BASAL LVS in rosette

J F M A M J J A S O N D

● **Gentians** | A group of plants that inspire almost as much passion as orchids. Gentians (*Gentiana, Gentianopsis* and *Gentianella*) usually have blue (often vividly so) or purple flowers, with a longer petal-tube than others in the family. The most widespread is the variable **Autumn Gentian** *Gentianella amarella* (*right*), found especially on sand dunes and calcareous grassland.

AUTUMN GENTIAN

189

63 Phrymaceae | **Monkeyflower** family
1 sp. | 5 spp. B&I

Form waterside plants spreading by leafy runners. **Fls** showy; yellow or red; PETALS fused into a tube with 2-lobed upper lip and 3-lobed lower lip (as with several related families).

P Monkeyflower *Erythranthe guttata*

H to 80 cm. **Form** erect, creeping. **Fls** D 25–45 mm; **bright yellow**, some with **small red spots** in the 'throat', which is closed by 2 swellings of the lower lip; upper tooth of sepal-tube longer than the rest. **Lvs** L to 70 mm; broadly oval to triangular; coarsely toothed; opposite.

Hab stream and pond edges, river shingle, ditches and marshes. **SS** Hybrids with the scarcer **Blood-drop-emlets** *E. luteus* (N/I) [more red on flower] are probably under-recorded.

FLS distinctive shape; bright yellow; can have small red spots

LF

64 Orobanchaceae | **Broomrape** family
10 spp. | 42 spp. B&I

Form root parasites; either total (lacking chlorophyll and consisting principally of aerial flowering shoots and leaves reduced to scales) or partial (green leaves in opposite pairs and a reduced root system). **Fls** bilaterally symmetrical; 2-lipped; SEPALS 4, fused; STAMENS 4.

FORM **partial root parasites with chlorophyll**   1/2

A **Red Bartsia**
Odontites vernus

H to 50 cm. **Form** erect, branched. **Fls** L 8–10 mm; **pink**; in **drooping spike**. **Lvs** L to 40 mm; opposite; narrowly triangular; margin sparsely toothed. **Hab** brownfield, grassland, upper saltmarsh.

INFL distinctive

A **Common Cow-wheat**
Melampyrum pratense

H to 60 cm. **Form** erect. **Fls** L 10–20 mm; **yellow, some with with purple marks**; mouth partially closed by swellings on lower lip. **Lvs** opposite. **Fr** sepal-tube not inflated. **Hab** woods, scrub, heathland.

SEPAL-TUBE not inflated

A **Yellow-rattle**
Rhinanthus minor

H to 50 cm. **Form** erect. **Fls** L 12–15 mm; **yellow**; UPPER LIP with two **triangular, often blue, downward-pointing teeth** at tip. **Lvs** opposite. **Fr** sepal-tube inflated. **Hab** grassland, fens, sand dunes.

SEPAL-TUBE inflated

● **Louseworts** | **Fls** pink. **Lvs** alternate; pinnate with lobed segments.

Lousewort *Pedicularis sylvatica*

H to 24 cm. **Form** spreading with erect flowering shoots. **Fls** PETAL-TUBE L 20–25 mm; upper lip with 1 terminal and lateral tooth on each side; SEPAL-TUBE with **4 dissected lobes**. **Fr** capsule L ≤ sepals. **Hab** usually drier parts of bogs, wet heaths.

J F M A M J J A S O N D

Marsh Lousewort *Pedicularis palustris*

H to 60 cm. **Form** erect. **Fls** PETAL-TUBE L 20–25 mm; upper lip with 1 terminal and 2 lateral teeth on each side; SEPAL-TUBE with **2 broad lobes**, dissected in some. **Fr** capsule L > sepals. **Hab** bogs, wet heaths.

J F M A M J J A S O N D

1 lateral tooth

LF L to 20 mm; pinnate; LFLTS lobed

2 lateral teeth

LF L to 40 mm; pinnate; LFLTS deeply lobed

● **Eyebrights** | **Fls** distinctive white or mauve with darker veins and usually yellow in the throat; lower lip with three notched lobes.

SS Although the genus is distinctive, the 19 species and numerous hybrids that occur are difficult to identify with confidence. They are all differentiated by critical characters, many of which require multiple specimens to be examined in order to establish a 'character average'. Most botanists are content simply to call them eyebrights (*Euphrasia* spp.). Only the two that are regarded as the most widespread, based on current knowledge, are covered here.

❹ Common Eyebright
Euphrasia nemorosa

H to 35 cm. **Form** erect, variably branched. **Fls** L 5–8 mm; D 7.0–12.5 mm. **Lvs** L to 12 mm; broadly oval to triangular; toothed. **Fr** capsule L ≤ 2× W. **Hab** grassland (often damp), woodland rides, dunes.

❹ Arctic Eyebright
Euphrasia arctica

H to 30 cm. **Form** erect. **Fls** L 6–11 mm; D 6–9 mm. **Lvs** L to 12 mm; broadly oval to triangular; toothed. **Fr** capsule L > 2× W. **Hab** mostly upland grassland.

COMMON
Eyebright agg.

ARCTIC
Eyebright agg.

J F M A M J J A S O N D   J F M A M J J A S O N D

EYEBRIGHT SP.

FL lobes with mauve or purple areas are frequent

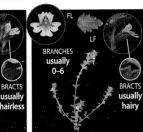

FL

LF

BRANCHES usually **numerous**

BRACTS usually **hairless**

FL

LF

BRANCHES usually **0–6**

BRACTS usually **hairy**

FORM total root parasites lacking chlorophyll

℗ Toothwort *Lathraea squamaria*

H to 30 cm. **Form** erect; **whitish with white scales**; underground runners. **Fls** L 14–22 mm on short stalks in 1-sided spike; PETALS **creamy-white with pink tinge**; glandular hairy. **Hab** woods, hedgebanks, parasitic on roots of Hazel, Ash and elms in

FLS creamy-white with tinge of pink in **1-sided spike**

particular. **SS Purple Toothwort** *L. clandestina* (N/I) [FORM larger; FL purple more numerous; borne at ground level].

● Broomrapes | **Form** erect. **Fls** PETALS 4; fused into a tube; 2-lipped, the lower lip 3-lobed and bent downwards at the tip. **Lvs** absent.

SS Other broomrapes (some very rare). Identification needs very careful examination of the flowers. Other plants without chlorophyll (outside the scope of this book) are the saprophytic **Bird's-nest Orchid** *Neottia nidus-avis* (N/I) and **Yellow Bird's-nest** *Hypopitys monotropa* (N/I).

℗ Common Broomrape *Orobanche minor*

H to 60 cm. **Form** STEM **yellowish, often flushed reddish. Fls** L 10–18 mm; lower fls may be stalked; PETALS dull yellow, suffused with purple, **soon turning brown**; STIGMA usually purple; rarely yellow. **Hab** grassland, gardens, on a wide range of hosts but especially Fabaceae and Asteraceae.

STIGMA usually **purple**

℗ Ivy Broomrape *Orobanche hederae*

H to 60 cm. **Form** STEM **typically brownish-purple**; a few yellow. **Fls** L 10–22 mm; lower fls may be stalked; PETALS **cream; tinged with reddish-purple**; STIGMA usually yellow; a few purple. **Hab** hedges, woodland, gardens, cliff slopes, **parasitic on ivy** species and close relatives.

STIGMA usually yellow

FLOWER SPIKE less crowded than that of Common Broomrape

65 Lentibulariaceae | **Bladderwort + Butterwort** family `1 sp. | 11 spp. B&I`

Form insectivorous; two genera with very different feeding methods (sticky leaves; underwater suction bladders). **Fls** 2-lipped with a backward-pointing spur; STAMENS 2.

● **Butterworts** | **Form** rosette-forming. **Fls** solitary on a leafless stalk. **Lvs** yellow-green; entire; margins inrolled; covered in sticky secretions from their many glands.

℗ **Common Butterwort** *Pinguicula vulgaris*

H to 18 cm. **Fls** L 14–22 mm (incl. SPUR L 4–7 mm); **violet with white in throat**; 2-lipped, LOWER LIP 3-lobed (lobes L>W). **Lvs** L to 80 mm; broadly oval; margins curled upwards. **Hab** wet heaths, bogs, springs on limestone. **SS** flower form and colour are superficially similar to **violets** (*p. 118*) [LVS not sticky or in rosette].

J F M A M J J A S O N D

FL distinctive

LF margins curved upwards

● **Bladderworts** *Utricularia* spp. `AQ` | **H** to 15 cm.
Form submerged aquatic with emergent erect Fl-stalks or creeping over exposed mud. **Fls** L 12–18 mm; **yellow**; with conical spur; UPPER LIP broadly oval, not lobed. **Lvs** hair-like; some segments with bladders (D < 4 mm). **Hab** in standing or very slow-moving, unpolluted water – species have differing preferences for calcareous or acid water.

J F M A M J J A S O N D

SS There are 9 species in B&I (some very rare). Identification is not easy and depends in part on leaf features which are especially hard to discern in the collapsed mass of greenery the leaves become when removed from water.

BLADDERS

LESSER BLADDERWORT

Yellow, 2-lipped fls and bladders distinguish bladderworts from all other superficially similar aquatic plants with hair-like leaf-segments

GREATER BLADDERWORT

66 Adoxaceae | **Moschatel** family `1 sp. B&I`

The sole member of its family worldwide, its unique cubical flowerheads comprising 5-lobed flowers on the faces and a 4-lobed flower on top, hence the alternative name Town Hall Clock.

℗ **Moschatel** *Adoxa moschatellina*

H to 15 cm. **Infl** cubical flowerheads (D 8–10 mm) with a **pale green 5-lobed flower on each horizontal face** and a **4-lobed flower on top**. **Lvs** BASAL trifoliate; LFLTS trifoliate, segments lobed or further trifoliate; ultimate lflts rounded at tip; STEM similar but smaller; opposite. **Hab** woods, hedges, shady places in mountains. **SS** leaves – **Wood Anemone** (*p. 72*) [ultimate lflts pointed]; **Sanicle** (*p. 228*) [LVS underside glossy].

J F M A M J J A S O N D

INFL unmistakable cubical 'town-hall clock'

193

67 Asteraceae | **Daisy + Thistle** family   **76 spp.** | 188 spp. B&I

Infl looks like a single flower but is actually a flowerhead (**capitulum** (see p.17)) consisting of, in most cases, numerous tiny flowers that sit on an enlarged stem-apex (**receptacle**). The 'single-flower' appearance is most noticeable in those with elongated, flat, petal-like outer flowers (**ray-flowers**). The receptacle is clothed in sepal-like bracts, and the tiny flowers may be interspersed with '**receptacle-scales**'. The flowerheads may themselves be clustered into a (typically branched) inflorescence.

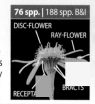

DISC-FLOWER / RAY-FLOWER / BRACTS / RECEPTACLE

IDENTIFY TO GROUP BY: ► **inflorescence shape + branching;** ► **flower + receptacle details**

Daisy family group and species identification

1 FLOWERHEADS **consisting of tubular (disc-) flowers only** (see box opposite, top)

FLOWERHEADS thistle-like; straw-coloured	FLOWERHEADS thistle-like; purplish		
	LEAVES + BRACTS spiny	LEAVES + BRACTS not spiny	LEAVES not spiny; BRACTS hooked bristles
Carline Thistle *p.196*	**THISTLES** *pp.196–197*	**KNAPWEEDS** *p.198*	**BURDOCKS** *p.199*

FLOWERHEADS **in tight buttons; typically lacking rays**

FLOWERHEADS in flat-topped umbel	FLOWERHEADS in loosely branched inflorescence – see also **3** opposite			
Tansy *p.198*	**BUR-MARIGOLDS** *p.199*	**Pineappleweed** *p.200*	**Groundsel** *p.200*	**Sea-aster** (rayless form) *p.200*

FLOWERHEADS **inconspicuous; aggregated into clusters or spikes**

FLOWERS brownish		FLOWERS pinkish/whitish		
CUDWEEDS *p.200*	**WORMWOODS + MUGWORTS** *p.201*	**Hemp-agrimony** *p.201*	**Mountain-everlasting** *p.202*	**'HELIOTROPES'** *p.202*

DAISY FAMILY Flowerhead STRUCTURE EXPLAINED (see also pp. 24, 52)

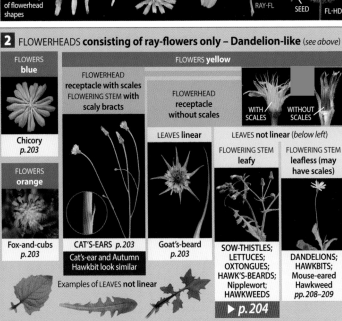

1 ALL DISC-FLOWERS: tubular with 4 or 5 apical tooth-like 'petals'

2 ALL RAY-FLOWERS: petal-tube flattened and strap-like

3 DISC- + RAY-FLOWERS COMBINED

IN FRUIT

PAPPUS

'Petals' range in length from very short to long, making for a diversity of flowerhead shapes

YARROW   DISC-FL

RAY-FL   SEED   FL-HD

---

2 FLOWERHEADS consisting of ray-flowers only – Dandelion-like (see above)

FLOWERS **blue**

Chicory
p. 203

FLOWERS **orange**

Fox-and-cubs
p. 203

FLOWERS **yellow**

FLOWERHEAD receptacle with scales
FLOWERING STEM **with scaly bracts**

FLOWERHEAD receptacle **without scales**

WITH SCALES   WITHOUT SCALES

CAT'S-EARS p. 203

Cat's-ear and Autumn Hawkbit look similar

Examples of LEAVES **not linear**

LEAVES **linear**

Goat's-beard
p. 203

LEAVES **not linear** (below left)

FLOWERING STEM **leafy**

SOW-THISTLES; LETTUCES; OXTONGUES; HAWK'S-BEARDS; Nipplewort; HAWKWEEDS

▶ p. 204

FLOWERING STEM **leafless (may have scales)**

DANDELIONS; HAWKBITS; Mouse-eared Hawkweed
pp. 208–209

---

3 FLOWERHEADS comprising both disc- and ray-flowers – Daisy-like

DISC-FLOWERS + RAY-FLOWERS **grey**

Yarrow + Sneezewort
p. 213

some resemblance to members of the carrot family p. 222

DISC-FLOWERS **yellow**

RAY-FLOWERS **yellow/orange**

FLS single

MARIGOLDS; GROUNDSELS; RAGWORTS; FLEABANES; Colt's-foot
pp. 210–212

FLS in terminal inflorescence

GOLDENRODS
p. 213

RAY-FLOWERS **blue/purple**

MICHAELMAS-DAISIES; Blue Fleabane; Sea-aster
p. 209

RAY-FLOWERS **white**

DAISIES; MAYWEEDS; SOLDIERS; Feverfew
pp. 214–215

RAY-FLOWERS **white (pink tinge)**

Daisy; Mexican Fleabane; *ERIGERON* FLEABANES
p. 214, p. 216

 Blue Fleabane (*left*) can, and *Erigeron* fleabanes (p. 216) (*right*) do, have very short rays

1

FLOWERHEADS **consisting of tubular (disc-) flowers only**

● **Thistles** | **Form** erect. **Lvs + bracts** spiny.

IDENTIFY BY: ▶ combination of stem detail; leaf shape and inflorescence/flowerhead. During late season details of the pappus provide a distinction between most.

FLOWERS **straw-coloured**

B Carline Thistle *Carlina vulgaris*

H to 60 cm. **Form** can be squat. **Fl-hds** D 15–40 mm; solitary or in clusters of 2–5. **Fls** pinkish-purple when fresh; **turning yellow**. **Lvs** pinnately lobed, with spiny margins, cottony beneath. **Fr** L 2–4 mm, PAPPUS feathery; brownish-cream. **Hab** dry calcareous grassland.

INNER BRACTS
**spreading,
straw-coloured** –
resembling ray-florets

FLOWERS **purple**; PAPPUS **feathery**

B Spear Thistle *Cirsium vulgare*

H to 150 cm. **Form** STEM with sharp spiny wings. **Fl-hds** rounded; D 25–50 mm; solitary or in open cluster. **Fls** purple. **Lvs** BASAL ROSETTE unlobed; almost spineless at first – spinier with age; STEM somewhat clasping; slightly spreading down stem; pinnately lobed. **Fr** L 3·5–5·0 mm; PAPPUS pale silver-white; SEED pale brown (can have black streaks). **Hab** grassland, disturbed ground.

BRACTS
**spines
yellow;
straight**

LF UPP
prickly hairs;
MARGIN spines
sharp, unequal

STEM with
**sharp, spiny
wings**

B Marsh Thistle *Cirsium palustre*

H to 200 cm. **Form** STEM with **wings and spines**. **Fl-hds** D 10–20 mm; in clusters. **Fls** purple, occasionally white. **Lvs** BASAL ROSETTE present; STEM deeply pinnately lobed, very spiny; upperside hairy. **Fr** L 3–4 mm, PAPPUS dirty-white; SEED pale fawn. **Hab** marshes, damp grassland, woodland rides.

BRACTS
**spines
purplish;
small;
appressed**

LF UPP hairy;
MARGIN with
white hairs
and spines

STEM with
wings and spines

P Creeping Thistle *Cirsium arvense*

H to 150 cm. **Form** STEM **without wings or spines**. **Fl-hds** egg-shaped; D 15–25 mm; solitary or in small clusters, on cobwebby stalks. **Fls** lilac-mauve, honey-scented. **Lvs** BASAL ROSETTE **absent**; STEM pinnately lobed; higher lvs stalkless. **Fr** L 3–4 mm, PAPPUS pale brown; SEED dark brown. **Hab** grassland, disturbed ground, road verges.

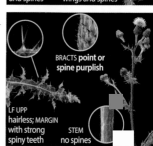

BRACTS **point or
spine purplish**

LF UPP
hairless; MARGIN
with strong
spiny teeth

STEM
no spines

SEPARATE BY: ► flower/plant size ► leaf underside ► stem details ► distribution

ⓟ Meadow Thistle *Cirsium dissectum*

H to 80 cm. **Form** creeping; STEM downy; unbranched without wings. **Fl-hds** solitary; D 20–30 mm. **Fls** purple. **Lvs** narrowly oval, usually ± **unlobed**, soft prickles on margins. **Fr** L 3–4 mm; PAPPUS pure white; SEED oval. **Hab** peaty wet meadows, fens and less acid bogs.

ⓟ Melancholy Thistle *Cirsium heterophyllum*

H to 100 cm. **Form** creeping; STEM ridged and cobwebby, sparsely branched, without wings. **Fl-hds** solitary; D 30–50 mm on cobweb-hairy stalk. **Fls** purple. **Lvs** narrowly oval, toothed, can be pinnately lobed; UPPER small, clasping stem; LOWER stalked. **Fr** L 4–5 mm; PAPPUS pure white; SEED elongate oval. **Hab** upland hay meadows, damp grassland, scrub and streamsides.

STEM white downy

LF **soft prickles on margins**

LF UND downy

STEM ridged; cobwebby

LF UND cottony

D

FLOWERS **purple**; PAPPUS **non-feathery**

ⓑ Welted Thistle *Carduus crispus*

H to 150 cm. **Form** STEM branched with **continuous spiny wing (welt)** to just below fl-heads. **Fl-hds** rounded; D 10–25 mm; solitary or in small clusters. **Fls** reddish-purple. **Lvs** pinnately lobed, weakly spiny, somewhat cobwebby. **Fr** L 3–4 mm; PAPPUS L<12 mm, with scaly hairs; SEED with prominent projection at tip. **Hab** hedgerows, waysides, streamsides, disturbed areas.

ⓑ Musk Thistle *Carduus nutans*

H to 150 cm. **Form** STEM usually branched; wingless and cobwebby below fl-hds, winged lower down. **Fl-hds** large, nodding; D 30–50 mm; solitary or in clusters on non-spiny stalks. **Fls** bright reddish-purple; musky scented. **Lvs** pinnately lobed, weakly spiny, somewhat cobwebby. **Fr** L 3–4 mm; PAPPUS L<12 mm, with toothed hairs. **Hab** pastures, rough grassland.

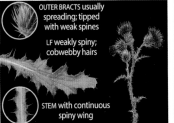

OUTER BRACTS usually spreading; tipped with weak spines

LF weakly spiny; cobwebby hairs

STEM with continuous spiny wing

FL-HDS **large; nodding**

BRACTS **erect to strongly recurved; deep purplish-brown**

D

197

1

FLOWERHEADS **consisting of tubular (disc-) flowers only**

● **Knapweeds** | **Form** erect; **STEM** thickened below fl-hd. **Fls** thistle-like. **Bracts** non-spiny.

IDENTIFY BY: ▶ leaf shape ▶ stem hairs
▶ bract and bract fringe details

SS Chalk Knapweed *C. debeauxii* (N/I) [most similar to **Common Knapweed**; BRACT central undivided part L 2× W; STEM only slightly swollen below FL- head; FL-heads smaller].

COMMON   GREATER   PERENNIAL

J F M A M J J A S O N D   J F M A M J J A S O N D   J F M A M J J A S O N D

℗ **Common Knapweed**
Centaurea nigra

H to 100 cm. **Form** erect, can be well-branched. **Fl-hds** D 20–40 mm; solitary or in branched clusters.
Lvs narrowly oval; can have basal lobes; rough.
Fr seeds L 3 mm; pale brown; compressed. **Hab** grassland.

FLS **L all ± equal**

BRACTS fringe blackish; broad; central part L<2×W

LF basal lobes at most

℗ **Greater Knapweed**
Centaurea scabiosa

H to 120 cm. **Form** erect, branched in upper half.
Fl-hds D 20–40 mm; solitary or in branched clusters.
Lvs pinnately lobed.
Fr seeds L 3 mm; pale brown; compressed. **Hab** rough calcareous grassland.

FLS L outer > inner

BRACTS fringe dark brown; horseshoe-shaped

LF pinnately lobed

℗ **Perennial Cornflower**
Centaurea montana

H to 80 cm. **Form** clump-forming. **Fl-hds** D 60–80 mm; solitary. **Lvs** narrowly oval; soft. **Fr** seeds L 5–7 mm; brown. **Hab** brownfield, rough grassland. **SS Cornflower** *C. cyanus* (N/I) [LOWER LVS pinnately lobed].

FLS L outer > inner

BRACTS fringe dark brown; pointed

LF elongate; undivided

FLOWERHEADS **tight buttons; typically lacking rays**

INFL **flat top umbels; FL yellow**

℗ **Tansy** *Tanacetum vulgare*

J F M A M J J A S O N D

H to 150 cm. **Form** erect, patch-forming, almost hairless, strongly aromatic when crushed. **Fl-hds** D 8–12 mm; **lacking rays**; in **large, flat-topped clusters. Lvs** L to 25 cm, **deeply pinnately lobed** with toothed segments.
Hab brownfield, riverbanks, road verges.

FL-HD

FL-HD

FL

LF deeply pinnately lobed; LFLTS sharply toothed

● **Burdocks** | **Form** large (H to 150 cm); erect; branched. **Fls** thistle-like; usually purplish.
Lvs large (Lesser L to 50 cm; Greater L to 80 cm), triangular. **Bracts** hooked spines.

IDENTIFY BY: ► flowerhead shape, stalk length + bract size ► leaf-stalk hollow or not

B Lesser Burdock
Arctium minus

Form hairy. **Fl-hds** egg-shaped; D 15–25 mm (larger in fruit) unstalked or almost so. **Lvs** LF-STEM hollow.
Fr seeds; L 5–6 mm. **Hab** field borders, woodland rides, rough grassland. **SS Wood Burdock** *A. nemorosum* (N/I) [FL-HDS D>25 mm – but beware larger-flowered forms (perhaps hybrids) of Lesser].

B Greater Burdock
Arctium lappa

Form hairy or hairless.
Fl-hds globular; D 30–42 mm (larger in fruit) unstalked or almost so. **Lvs** LF-STEM solid.
Fr seeds; L 6·0–7·7 mm.
Hab unshaded field borders, rough grassland.

LESSER　　GREATER

J F M A M J J A S O N D　J F M A M J J A S O N D

FL

D

LF both species similar

FL-HDS **unstalked** (or almost so)　FL-HDS **stalked** (can be short)

LF-STEM **hollow**　　LF-STEM **solid**

INFL **loosely branched**; FL **yellow**; solitary at tip of branches on short stalks

The bur-marigolds often grow in proximity; **Trifid** tends to prefer more acid sites.

A Trifid Bur-marigold *Bidens tripartita*

H to 75 cm. **Form** erect, branched, hairless.
Fl-hds D 15–25 mm.
Lvs typically trifoliate; end lobe largest; coarsely toothed; on winged stalk.
Fr with barbed bristles.
Hab beside standing or slow-moving fresh water.

J F M A M J J A S O N D

FL-HD **erect**;
receptacle-scales
L>8 mm

LF **3–5-lobed**

A Nodding Bur-marigold *Bidens cernua*

H to 75 cm. **Form** erect, branched, hairless. **Fl-hds** D 15–25 mm. **Lvs** narrowly oval; toothed; undivided; unstalked. **Fr** with barbed bristles. **Hab** beside standing or slow-moving fresh water.

J F M A M J J A S O N D

FL-HD **nodding**;
receptacle-scales
L<8 mm

LF **undivided**; toothed

199

1

FLOWERHEADS **consisting of tubular (disc-) flowers only**

FLOWERHEADS **tight buttons; typically lacking rays**

ⒶPineappleweed *Matricaria discoidea*

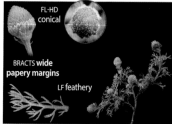

H to 35 cm. **Form** erect; branched; hairless; AROMA **pineapple-like** when crushed. **Fl-hds** conical; D 5–8 mm; solitary on short stalks. **Lvs** 1–2 pinnate; bright green; slightly fleshy. **Fr** L to 1·5 mm; ribbed. **Hab** arable margins, compacted tracksides, brownfield.

FL-HD conical

BRACTS **wide papery margins**

LF feathery

ⒶCommon Groundsel *Senecio vulgaris*

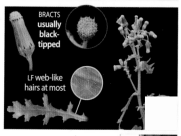

H to 30 cm. **Form** erect, irregularly branched. **Fl-hds** D 4–5 mm; RAYS **usually lacking** (can be up to 11, short (L < 5 mm)). **Lvs** L to 50 mm; pinnately lobed; LOWER LVS stalked **Fr** L<2·5 mm; ribbed; hairs between ribs. **Hab** brownfield, gardens, arable margins. **SS** other groundsels (*p. 211*) [typically rayed].

BRACTS **usually black-tipped**

LF web-like hairs at most

Sea Aster *Tripolium pannonicum* ssp. *discoidea* (*right*) lacks the rays found in the typical rayed form (ssp. *pannonicum*) – see *p. 209*; the mauve rays of **Blue Fleabane** (*p. 209*) can be so short it can appear rayless at first glance.

FLOWERHEADS **inconspicuous; aggregated into clusters or spikes**

● **Cudweeds** | **Form** low-growing, white-woolly. **Fl-hds** ± conical, in tight clusters.

ⒶMarsh Cudweed *Gnaphalium uliginosum*

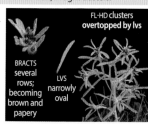

H to 20 cm. **Form** sprawling to bushy; basally branched; STEM **silvery-grey**. **Fl-hds** L 4 mm; in tight terminal clusters of 2–10; RECEP-SCALES absent. **Lvs** linear to narrowly oval; UPPERSIDE green; UNDERSIDE woolly-grey. **Fr** L to 0·7 mm, cylindrical. **Hab** trampled, compacted areas subject to winter flooding.

FL-HD clusters **overtopped by lvs**

BRACTS several rows; becoming brown and papery

LVS narrowly oval

ⒶCommon Cudweed *Filago germanica*

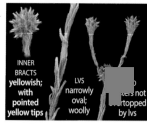

H to 40 cm. **Form** erect; woolly; branched from middle. **Fl-hds** L 5 mm; in **globular clusters** of 20+ (D 10–14 mm); BRACTS 3 rows; yellowish (can be reddish); RECEP-SCALES present; grading into bracts. **Lvs** narrowly oval; woolly; can be wavy. **Fr** L to 0·8 mm; egg-shaped. **Hab** sandy arable margins, open grassland, sand dunes, heaths.

INNER BRACTS **yellowish; with pointed yellow tips**

LVS narrowly oval; woolly

...ers not overtopped by lvs

FLOWERHEADS **inconspicuous; aggregated into clusters or spikes**

FL **brownish to yellow/orange; protruding stamens typically yellowish**

● **Mugwort /Wormwoods** | Form erect; tufted/bushy; aromatic; wormwoods with woody base. **Fl-hds** clustered on branches. **Lvs** deeply pinnately lobed.

Ⓟ Mugwort
Artemisia vulgaris

H to 150 cm.
Form erect,
tufted; STEM
white central
pith; AROMA
slight. **Fl-hds**
L 4 mm; oval.
Lvs L to 80 mm;
1–2-pinnately
lobed. **Fr** L 1 mm; oblong. **Hab**
brownfield, rough grassland,
roadsides.

Ⓟ Wormwood
Artemisia absinthium

H to 100 cm.
Form erect,
tufted; AROMA
strong. **Fl-hds**
L 5 mm;
globular; on
nodding stalks.
Lvs L to 80 mm;
2–3-pinnately
lobed. **Fr** L 1·6 mm; cylindrical.
Hab brownfield, rough
grassland.

Ⓟ Sea Wormwood
Artemisia maritimum

H to 50 cm.
Form bushy,
patch-forming;
AROMA strong.
Fl-hds W 2 mm;
egg-shaped;
on short stalks.
Lvs L to 50 mm;
2–3-pinnately
lobed. **Fr** L 2·5 mm; cylindrical.
Hab upper saltmarshes, sea
cliffs, shingle, salted roadsides.

FL tinged
yellow or
purple

LF
hairless
above

LF downy
beneath

FL
yellowish

LF silky-white
both sides

LF terminal
lobes oval

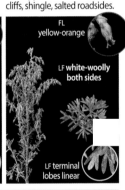

FL
yellow-orange

LF white-woolly
both sides

LF terminal
lobes linear

FLOWERHEADS **inconspicuous; aggregated into clusters or spikes**

FL **whitish/pinkish**

Ⓟ Hemp-agrimony
Eupatorium cannabinum

H to 150 cm. **Form** erect;
robust; clump-forming;
STEM usually purplish.
Fl-hds D to 5 mm;
densely clustered in
flat or rounded infl.
Lvs palmate or
palmately 3- to 5-lobed;
coarsely toothed. **Fr** L
3 mm; PAPPUS unbranched white hairs
(L 5 mm). **Hab** marshes, fens, on banks
of waterbodies; also drier habitats,
including woods and brownfield.

FL-HD

FL purplish-
pink; styles
protruding,
pale, forked

LF
palmate

201

1

FLOWERHEADS **inconspicuous; aggregated into clusters or spikes**

FL **whitish/pinkish**

℗ Mountain-everlasting *Antennaria dioica*

H to 20 cm. **Form** patch-forming; woody at base. **Fl-hds** D ♀ 10–12 mm, ♂ 5–6 mm, in terminal clusters of 2–8 on whitish stem arising from rosette. **Lvs** L to 40 mm; narrowly oval; entire; UPPERSIDE dark green; UNDERSIDE downy white. **Fr** L to 1 mm; PAPPUS unbranched white hairs, (L to 8 mm). **Hab** montane grassland and rocks, calcareous grassland, sand dunes.

♂ FLS pale pink; BRACTS whitish, **spreading**

♀ FLS pale to deep pink; BRACTS **erect**

℗ Winter Heliotrope *Petasites pyrenaicus*

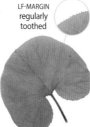

H to 150 cm. **Form** densely patch-forming; only ♀ plants found in B&I. **Fl-hds** L to 30 cm; up to 20 in broad terminal cluster; AROMA **almond/vanilla**. **Lvs** all basal; kidney- or heart-shaped; **appearing with fls** on stalks L to 30 cm. **Fr** L to 3 mm; cylindrical. **Hab** brownfield, streamsides, road verges.

LF-MARGIN regularly toothed

℗ Butterbur *Petasites hybridus*

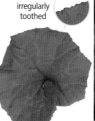

H to 40 cm at flowering. **Form** densely patch-forming; ♂ and ♀ fls on separate plants. **Fl-hds** L to 30 cm; numerous in leafless terminal clusters. **Lvs** all basal; ± round; lf-bases overlap; **appearing after fls** and becoming very large (D to 100 cm) on stalks L to 150 cm. **Fr** L to 3 mm; cylindrical. **Hab** banks of watercourses, wet meadows.

LF-MARGIN irregularly toothed

Butterbur can form extensive patches; female plants (inset) are far more common than males.

♀ FLS

FLOWERHEADS **consisting of ray-flowers only** · 1/4

FLOWERS **blue**

Ⓟ Chicory · FL-HD
Cichorium intybus

H to 120 cm. **Form** erect, branched. **Fl-hds** D 25–40 mm, in elongated leafy spikes, **usually closing by midday.** **Lvs** pinnately lobed. **Fr** L to 3 mm; PAPPUS short scales. **Hab** arable margins, road verges.

J F M A M J J A S O N D

FLOWERS **orange**

Ⓟ Fox-and-cubs · FL-HD
Pilosella aurantiaca

H to 40 cm. **Form** rosette-forming, with **blackish hairs.** **Fl-hds** D 10–20 mm; up to 12 on branches at top of unbranched stem. **Lvs** narrowly oval; entire; bluish-green. **Fr** L to 2·5 mm. **Hab** grassland, brownfield.

J F M A M J J A S O N D

FLOWERS **yellow** | **receptacle-scales present** | FL-STEM **leafless (small bracts present)**

Ⓟ Cat's-ear *Hypochaeris radicata*

H to 60 cm. **Form** erect, rosette-forming. **Fl-hds** D 20–40 mm; usually >1 in widely branched infl. **Lvs** pinnately lobed; usually hairy. **Fr** ribbed, PAPPUS (where present) 2 rows of unbranched off-white hairs. **Hab** grassland. **SS** hawkbits (*p. 208*) [RECEP-SCALES absent].

J F M A M J J A S O N D

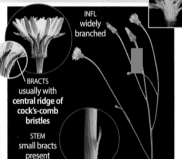

INFL **widely branched**

BRACTS usually with **central ridge of cock's-comb bristles**

STEM small bracts present

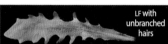

LF with unbranched hairs

FLOWERS **yellow** | **receptacle-scales absent** | LVS **linear**

ⓐⓟ Goat's-beard *Tragopogon pratensis*

H to 70 cm. **Form** erect, unbranched. **Fl-hds** D 25–40 mm solitary; bracts in 1 row; open in the morning. **Lvs** L to 35 cm; **linear; channelled; base sheathing.** **Fr** L to 22 mm; PAPPUS white; **feathery** in conspicuous fruiting head.

J F M A M J J A S O N D

Hab dry grassland, road verges. **SS** Salsify *T. porrifolius* (N/I) [FLS purplish].

BRACTS much longer than outer rays

FL-HDS closed by early afternoon

LF very narrow; conspicuous white midrib

203

2

FLOWERHEADS **consisting of ray-flowers only**

FLOWERS **yellow** | **receptacle-scales absent** | FL-STEM **leafy** | LVS **not linear**

Sow-thistles, lettuces, oxtongues, hawk's-beards and Nipplewort

UPPER LVS **basal lobe broadly clasping stem**

FL-HDS **smaller** (D 8–20mm)	FL-HDS **larger** (D 15–35mm)
LACTUCA LETTUCES *p. 206*	SOW-THISTLES *p. 207*

UPPER LVS **basal lobe not or only narrowly clasping stem**

	BRACTS **in 2 rows**		BRACTS **in >2 rows**	
RAYS **5**	RAYS **>5**			
	BRACTS **all erect**	BRACTS **outer spreading**	BRACTS **sharply bristly**	BRACTS **hairy**
			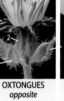	
Wall Lettuce *below*	**Nipplewort** *below*	**HAWK'S-BEARDS** *opposite*	**OXTONGUES** *opposite*	**HAWKWEEDS** *p. 206*

BRACTS **in 2 rows** | RAYS **=5**　　　　　BRACTS **in 2 rows** | RAYS **>5** | BRACTS **all erect**

P **Wall Lettuce** *Mycelis muralis*　　　　**A** **Nipplewort** *Lapsana communis*

H to 100cm. **Fl-hds** D 10–15mm on **wide-spreading** branches. **Lvs** L to 22cm, pinnately lobed. **Fr** L to 5mm; dark with pale beak; PAPPUS 2 rows of unbranched white hairs. **Hab** hedgebanks, rocky habitats on calcareous soil.

J F M A M J J A S O N D

H to 100cm. **Form** erect. **Fl-hds** D 15–20mm; several in widely branched INFL; FLS up to 15 per head. **Lvs** pinnately lobed; TERMINAL LOBE **very large; triangular; sharply** toothed. **Fr** ribbed; PAPPUS **absent.** **Hab** open woodland, hedges.

J F M A M J J A S O N D

FL-HD **just 5 flowers**

FR dark with pale beak　　LF lobes angular

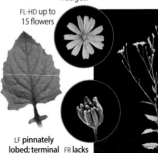

FL-HD up to 15 flowers

LF **pinnately lobed; terminal lobe very large**　FR **lacks pappus**

● **Hawk's-beards** | **Form** erect; rosette-forming. **Fls** rays >5; BRACTS in two rows; inner row erect; outer spreading. **Lvs** STEM narrowly clasping; variable, pinnately lobed to toothed.

IDENTIFY BY: ► bract details ► fruit details

SS Other rare hawk's-beards (3 spp.) (N/I) [various characters FR; STEM-LVS; AROMA].

Ⓑ Beaked Hawk's-beard *Crepis vesicaria*

H to 80 cm. **Fl-hds** D 15–25 mm. **Fls** RAYS underside usually with red-brown stripe. **Lvs** pinnately lobe; lobes toothed or lobed. **Fr** SEED L to 5 mm; tapering to beak (L = seed); PAPPUS 1–2 rows of soft white hairs. **Hab** rough grassland, brownfield.

OUTER BRACTS **spreading**

STEM-LF typically pinnately lobed

SEED **beaked**

NOTE: stem and leaf characters are variable and cannot be relied upon alone for identification

Ⓐ Smooth Hawk's-beard *Crepis capillaris*

H to 75 cm. **Form** often branched and bushy. **Fl-hds** D 15–25 mm. **Lvs** pinnately lobed or toothed. **Fr** SEED L to 2·5 mm, ribbed; PAPPUS 2 rows of white hairs. **Hab** grassland, brownfield, arable margins.

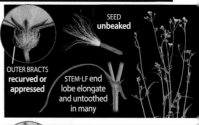

SEED **unbeaked**

OUTER BRACTS **recurved or appressed**

STEM-LF end lobe elongate and untoothed in many

Ⓟ Marsh Hawk's-beard *Crepis paludosa*

H to 90 cm. **Fl-hds** D 15–25 mm in loose clusters on leafy stem. **Lvs** narrowly triangular; sharply toothed; BASAL with short, winged stalk; UPPER unstalked, clasping stem. **Fr** SEED L to 5 mm, ribbed; PAPPUS yellow-brown, brittle, 1–2 rows of soft white hairs. **Hab** marshes, fens, damp meadows; some tolerance for shade.

STEM-LF lower lobes long; clasping

BRACTS woolly, with sticky black glands

SEED slightly tapering

BRACTS **in more than 2 rows; sharply bristly**

ⓐⒷ Bristly Oxtongue *Helminthotheca echioides*

H to 80 cm. **Form** erect, branched, bristly. **Fl-hds** D 20–25 mm; on branched stem. **Lvs** narrowly oval, with wavy edges and clasping stem. **Fr** SEED beaked; PAPPUS feathery white. **Hab** open grassland, river banks, sea walls. **SS** Hawkweed Oxtongue *Picris hieracioides* (N/I) [bristles not from pimples; all INFL-bracts linear]; leaves – **Green Alkanet** (*p. 171*) [LF bristle pustules smaller].

OUTER BRACTS **heart-shaped; spreading**

LVS **bristles usually arising from a pimple**

2

FLOWERHEADS **consisting of ray-flowers only**

FLOWERS **yellow** | **receptacle-scales absent** | FL-STEM **leafy** | LVS **not linear**

UPPER LEAVES **not , or only narrowly, clasping stem**

BRACTS **in more than 2 rows; hairy**

● **Hawkweeds** *Hieracium* spp. |
a genus (like dandelions – *p. 209*)
that produces seed without
sexual reproduction, so that
small differences are perpetuated.
400+ such microspecies are
recognized, many are rare.
**Identification requires
specialist keys.**
H to 60 cm (montane forms mostly smaller).
Form erect, usually with leafy stem and/
or rosette-forming. **Fl-hds** D 20–30 mm; on
branched stem; FL-HEAD BRACTS in overlapping
rows unequal in length. **Lvs** oval; ± entire, but
variable. **Fr** L to 4·5 mm; PAPPUS simple, pale
brown; SEEDS ribs merge into a swollen ring.
Hab grassland, open woodland, uplands.

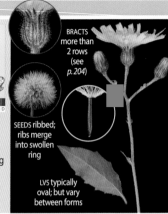

BRACTS
more than
2 rows
(see
p. 204)

SEEDS ribbed;
ribs merge
into swollen
ring

LVS typically
oval; but vary
between forms

UPPER LEAVES **basal lobes broadly clasping stem**

● *Lactuca* **lettuces** | **Form** Tall (H to 200 cm); erect. **Fl-hds**
numerous; closed by midday; closely spaced on spreading
branches. **Fr** L to 4 mm; PAPPUS unbranched, white hairs.
Hab disturbed ground, brownfield, coastal sea walls and dunes.

PRICKLY　　GREAT

PRICKLY LETTUCE

ⓐⓑ **Prickly Lettuce**
Lactuca serriola

Fl-hds D 8–13 mm. **Form**
STEM ± smooth. **Lvs** L to 15 cm;
UPPER **held vertical and
orientated N–S in full sun**;
STEM-LVS broadly oval; basal
lobes, pointed; MARGINS with
rigid prickles; UNDERSIDE
midrib whitish with rigid
prickles. **Fr** pale brown.

ⓐⓑ **Great Lettuce**
Lactuca virosa

Fl-hds D 15–20 mm. **Form**
STEM **prickly**. **Lvs** L to 23 cm;
UPPER not orientated towards
sun; STEM-LVS broadly oval to
pinnately lobed, basal lobes
rounded; MARGINS with rigid
prickles; UNDERSIDE **midrib
purplish** with dense prickles.
Fr purplish-black.

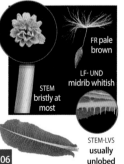

FR pale
brown

LF- UND
midrib whitish

STEM
bristly at
most

STEM-LVS
usually
unlobed

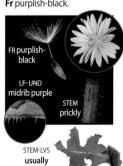

FR purplish-
black

LF- UND
midrib purple

STEM
prickly

STEM-LVS
usually
pinnately lobed

UPPER LEAVES **basal lobes broadly clasping stem**

● **Sow-thistles** | Form erect; branched. **Fl-hds** yellow; in loose clusters.

IDENTIFY BY: ▶ presence/absence of yellow glandular hairs then
▶ stem-leaf basal lobe shape ▶ fruit details

A Prickly Sow-thistle *Sonchus asper*

H to 120 cm. **Form** hairless. **Fl-hds** D 15–35 mm in irregular clusters. **Lvs** glossy, dark-green, pinnately lobed; MARGINS spiny. **Fr** L to 2·5 mm; PAPPUS unbranched, white. **Hab** field margins, brownfield, disturbed areas.

J F M A M J J A S O N D

STEM-LVS glossy; dark-green; relatively stiff; LF-MARGINS spiny

STEM-LVS **basal lobes rounded; spiny;** appressed

BRACTS hairless

SEEDS **smooth** between ribs

NOTE: Sow-thistle leaves are highly variable in shape

A Smooth Sow-thistle *Sonchus oleraceus*

H to 150 cm. **Form** largely hairless. **Fl-hds** D 15–35 mm; often paler than other sow-thistles. **Lvs** dull, greyish-green, pinnately lobed; MARGINS not spiny. **Fr** L to 3 mm; PAPPUS unbranched, white. **Hab** brownfield, disturbed areas, field margins.

J F M A M J J A S O N D

STEM-LVS dull; greyish-green; relatively soft; LF-MARGINS not spiny

STEM-LVS basal lobes pointed; toothed; **spreading**

BRACTS can have a few white glandular hairs

SEEDS **wrinkled** at right-angles to ribs

P Perennial Sow-thistle *Sonchus arvensis*

H to 150 cm. **Form yellow glandular hairs;** patch-forming. **Fl-hds** D 40–50 mm in loose clusters. **Lvs** elongate, pinnately lobed; MARGINS with spiny teeth. **Fr** L to 3·5 mm; PAPPUS unbranched, white. **Hab** field margins, river banks, strandlines.

J F M A M J J A S O N D

STEM-LVS dull; green; LF-MARGINS softish spiny teeth

STEM-LVS basal lobes rounded; toothed; **appressed**

BRACTS with yellow glandular hairs

SEEDS **ribbed;** dark brown

2

FLOWERHEADS **consisting of ray-flowers only**

FLOWERS **yellow** | **receptacle-scales absent** | FL-STEM **leafless (may have scales)** | LVS **not linear**

FL-STEM **solid or narrowly hollow at most**; PLANT **with leafy runners**; LVS **whitish underneath**

℗ Mouse-eared Hawkweed
Pilosella officinarum

H to 30 cm. **Form** rosette-forming; in patches. **Fl-hds** D 15–25 mm; singly on unbranched leafless stem. **Lvs** L to 80 mm; broadly oval; entire. **Fr** SEED purple-black; ribbed; PAPPUS 1 row of feathery, off-white hairs. **Hab** dry grassland, heaths, dunes.

J F M A M J J A S O N D

FLS **pale lemon-yellow**; outer rays usually red underneath

LVS with **long, white, whiskery hairs**; whitish underneath

FL-STEM **solid or narrowly hollow at most**; PLANT **without leafy runners**; LVS **green underneath**

● Hawkbits |
Form erect; rosette-forming; FL-STEM small bracts present. **Fl-hds** clustered on branches. **Lvs** typically pinnately lobed but very variable; UNDERSIDE green. **SS** Cat's-ear (*p. 203*) [RECEP-SCALES present; SEED beaked].

SCALES absent

STEM X-SECT.

AUTUMN　LESSER　ROUGH

J F M A M J J A S O N D　J F M A M J J A S O N D　J F M A M J J A S O N D

℗ Autumn Hawkbit
Scorzoneroides autumnalis

H to 60 cm. **Fl-hds** D 20–35 mm; **usually >1 in widely branched infl**; FLS up to 15 per head. **Lvs** deeply pinnately lobed to almost unlobed; sparse unforked hairs at most. **Fr** ribbed; PAPPUS (ALL FRUIT) 1 row of feathery, off-white hairs. **Hab** grassland.

℗ Lesser Hawkbit
Leontodon saxatilis

H to 30 cm. **Fl-hds** D 12–20 mm; solitary on unbranched stem. **Lvs** variably pinnately lobed; MARGIN with forked hairs. **Fr** finely ribbed; INNER FRUIT PAPPUS off-white hairs, OUTER FRUIT PAPPUS absent. **Hab** well-drained grassland.

℗ Rough Hawkbit
Leontodon hispidus

H to 60 cm. **Form** usually very hairy. **Fl-hds** D 25–40 mm; solitary on unbranched stem. **Lvs** pinnately lobed or toothed, with some forked hairs. **Fr** finely ribbed; PAPPUS (ALL FRUIT) 2 rows of off-white hairs. **Hab** well-drained grassland.

LVS shape highly variable; sparse unforked hairs at most

RAYS usually **reddish beneath**

FR all seeds beakless; with pappus

INFL usually branched

LVS forked hairs on margin

RAYS usually **greyish-violet beneath**

INFL unbranched

FR outer seeds lack pappus

INNER　OUTER

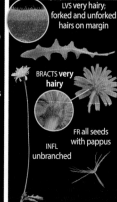

LVS very hairy; forked and unforked hairs on margin

BRACTS **very hairy**

FR all seeds with pappus

INFL unbranched

ⓟ Dandelions *Taraxacum* spp. | a genus (like hawkweeds – *p. 206*) that produces seed without sexual reproduction, so that small differences are perpetuated. More than 232 such microspecies are recognized, divided into 9 sections. **Identification requires specialist keys. H** to 40 cm. **Form** rosette-forming. **Fl-hds** D 20–50 mm; solitary; FL-STALK hollow, leafless; OUTER RAYS can have brownish stripe on underside; BRACTS typically 'untidy', bent back, greyish-green, papery margin. **Lvs** soft; variable; entire to pinnately lobed. **Fr** ribbed, may have tiny spines, long beak; PAPPUS several rows of unbranched white hairs. **Hab** grassland, lawns, brownfield.

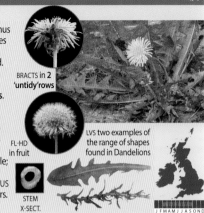

BRACTS in **2** 'untidy' rows

FL-HD in fruit

LVS two examples of the range of shapes found in Dandelions

STEM X-SECT.

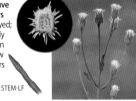

J F M A M J J A S O N D

FLOWERHEADS consisting of both ray- and disc-flowers — 1/6

DISC **yellow**	RAYS **bluish/mauve/purple**

ⓐⓑ Blue Fleabane *Erigeron acris*

H to 60 cm. **Form** erect; branched, greyish-hairy, often red-tinged. **Fl-hds** D 10–18 mm; in clusters. **Lvs** BASAL narrowly oval, stalked; STEM ± linear; unstalked. **Fr** seeds; L 2–3 mm; yellowish, cylindrical, hairy; PAPPUS 1 row of reddish hairs. **Hab** open, dry grassland.

J F M A M J J A S O N D

FL grey-mauve ray-flowers scarcely splayed; only slightly longer than pale yellow disc-flowers

STEM-LF

ⓟ Common Michaelmas-daisy

Symphyotrichum × *salignum*.

H to 130 cm. **Form** erect; STEM leafy; ridged. **Fl-hds** D 15–25 mm; in loose clusters. **Lvs** ± broadly oval; narrowing to clasping base. **Fr** seeds; L 1·5–2·0 mm; PAPPUS several rows of finely toothed white hairs. **Hab** rough grassland, brownfield. **SS** many cultivated species and hybrids [BRACT form, RAY colour].

J F M A M J J A S O N D

FL

BRACTS 'neat'; tips reddish

LVS long-oval

ⓟ Sea Aster *Tripolium pannonicum*

H to 100 cm. **Form** erect, hairless, somewhat **succulent**. **Fl-hds** D 8–30 mm; **rayed, partly rayed or rayless**; in loose clusters of heads. **Lvs** narrowly to broadly oval, half-clasping stem. **Fr** L 5–6 mm; PAPPUS 1 or 2 rows of brownish hairs. **Hab** muddy saltmarshes, sea cliffs, brackish ditches.

J F M A M J J A S O N D

FL

BRACTS 'untidy'; tips not red

LVS similar to those of Common Sea-lavender (*p. 140*) [usually less fleshy with extending terminal point]

3

FLOWERHEADS **consisting of both ray- and disc-flowers**

DISC **yellow or orange**	RAYS **yellow or orange**

FLOWERHEAD **bracts in 2 rows**

FLOWERHEADS **larger (typically D >25 mm); ± solitary**

⊕ Pot Marigold *Calendula officinalis*

H to 50 cm. **Form** branched, bushy, often roughly hairy. **Fl-hds** D 30–70 mm, solitary **Lvs** narrowly to broadly oval; entire; stalkless. **Fr** L to 25 mm; curved; warty. **Hab** brownfield and disturbed sites.

J F M A M J J A S O N D

FL rays orange or yellow; broad

BRACTS in 2 rows; greyish-green with narrow papery margins

LF entire; stalkless

FLOWERHEADS **smaller (D ≤25 mm); in branched inflorescence**   1/2

⊕ Common Ragwort *Jacobaea vulgaris*

H to 100 cm. **Form** erect; STEM ridged; basal rosette. **Fl-hds** D 15–25 mm in dense clusters. **Lvs** BASAL + STEM deeply pinnately lobed; MARGINS not downrolled; UNDERSIDE veins hairy. **Fr** L to 2·2 mm. **Hab** grassland, brownfield, sand dunes.

J F M A M J J A S O N D

LF UNDERSIDE only veins hairy LF TERMINAL LOBE a little wider than lateral lobes

INNER BRACTS **black tips**

INFL **dense clusters**

⊕ Hoary Ragwort *Jacobaea erucifolia*

H to 120 cm. **Form** erect; patch-forming. **Fl-hds** D 15–20 mm in loose, flat-topped clusters. **Lvs** BASAL + STEM irregularly deeply pinnately lobed; MARGINS **downrolled**; UNDERSIDE hairless or sparsely grey-hairy underneath. **Fr** L to 2 mm; hairy on ribs. **Hab** meadows, pastures, sand dunes.

J F M A M J J A S O N D

LF UNDERSIDE sparsely hairy at most LF TERMINAL LOBE narrow; width ± same as lateral lobes

INNER BRACTS pale brown tips

INFL **loose; flat-topped clusters**

⊕ Marsh Ragwort *Jacobaea aquatica*

H to 80 cm. **Form** erect; branched. **Fl-hds** D 15–20 mm; in spreading clusters. **Lvs** BASAL unlobed; STEM irregularly pinnately lobed; MARGINS not downrolled; UNDERSIDE sparsely hairy at most. **Fr** L to 3 mm, ribbed. **Hab** damp grassland, marshes, river margins.

J F M A M J J A S O N D

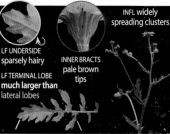

LF UNDERSIDE sparsely hairy LF TERMINAL LOBE **much larger than** lateral lobes

INNER BRACTS pale brown tips

INFL **widely spreading clusters**

FLOWERHEADS **smaller (D ≤25 mm); in branched inflorescence**   **2/2**

Oxford Ragwort *Senecio squalidus*

H to 50 cm. **Form** erect or sprawling; hairless; BASE woody in many. **Fl-hds** D 16–20 mm; RAYS 12–15; in open clusters. **Lvs** L to 120 mm; pinnately lobed or sharply toothed with pointed tip; LOWER LVS stalk winged. **Fr** L to 3 mm. **Hab** brownfield, walls.

J F M A M J J A S O N D

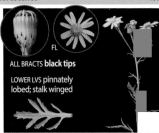

ALL BRACTS **black tips**

LOWER LVS pinnately lobed; stalk winged

Narrow-leaved Ragwort *Senecio inaequidens*

H to 80 cm. **Form** erect or sprawling; hairless; BASE woody. **Fl-hds** D 10–25 mm; RAYS 7–15; in open clusters. **Lvs** L to 60 mm; stalkless; linear; MARGINS incurved. **Fr** L to 2·5 mm, ribbed. **Hab** brownfield, tracksides, sandy areas.

J F M A M J J A S O N D

ALL BRACTS **tips black with tuft of hairs**

FL

LF **linear**

Sticky Groundsel *Senecio viscosus*

H to 60 cm. **Form** erect, branched, **sticky**. **Fl-hds** D 10–15 mm; RAYS L to 8 mm, soon bending back. **Lvs** L to 60 mm, greyish; pinnately lobed; LOWER LVS stalked. **Fr** L >4 mm; ribbed; **hairless**. **Hab** sand and shingle, especially by the coast.

J F M A M J J A S O N D

FL D 10–15 mm

INNER BRACTS blackish tips

LF glandular hairs only

Heath Groundsel *Senecio sylvaticus*

H to 70 cm. **Form** erect; well-branched; **tacky**. **Fl-hds** D 5–6 mm; L to 6 mm; soon bending back. **Lvs** L to 90 mm; greyish; oblong; pinnately lobed; LOWER LVS **some clasping stem**. **Fr** L <2·5 mm; ribbed; **hairs on ribs**. **Hab** heaths and sandy places.

J F M A M J J A S O N D

FL D <6 mm

INNER BRACTS green or pale brown tips

LF both web-like and glandular hairs

SS Both Heath and Sticky Groundsel could be confused with the rayed form of **Common Groundsel** (typical form – *p. 200*) [RAYS up to 11 (L <5 mm); BRACTS usually black-tipped; LVS hairless or with web-like hairs only].

211

3

FLOWERHEADS **consisting of both ray- and disc-flowers**

DISC **yellow**	RAYS **yellow**

FLOWERHEAD **bracts in >2 rows**

FLOWERHEADS **in widely branched inflorescence** | RAYS **L ≥ disc diameter**

A Corn Marigold *Glebionis segetum*

FL

H to 60 cm. **Form** erect; can be branched; hairless. **Fl-hds** D 30–65 mm; solitary, on long stalks. **Lvs** L to 80 mm; narrowly to broadly oval; toothed or lobed. **Fr** L to 3 mm, ridged or 3-angled. **Hab** arable land, brownfield and disturbed sites.

STEM-LVS oblong; toothed or lobed

BRACTS rounded with a **wide, brownish papery margin**

FLOWERHEADS **in widely branched inflorescence** | RAYS **L < disc diameter**

P Common Fleabane *Pulicaria dysenterica*

FL

H to 100 cm. **Form** erect; branched; clump-forming. **Fl-hds** D 15–30 mm in loose clusters. **Lvs** BASAL L to 80 mm; broadly oval; withered by flowering; STEM narrowly oval, **basal lobes clasping stem**; UNDERSIDE downy grey. **Fr** L to 1·5 mm; hairy. **Hab** damp meadows, by rivers and streams.

STEM-LVS lobes clasp stem

BRACTS short; linear; downy

FLOWERHEADS **on scaly stem before lvs, borne singly**

P Colt's-foot *Tussilago farfara*

FL

FL-STALK covered in scales

H to 15 cm. **Form** erect at first, then nodding; patch-forming. **Fl-hds** D 15–35 mm; solitary, **appearing before LVS**, on stem (L to 15 cm). **Lvs** basal; rounded; W to 30 cm. **Fr** L to 10 mm; pappus of unbranched white hairs (L to 14 mm). **Hab** cliffs, screes, damp grassland and road verges.

LF-MARGIN ragged, undulating; LF-UNDERSIDE dense white felt

FLOWERHEADS **in terminal branched inflorescence; often congested**

🅟 Goldenrod *Solidago virgaurea*

H to 70 cm. **Form** erect; variable in stature; stem leafy. **Fl-hds** D 6–12 mm in **irregular stalked clusters up stem**. **Lvs** BASAL L to 100 mm; oval, stalked, crinkled; STEM narrower, pointed, stalkless, decreasing in size up stem. **Fr** L to 3·2 mm, brown; PAPPUS white hairs; (L to 8 mm). **Hab** heaths, dry woods, cliff tops, limestone pavement.

FL-HEADS up to 12 bright yellow rays (L to 9 mm); up to 30 disc-flowers

🅟 Canadian Goldenrod *Solidago canadensis*

H to 200 cm. **Form** erect, densely patch-forming. **Fl-hds** D 3–4 mm in **terminal conical clusters of curved spikes** (W to 15 cm). **Lvs** narrowly oval, pointed, toothed, roughly hairy. **Fr** L to 1·2 mm, pale brown; PAPPUS very short dirty white hairs (L to 8 mm). **Hab** brownfield, roadsides, railway embankments, riverbanks.

FL-HEADS up to 17 short, golden-yellow rays (L to 1·5 mm); up to 8 disc-flowers

FLOWERHEADS **consisting of both ray- and disc-flowers**　　　4/6

DISC **greyish**	RAYS **greyish**

🅟 Yarrow *Achillea millefolium*

H to 70 cm. **Form** patch-forming, softly hairy, aromatic. **Fl-hds** D 4–6 mm in dense, branched, flat-topped infl; can be tinged pink. **Lvs** L to 150 mm; **2-pinnately lobed; spreading in 3 dimensions**. **Fr** seeds L to 2 mm, compressed, shiny. **Hab** grasslands, lawns.

🅟 Sneezewort *Achillea ptarmica*

H to 60 cm. **Form** erect; creeping; woody rootstock. **Fl-hds** D 12–18 mm in loose, branched infl. **Lvs** L to 80 mm; ± linear; stalkless; pointed; margins with sharp teeth. **Fr** seeds L to 2 mm, compressed, pale grey. **Hab** damp acidic grassland, upland springs.

DISC-FLOWERS dirty-white/cream; RAY-FLOWERS usually 5

LF 'ferny'

STEM furrowed

DISC-FLOWERS greenish-white; RAY-FLOWERS up to 13

LF ± linear; margins with sharp teeth

STEM angled

213

3

FLOWERHEADS **consisting of both ray- and disc-flowers**

| DISC **yellow** | RAYS **white; usually with some pink tinge** |

Ⓟ Daisy *Bellis perennis*

H to 15 cm. **Form** low-growing; rosette-forming; hairy. **Fl-hds** D 15–25 mm, solitary on long stalks. **Lvs** L to 40 mm; broadly oval; irregularly toothed. **Fr** compressed, hairy, no pappus; seeds L 1–2 mm. **Hab** mown or grazed grassland.

J F M A M J J A S O N D

FL-HD rays can be tinged pink

LVS spoon-shaped

Ⓟ Mexican Fleabane *Erigeron karvinskianus*

H to 25 cm. **Form** loosely trailing; branched; slender. **Fl-hds** D 15–20 mm, solitary on short stalks. **Lvs** small; L to 15 mm; linear to diamond-shaped; can be lobed. **Fr** seeds L 1·5 mm, shiny. **Hab** rocks, walls and gardens.

J F M A M J J A S O N D

FL-HD rays usually with some crimson flush

LVS linear to lobed

| DISC **yellow** | RAYS **white** |

RAYS **length > disc diameter**   1/2

Ⓜ Mayweeds | **Form** erect or sprawling. **H** to 60 cm. **Lvs** finely divided.

IDENTIFY BY: ► leaf-tip shape ► bract margins ► seed details

SS Stinking Chamomile *Anthemis cotula* (N/I) [most like **Scented Mayweed**; SCENT unpleasant when crushed; RECEPTACLE with scales].

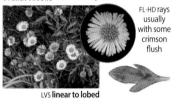

SCENTED   SCENTLESS   SEA

J F M A M J J A S O N D

Mayweed ID	Ⓐ **Scented Mayweed** *Matricaria chamomilla*	Ⓐ **Scentless Mayweed** *Tripleurospermum inodorum*	Ⓟ **Sea Mayweed** *Tripleurospermum maritimum*
SCENT	**pleasantly aromatic**	± scentless	
FL-HEADS	D 10–25 mm; DISC domed, becoming strongly so	D 30–50 mm; DISC flat, becoming domed	
RECEPTACLE	**conical; hollow**	rounded; solid	
BRACTS	**linear;** MARGIN **narrow, papery**	narrowly oblong; MARGIN **blackish, papery**	oblong to triangular; MARGIN brown, papery
LEAF SEGS.	contracted to abrupt tip	bristle-tipped or pointed	fleshy; tip blunt or pointed
FRUIT FACE 1	**4–5 ribbed**	3-ribbed	2-ribbed (almost touching)
FRUIT FACE 2	OIL GLANDS **absent**	OIL GLANDS (➡) rounded	OIL GLANDS (➡) elongated
HABITAT	arable, brownfield (including coastal)		coastal; sand, shingle, cliffs

conical / hollow   blackish / rounded / solid   brown   SEA MAYWEED

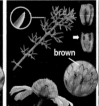

DISC **yellow**	RAYS **white**

RAYS **length > disc diameter** 2/2

ⓟ Feverfew *Tanacetum parthenium*

H to 70cm. **Form** erect, woody-based. **Fl-hds** D 15–25mm in loose, branched infl. **Lvs** L to 100mm; 1–2-pinnate, with broad, rounded lflts; long-stalked (L to 80mm); strongly aromatic. **Fr** seeds L to 1·5mm, ribbed. **Hab** gardens, brownfield and urban sites.

LF 'ferny'

ⓟ Oxeye Daisy *Leucanthemum vulgare*

FL-HD

H to 70cm. **Form** erect, clump-forming. **Fl-hds** D 20–60mm, solitary on long stalks; BRACTS L 6–8mm. **Lvs** BASAL L to 80mm, broadly oval, toothed; STEM unstalked, clasping stem. **Fr** seeds L 2–3mm, ribbed, pale grey. **Hab** grassland. **SS** Shasta Daisy *L. × superbum* (N/i) [planted/garden escape FL-HD D 60–100mm; BASAL LVS narrowly oval, lacking distinct stalk].

BASAL LF stalked

RAYS **length < disc diameter** 1/2

● **Soldiers** | **H** to 80cm. **Form** erect, branched. **Fl-hds** D 4–7mm; with 5 short white rays. **Lvs** broadly oval; opposite, stalked, toothed. **Hab** brownfield, arable.

GALLANT- SHAGGY-

IDENTIFY BY: ► receptacle-scale shape ► pappus details ► stem hairs

Soldier ID	Ⓐ Gallant-soldier *Galinsoga parviflora*	Ⓐ Shaggy-soldier *Galinsoga quadriradiata*
STEM HAIRS	hairless; sparsely downy at most	densely downy; some hairs long and glandular
RECEPTACLE-SCALES	3-lobed	mostly unlobed
PAPPUS SCALE	PROJECTIONS **absent**	PROJECTIONS **present**

SHAGGY-SOLDIER

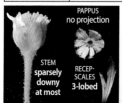

PAPPUS no projection — STEM sparsely downy at most — RECEP-SCALES 3-lobed

PAPPUS projections present — STEM some long hairs — RECEP-SCALES mostly unlobed

3

FLOWERHEADS **consisting of both ray- and disc-flowers**		6/6
DISC **yellow**		RAYS **white (can have reddish tinge)**
RAYS **length < disc diameter**		2/2

● **Erigeron** fleabanes | **Form** erect, branched. **Fl-hds** many (200+) in branched leafy inflorescence. **Fls** RAYS in 2 rows. **Lvs** ROSETTE withering early; STEM narrowly oval to linear, toothed or lobed, stalkless. **Hab** dry, open, disturbed habitats, brownfield, sand and shingle.

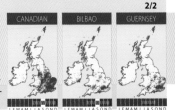

CANADIAN　BILBAO　GUERNSEY

J F M A M J J A S O N D　J F M A M J J A S O N D　J F M A M J J A S O N D

IDENTIFY BY: ▶ flower details + bract colour + bract shape + leaf-margin hairs near stem

Erigeron fleabane ID	**Ⓐ Canadian Fleabane** *Erigeron canadensis*	**Ⓐ Bilbao Fleabane** *Erigeron floribundus*	**Ⓐ Guernsey Fleabane** *Erigeron sumatrensis*
H	to 100 cm	to 150 cm	to 200 cm
FL-HEADS	D 3–5 mm		D 5–8 mm
INFL*	usually narrow; ± cylindrical	usually widely spreading with long branches	± pyramidal
FLOWERS	DISC **4-lobed** (can be 5)	DISC 5-lobed	
	RAY **L 0·5–1·0 mm**	RAY L ≤0·5 mm	
BRACTS	yellowish-green; not hairy		**greyish-green; hairy**
UPPER	**narrowly triangular;** TIP **pointed**	± **strap-shaped;** TIP **blunt**	
LOWER	**all short** (L <½ upper)	**some longer** (L >½ upper)	
STEM	sparsely hairy	**stiffly hairy**	very hairy
LEAF	L to 100 mm	L to 150 mm	L to 90 mm
LF-HAIRS NEAR BASE OF LEAF	± straight		**stiff curved**
FRUIT	L 1·5 mm	L 1·3 mm	L 1·0 mm
PAPPUS	**yellowish to off-white**		off-white

* inflorescence shape is variable; narrow and ± cylindrical suggests Canadian; but all plants need careful checking of the features described in the table

CANADIAN FLEABANE

BILBAO FLEABANE

CANADIAN

RAYS **relatively conspicuous**

BRACTS **'thin'; lower short**

IN FRUIT

STEM **sparsely hairy**

BILBAO

RAYS **relatively short; can be reddish**

BRACTS **'fat'; some lower long**

IN FRUIT

STEM **stiffly hairy**

GUERNSEY

BRACTS **greyish-green; hairy**

LF-HAIRS **stiff; curved**

STEM **very hairy**

68 Viburnaceae | **Viburnum** family

2 spp. | 5 spp. B&I

Form woody. **Infl** flat-topped flowerhead. **Fls** PETALS fused;
OVARY inferior; SYMMETRY radial. **Lvs** oval or lobed.

℗ **Guelder-rose** *Viburnum opulus* **WS**

J F M A M J J A S O N D

H to 4 m. **Infl** flat-topped
flowerheads (D to 100 mm).
Fls INNER small (D 6 mm),
greenish-white, fertile;
OUTER large (D 15–20 mm),
white, sterile. Lvs L to 80 mm;
palmately 3- lobed; margins
toothed. **Fr** globular berries
(D 8–11 mm); **bright red** (some
cultivars orange or yellow). **Hab** woods, scrub,
hedges; a very frequent species of drier fen
edges. **SS** Asian Guelder-rose *V. sargentii*
(planted; N/I) [ANTHERS purple; FRUITS L > W].

FR
globular;
red

LVS 3-lobed

INNER (FERTILE) FLS

℗ **Wayfaring-tree** *Viburnum lantana* **WS**

J F M A M J J A S O N D

H to 6 m. **Infl** flat head (D
to 100 mm). **Fls** D 5–6 mm;
**creamy-white; all fertile.
Lvs** L to 100 mm; broadly oval;
regularly, finely toothed; **very
hairy underneath;** veins deeply
impressed. **Fr** compressed
(L to 8 mm); red, **ripening
to black. Hab** woods, scrub,
and hedges on dry calcareous soils; planted
elsewhere. **SS** Chinese Wayfaring-tree
V. veitchii (planted; N/I) [SEPALS densely hairy].

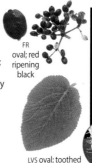

FR
oval; red
ripening
black

LVS oval; toothed

FLS

69 Sambucaceae | **Elder** family

1 sp. | 4 spp. B&I

Form woody. **Infl** small flowers in large, flat-topped to concave cluster.
Fls PETALS fused; OVARY inferior; SYMMETRY radial. **Lvs** pinnate.

℗ **Elder** *Sambucus nigra* **WS**

FLS

J F M A M J J A S O N D

H to 10 m. **Form** BARK deeply
fissured; corky, often well-
covered by mosses; with
numerous suckers and twigs
with prominent lenticels.
Infl flat or concave head (D
10–20 cm). **Fls** D 5 mm; creamy-
white. **Lvs** pinnate; LFLTS 5–7,
L to 90 mm; AROMA **unpleasant**
when crushed. **Lvs** shiny, globular black
berries (D 6–8 mm), **in distinctively
drooping heads. Hab** woods, hedges, scrub,
especially on over-fertilized ground (*e.g.* under
communal bird roosts) and generally with
an understorey of Common Nettle (*p. 105*).

FR
black

LVS pinnate

217

70 Caprifoliaceae | **Honeysuckle** family | 4 spp. | 14 spp. B&I

Form woody. **Infl** various, but not flat-topped. **Fls** PETALS fused; OVARY inferior; SYMMETRY bilateral. **Lvs** oval or broadly 3-lobed.

ⓟ Honeysuckle *Lonicera periclymenum* **WC**

J F M A M J J A S O N D

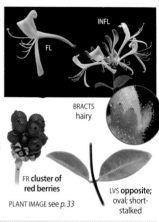

INFL

FL

BRACTS hairy

L to 8 m. **Form** deciduous **climber** or ground-covering; BARK peeling; STEM twining. **Infl** dense terminal whorl. **Fls** L 40–50 mm; SYMMETRY **bilateral; yellow**, can be tinged pink-purple; PETALS fused into a tube; UPPER LIP 4-toothed; LOWER LIP single lobe; AROMA **sweetly fragrant, especially at night.** NOTE: may not flower in shade. **Lvs** broadly oval (L to 70 mm) on short stalks; opposite; usually softly downy. **Fr** red berry; D to 8 mm (though often some much smaller ones in the cluster). **Hab** woods, scrub, hedges. **SS Garden Honeysuckle** *L. × italica* (N/I) [BRACTS hairless]; **Traveller's-joy** (*p. 73*) [FRUIT characteristic plumes persist into midwinter; STEM draping and festooning].

FR cluster of red berries

LVS **opposite**; oval; short-stalked

PLANT IMAGE see *p. 33*

● Distinctive-looking garden escape species.

ⓟ Himalayan Honeysuckle
Leycesteria formosa **WS**

H to 180 cm. **Form** erect; semi-herbaceous shrub. **Fls** L 10–20 mm; **pinkish; in drooping spike. Lvs** narrowly oval; L to 17 cm; long drawn-out tip. **Fr** purple-black rounded berry (D to 10 mm). **Hab** hedges, scrub, woodland.

J F M A M J J A S O N D

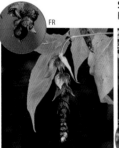

FR

ⓟ Wilson's Honeysuckle
Lonicera nitida **WS**

H to 180 cm. **Form** evergreen shrub with glandular-hairy stem. **Fls** L 5–7 mm; creamy-white; in pairs in leaf-axils. **Lvs** oval; L to 6 mm; in regular opposite pairs. **Fr** violet berry; D to 8 mm. **Hab** hedges, scrub, woodland. **SS** leaves – **Box** (*p. 75*) [STEM hairless, 4-angled].

J F M A M J J A S O N D

FL   FR

ⓟ Snowberry
Symphoricarpos albus **WS**

H to 200 cm. **Form** deciduous shrub; STEM arching. **Fls** bell-shaped; D 5–8 mm; PETALS **pink outside, white and hairy inside**; 4- or 5-lobed; in dense terminal clusters. **Lvs** broadly oval; can be 3-lobed; L to 50 mm. **Fr** pure white globular berry; D 8–15 mm. **Hab** woods, scrub.

J F M A M J J A S O N D

FR

71 Valerianaceae | **Valerian** family

5 spp. | 9 spp. B&I

Form erect. **Infl** complex, branched inflorescence of small flowers. **Fls** PETALS fused; spurred at base in some; SEPALS minute or absent. **Fr** SEEDS with feathery plumes in most.

Ⓟ Common Valerian
Valeriana officinalis

H to 150 cm.
Fls D 4–5 mm;
pale pink;
in compact
heads. **Lvs**
pinnate; LFLTS
narrowly
oval, can be
toothed. **Hab**
damp grassland, fens, marshes, wet woodland and dry calcareous grassland (a form with a narrow terminal leaflet).

Ⓟ Marsh Valerian
Valeriana dioica

H to 40 cm.
Fls pale pink
or white;
♂ D 5 mm;
♀ D 2 mm;
in compact
heads. **Lvs**
STEM pinnate;
BASAL **broadly
oval, entire. Hab** marshes,
fens, bogs.

STEM-LVS pinnate

Ⓟ Red Valerian
Centranthus ruber

H to 80 cm.
Fls red, pink
or white;
♂ D 5 mm;
♀ D 2 mm;
in compact
heads;
PETAL-TUBE
L 8–10 mm;
**extended backwards into
a spur**, L 4–10 mm. **Lvs** oval;
blue-green; waxy. **Hab**
walls, cliffs, old buildings,
brownfield, shingle.

LF pointed oval

FLS with
long spur

ALL LVS pinnate

INFL

BASAL LVS
oval

INFL

● **Cornsalads** | **Form** erect annuals with a repeatedly forked stem. **Fls** pale mauve; D 1–2 mm; in compact heads (D 10–20 mm) surrounded by ruff of bracts; STAMENS 3; STIGMAS 3; SEPALS tiny or absent. **Lvs** ± narrowly oval. **Hab** arable margins; bare patches in dunes and dry grassland.

SS The 5 **cornsalad** species (3 rare) are reliably distinguishable only on seed characters.

COMMON KEELED-FRUITED

Ⓐ Common Cornsalad
Valerianella locusta

H to 30 cm. **Fr** round and flattened D 1·8–2·5 mm; with a **shallow groove** on one face.

FR **rounded;
shallow groove**

Ⓐ Keeled-fruited Cornsalad
Valerianella carinata

H to 30 cm. **Fr** oblong; much longer than wide L×W 2·0–2·7 × 0·8–1·4 mm with a **deep groove** on one face.

FR **oblong;
deep groove**

KEELED-FRUITED

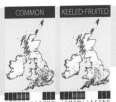

INFL

72 Dipsacaceae | Teasel + Scabious family `4 spp. | 8 spp. B&I`

Form erect. **Fls** numerous; in **dense heads** with a ruff of bracts at the base, borne on a common platform (receptacle); PETALS fused into a tube; SEPALS small, often extended as long bristles in fruit. **Fr** SEED enclosed in a ± inflated tube (epicalyx).

● **Scabiouses** | **Fl-heads** flat to gently rounded; STEM not prickly.

SS Sheep's-bit (p.229) [PETAL-TUBE split almost to the base; STAMENS fused].

Ⓟ Devil's-bit Scabious *Succisa pratensis*

H to 100 cm. **Form** erect; branched. **Fl-hds** D 15–25 mm. **Fls** PETAL-TUBE bluish-purple, **4-lobed**; SEPAL-TUBE with 4–5 bristles. **Lvs** oval; pointed; shallowly toothed. **Fr-tube** 4-angled. **Hab** grassland both wet and dry on a range of soils.

FL-HEAD

ALL LVS entire

Ⓟ Field Scabious *Knautia arvensis*

H to 100 cm. **Form** erect. **Fl-hds** D 15–40 mm; shallowly curved; OUTER FLS much larger than INNER. **Fls** PETAL-TUBE lilac-blue, **4 unequal lobes**. **Lvs** UPPER pinnate; LOWER simple to shallowly lobed. **Fr-tube** not papery; 4-ridged, with dense ring of short hairs at the mouth. **Hab** dry grassland away from heavy clay soils.

FL-HD BRACTS broad; in 2 rows

UPPER LVS pinnate; lflts relatively broad

Ⓟ Small Scabious *Scabiosa columbaria*

H to 70 cm. **Form** erect. **Fl-hds** D 15–35 mm; gently rounded; OUTER FLS longer than INNER. **Fls** PETAL-TUBE lilac-blue, **5 unequal lobes**. **Lvs** UPPER pinnate with narrow lflts; LOWER simple to pinnate. **Fr** expanded into a papery funnel; 8-ridged, with dense ring of short hairs at the mouth. **Hab** calcareous grassland.

FL-HD BRACTS narrow; in 1 row

UPPER LVS pinnate; lflts narrow

● **Teasels** | **Fl-heads** ovate to globular; STEM prickly.

Ⓑ Wild Teasel *Dipsacus fullonum*

H to 3 m. **Form** erect; STEM prickly. **Fl-hds** dense; egg-shaped (L to 9 cm); short **spine-tipped bracts** below each flower. **Fls** tiny; pink to lilac, opening sequentially in horizontal bands around the head. **Lvs** narrowly triangular; underside midrib with spines; STEM-LVS fused in pairs. **Fr** spiny seed-heads. **Hab** river banks, road verges, rough ground.

FL-HDS unmistakable

73 Araliaceae | **Ivy** family

2 spp. | 5 spp. B&l

Form woody climbers; usually buzzing with pollinators when in flower. **Infl** globular umbels of stalked flowers and berries. **Fls** AROMA strong, musky honey.

● **Ivies** | **L** to 30 m. **Fls** D6–8mm; PETALS 5, free; SEPALS 5, small. **Lvs** palmately 3–5-lobed on **climbing/creeping stem**; ± unlobed on flowering shoots. **Fr** globular berry D 6–8mm; green, ripening black. **Hab** growing up trees, rocks, cliffs; scrambling through hedges and along fences; trailing across woodland floor.

SS Both ivies can be separated using the detailed features below. Beware of much natural variation in leaf shape, size and colour (variegation) as a result of the escape of garden cultivars.

Ivy identification	● Ivy WC *Hedera helix*	● Atlantic Ivy WC *Hedera hibernica*
LEAF SIZE	W to 80mm	W to 120mm
LEAF-LOBES	W < L; divided > half way to base on many	W > L; divided < half way to base on many
LEAF- AND STEM HAIRS	whitish; star-shaped; **not closely pressed to surface**	yellowish; star-shaped; **closely pressed to surface**

IVY	ATLANTIC IVY

J F M A M J J A S O N D   J F M A M J J A S O N D

———— IVY ————   ———— ATLANTIC IVY ————

HAIRS whitish; not appressed   HAIRS yellowish; appressed

IVY

FL

INFL

FR

LVS variable in both species; those of **Atlantic Ivy** generally wider than those of **Ivy**

74 Hydrocotylaceae | **Pennywort** family

1 sp. | 2 spp. B&l

Form creeping. **Lvs** rounded; STALK attached centrally. **Fls** tiny; usually hidden by the leaves; in simple umbels with main spokes only.

● **Marsh Pennywort** *Hydrocotyle vulgaris*

J F M A M J J A S O N D

L to 30 cm. **Form** long creeping. **Fls** D 1 mm; PETALS 5; greenish- or pinkish-white in small umbels of 2–5; on stalks shorter than leaf-stalks. **Lvs round**, D 8–35mm; scarcely lobed; LF-STALK (L up to 20 cm) **attached centrally. Hab** bogs, fens, marshes and pond edges. **SS** leaves – Navelwort (*p. 78*) [dry habitats only; showy flower spikes]; **Floating Pennywort** *H. ranunculoides* (N/I) [LVS deeply lobed; much larger (D to 70mm); stalk not central].

FLS tiny

LF round; LF-STALK attached centrally

75 Apiaceae | **Carrot** family

19 spp. | 74 spp. B&I

Infl umbel, usually **compound** ('umbel of umbels'); usually **bracts** below main spokes and/or **bracteoles** below secondary spokes. **Fls** PETALS 5; STAMENS 5; STYLES 2, arising (in some spp.) from a swelling on top of ovary. **Lvs** usually with sheathing stalk-bases. **Fr** distinctively shaped seed.

IDENTIFY BY: ▶ **flower colour** ▶ **umbel structure** ▶ **number of bracts + bracteoles** ▶ **leaf shape** ▶ **fruit shape + size**

SS Other carrots (N/I) [separated using the identification features above].

BRACTEOLE | 2ʸ SPOKE
BRACT | SPOKE
FL-STALK

INFLORESCENCE **compound umbel;** LOWER LEAVES **'ferny'** | **1/3**

FORM **erect; at least shortly hairy** | **1/2**

Hab hedges, woodland margins, rough grassland.

IN FLOWER **April–May**	IN FLOWER **June–July**	IN FLOWER **July–August**

ⓟ Cow Parsley
Anthriscus sylvestris

H to 150 cm.
Form AROMA fls subtly musky. **Umbel** D to 60 mm.
Lvs OUTLINE ± triangular; 2–3-pinnate; LFLT-TIPS pointed. **Fr** L 6–10 mm.

J F M A M J J A S O N D

ⓑ Rough Chervil
Chaerophyllum temulum

H to 100 cm.
Form roughly hairy. **Umbel** D to 60 mm.
Lvs OUTLINE ± triangular; 2–3-pinnate; dull, flaccid; LFLT-TIPS rounded. **Fr** L 4·0–6·5 mm.

J F M A M J J A S O N D

ⓐ Upright Hedge-parsley
Torilis japonica

H to 120 cm.
Form STEM appressed hairs. **Umbel** D to 40 mm. **Lvs** 2–3-pinnate; OUTLINE rather narrowly triangular.
Fr L 2·0–2·5 mm.

J F M A M J J A S O N D

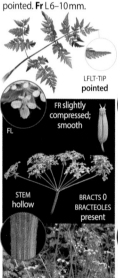

LFLT-TIP
pointed

FR slightly compressed; smooth

FL

STEM hollow

BRACTS 0
BRACTEOLES present

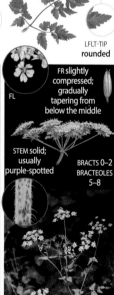

LFLT-TIP
rounded

FR slightly compressed; gradually tapering from below the middle

FL

STEM solid; usually purple-spotted

BRACTS 0–2
BRACTEOLES 5–8

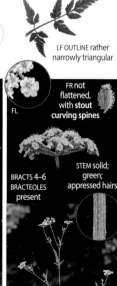

LF OUTLINE rather narrowly triangular

FR not flattened, with **stout curving spines**

FL

BRACTS 4–6
BRACTEOLES present

STEM solid; green; appressed hairs

INFLORESCENCE **compound umbel;** LOWER LEAVES **'ferny'** **2/3**

FORM **erect; at least shortly hairy** **2/2**

Ⓟ Sweet Cicely
Myrrhis odorata

J F M A M J J A S O N D

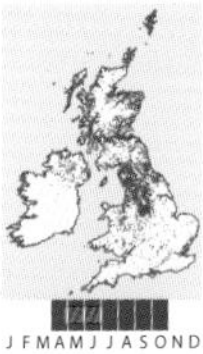

H to 180 cm. **Form** AROMA **strongly of aniseed when crushed. Umbel** D to 50 mm. **Lvs** 2–4-pinnate; softly hairy; LFLT-TIPS pointed. **Fr** L to 25 mm; L > 3 × W. **Hab** river banks, road verges, damp grassland; near gardens, especially in the south.

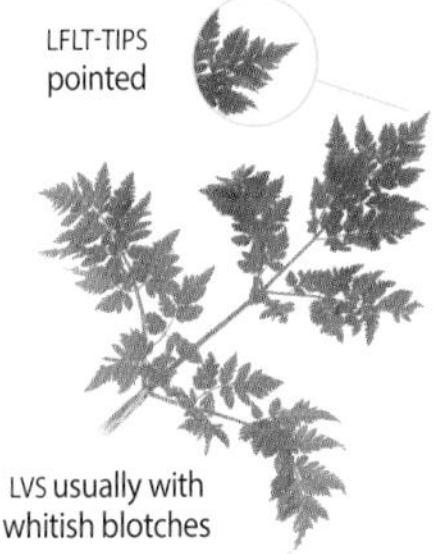

LFLT-TIPS
pointed
LVS usually with
whitish blotches

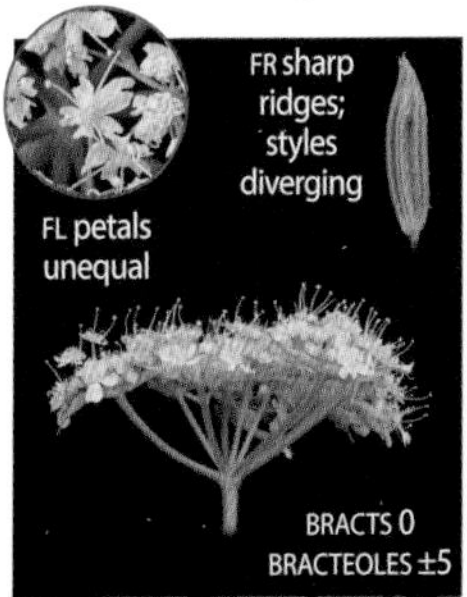

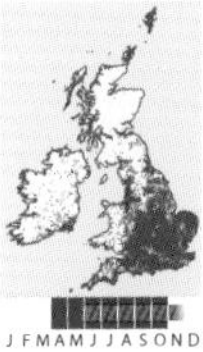

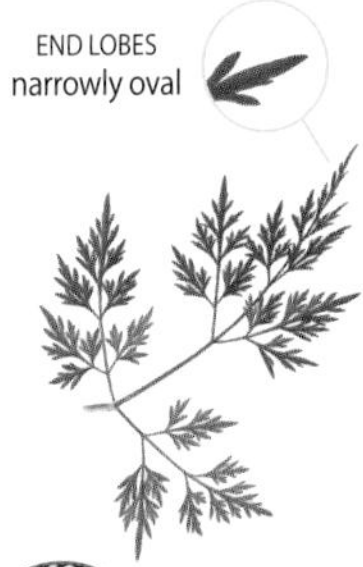

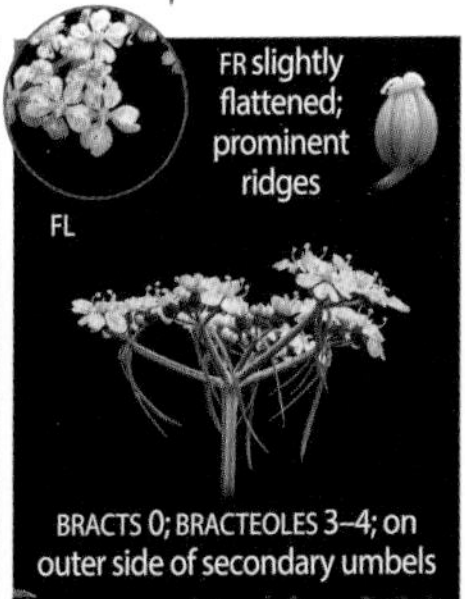

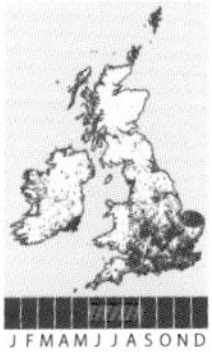

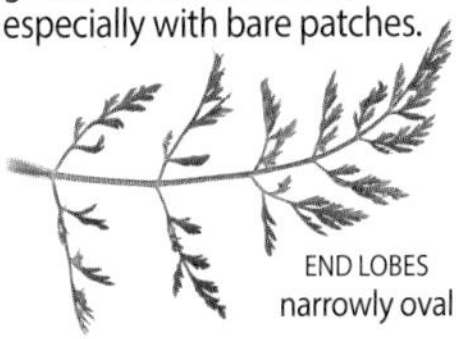

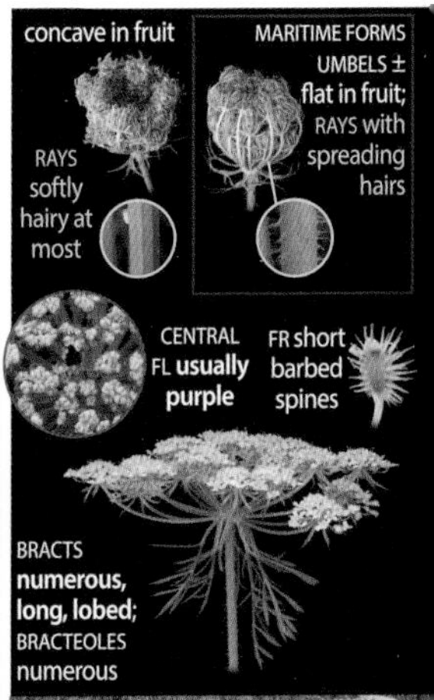

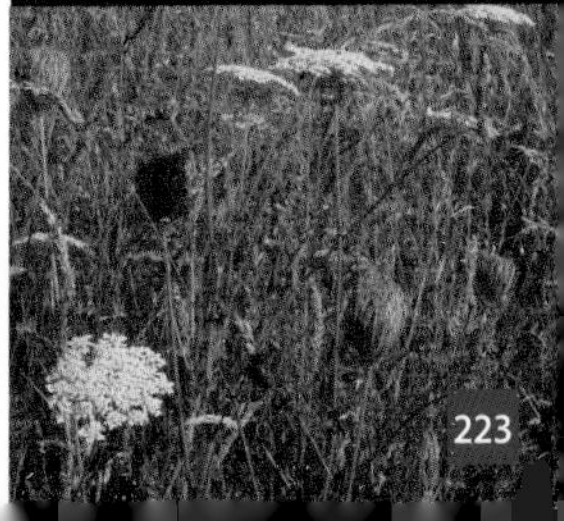

FR sharp
ridges;
styles
diverging
FL petals
unequal
BRACTS 0
BRACTEOLES ±5

Ⓐ Fool's Parsley ✕
Aethusa cynapium

J F M A M J J A S O N D

H to 80 cm. **Form** STEM hollow. **Umbel** D to 60 mm; BRACTEOLES L to 15 mm; strongly bent back, giving umbel a **'shaggy bottom'. Lvs** 2–3-pinnate; END LOBES narrowly oval. **Fr** L 2·5–3·5 mm. **Hab** gardens, cultivated ground, arable margins, brownfield.

END LOBES
narrowly oval

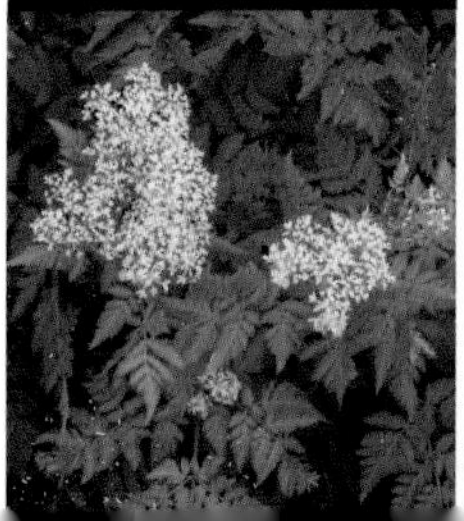

FR slightly
flattened;
prominent
ridges
FL
BRACTS 0; BRACTEOLES 3–4; on
outer side of secondary umbels

Ⓑ Wild Carrot
Daucus carota

J F M A M J J A S O N D

H to 100 cm. **Form** bristly hairy; AROMA of carrot when crushed. **Umbel** D to 70 mm; rays softly hairy at most. **Lvs** 3-pinnate; END LOBES narrowly oval. **Fr** L 2–4 mm; umbel concave in fruit. **Hab** grassland, often coastal, especially with bare patches.

END LOBES
narrowly oval

concave in fruit
RAYS
softly
hairy at
most
MARITIME FORMS
UMBELS ±
flat in fruit;
RAYS with
spreading
hairs
CENTRAL
FL usually
purple
FR short
barbed
spines
BRACTS
numerous,
long, lobed;
BRACTEOLES
numerous

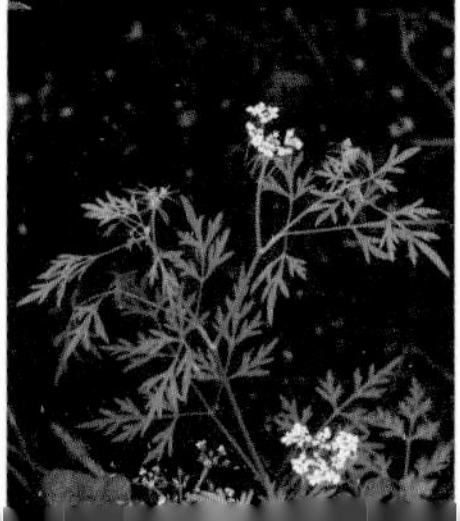

INFLORESCENCE **compound umbel;** LOWER LEAVES **'ferny'**

FORM **erect; hairless**

Ⓟ **Wild Angelica**
Angelica sylvestris

H to 2·5 m.
Form STEM
usually reddish;
leaf-veins and
umbel-rays
with short hairs
(lens best).
Umbel domed;
D to 150 mm.
Lvs 2–3-pinnate; LFLTS **broadly oval**, finely toothed. **Fr** L 4–6 mm. **Hab** fens, marshes, river banks, damp woodland.

Ⓟ **Hemlock Water-dropwort**
Oenanthe crocata ✗

H to 1·5 m.
Umbel
domed; D to
100 mm. **Lvs**
2–3-pinnate.
Fr L 4·0–
5·5 mm.
Hab ditches,
ponds, upper
estuarine areas, damp woodland. **SS** several other water-dropworts (N/I) [LEAF form + FRUIT details].

Ⓑ **Hemlock**
Conium maculatum ✗

H to 2·5 m.
Form STEM
purple
spotted;
AROMA **musty,**
mousy.
Umbel D to
50 mm. **Lvs**
2–4-pinnate.
Fr L 2·0–3·5 mm.
Hab riverbanks, dry marshland, road verges, brownfield.

LVS 2–3-pinnate
LFLTS broadly
oval

LVS upper
sheath
strongly
inflated

FLS can be
tinged pink
or green

BRACTS few or absent;
BRACTEOLES numerous

LVS 2–4-pinnate
LFLTS broadly diamond-shaped

FR oblong;
prominent
ridges; styles
persistent

FL

BRACTS +
BRACTEOLES
several

LVS 2–4-pinnate; LFLTS
narrowly oval; toothed

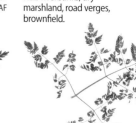

FR round;
several wavy
ridges

FLS

STEM
purple-spotted

BRACTS +
BRACTEOLES
several

FR strongly
compressed;
ribs with
papery wings

3/3

Two similar species. **Form** trailing to sprawling. **Lvs** 1-pinnate. **Hab** ditches, marshes, ponds and lake edges. Most easily **separated by leaf and leaf-stalk details**.

℗ Pignut *Conopodium majus*

J F M A M J J A S O N D

H to 50 cm.
Form tuberous.
Umbel D to 70 mm; **nodding in bud**. **Lvs** 2–3-pinnate; BASAL LVS withered at flowering. **Fr** L 3.0–3.5 mm. **Hab** grassland, open woodland; avoids calcareous soils.

℗ Lesser Water-parsnip *Berula erecta*

J F M A M J J A S O N D

L to 100 cm.
Umbel D to 60 mm.
Fls SEPALS present, but drop soon after fls open.
Lvs LFLTS 5–10 pairs; coarsely toothed.
Fr L 1.5–2.0 mm.

℗ Fool's Water-cress *Helosciadium nodiflorum*

J F M A M J J A S O N D

L to 100 cm.
Umbel D to 50 mm. **Fls** SEPALS absent.
Lvs LFLTS usually 2–4 pairs; finely toothed. **Fr** L 2.0–2.5 mm.
SS leaves – **Water-cress** (*p. 138*) [LF-STALK not ensheathing stem at base].

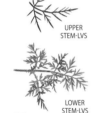

LVS variable; upper stem-lvs finer than lower stem-lvs

UPPER STEM-LVS

LOWER STEM-LVS

LF-STALK **whitish ring at base of leaf**

LF-STALK **whitish ring absent**

LVS usually 2–4 pairs of leaflets; finely toothed

LVS usually 5–7 pairs of leaflets; coarsely toothed

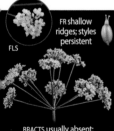

FR shallow ridges; styles persistent

FLS

BRACTS usually absent; BRACTEOLES few

FR slightly flattened; weak ridges

BRACTS several; BRACTEOLES several; can be lobed

FR egg-shaped; strongly flattened; prominent ridges

BRACTS none; BRACTEOLES several; undivided

225

INFLORESCENCE **compound umbel;** LOWER LEAVES **not 'ferny'**

ⓑ Hogweed
Heracleum sphondylium

H to 2·5 m. **Form** roughly hairy. **Umbel** D to 20 cm; MAIN SPOKES <20. **Lvs** L to 60 cm; pinnate; LFLTS broadly oval, **lobed**. **Fr** L 6–10 mm. **Hab** rough grassland, road verges, river banks, open woodland.

LF **large;** LFLTS **lobed**

ⓑⓟ Giant Hogweed
Heracleum mantegazzianum ✕

H to 5·5 m. **Form** softly hairy. **Umbel** D to 50 cm; MAIN SPOKES 50–120. **Lvs** L to 250 cm; pinnate; LFLTS deeply lobed, **jagged**. **Fr** L 9–14 mm. **Hab** river banks, road verges, brownfield.

LF **huge;** LFLTS **jagged**

ⓟ Ground-elder
Aegopodium podagraria

H to 1 m. **Form** patch-forming. **Umbel** D to 6 cm. **Lvs** 2-trifoliate (the lower lflts themselves trifoliate); LFLTS L to 70 mm, broadly oval, finely toothed. **Fr** L 3–4 mm. **Hab** gardens, waysides, woodland margins, brownfield.

LF **distinctive;** 2-trifoliate

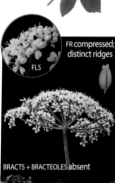

BRACTS 0 or few; BRACTEOLES several

FR

PET deeply cleft

FLS

BRACTS + BRACTEOLES several

FR

STEM very wide; blotched with purple

FR compressed; distinct ridges

FLS

BRACTS + BRACTEOLES absent

P Burnet-saxifrage
Pimpinella saxifraga

STEM-LVS

LVS variable

BASAL LF

H to 70 cm. **Umbel** D to 50 mm. **Lvs** STEM usually 2-pinnate, segments linear; LOWER 1-pinnate; lflts rounded. **Fr** L 2–3 mm. **Hab** grassland, especially calcareous.

J F M A M J J A S O N D

BRACTS + BRACTEOLES absent

FR slightly compressed; weak ridges

flowers yellow or greenish-yellow

bP Alexanders *Smyrnium olusatrum*

H to 150 cm. **Form** AROMA faintly of celery when crushed. **Umbel** domed; D to 80 mm. **Fls** greenish-yellow. **Lvs** 2–3-pinnate/ trifoliate; SEGMENTS broadly oval; toothed; glossy. **Fr** L 6·5–8·0 mm; black; rounded with prominent ribs. **Hab** cliffs, road verges, coastal grassland.

LVS 2–3-pinnate

J F M A M J J A S O N D

BRACTS + BRACTEOLES absent

P Fennel *Foeniculum vulgare*

H to 150 cm. **Form** AROMA strongly of **aniseed** when crushed. **Umbel** D to 80 mm. **Fls** bright yellow. **Lvs** divided repeatedly into **hair-like** ultimate segments. **Fr** L 4–5 mm; flattened; almost black when ripe. **Hab** open ground, road verges, disturbed grassland, brownfield, especially coastal. **SS** cultivar 'Purpureum' (N/I) has purplish stem and leaves.

J F M A M J J A S O N D

LF highly divided into hair-like segments

BRACTS + BRACTEOLES absent or very few at most

227

INFLORESCENCE **a branched aggregation of globular umbels**

ⓟ Sanicle *Sanicula europaea*

H to 50 cm. **Infl** W to 80 mm; UMBELS D to 5 mm in loose clusters. **Fls** white, or flushed pink. **Lvs** mostly basal; palmately 5-lobed; long-stalked; underside glossy.
Fr L 2–3 mm; laterally compressed; covered in **long, hooked** bristles.
Hab deciduous woodland.
SS leaves – **Wood Anemone** (*p. 72*) [LVS underside not glossy].

BRACTS + BRACTEOLES several, simple

FR hooked bristles

LF outline round; deeply lobed

---

76 Callitrichaceae | **Water-starwort** family

2 spp. | 7 spp. B&I

Lvs opposite pairs; entire, can be notched; the prostrate/floating shoots usually terminating in a rosette. **Fls** much reduced (♂ a single stamen; ♀ a single ovule with 2 styles); stalked or stalkless arising from the leaf-axils. Aquatic and semi-terrestrial specimens often appear very different. **Fr** in clusters of 4; distinctive in size, colour and form. **Hab** ponds, rivers, ditches, marshes – in fresh water or on drying mud (**Pedunculate Water-starwort** especially in acidic waters).

COMMON FL

SS All 6 water-starworts require detailed examination of mature fruits for definitive identification, which is not always possible. As a result, the distribution of all species is perhaps not fully known but the 2 species described are regarded as the most widespread. **SS New Zealand Pigmyweed** (*p. 78*) [LVS fleshy]; **Blinks** (*p. 154*) [PET + SEP present].

Water-starworts are different in their terrestrial (top) and aquatic forms (bottom).

Water-starwort ID		**ⓐⓟ Common Water-starwort** *Callitriche stagnalis* AQ	**ⓐⓟ Pedunculate Water-starwort** *Callitriche brutia* AQ		
Flower	POLLEN	bright yellow	translucent		
	STAMEN	L 0.5–2.0 mm	L 0.5–1.0 mm		
Leaves	TERRESTRIAL	± round (smaller than surface lvs)	oval		
	FLOATING	± round	oval		
	SUBMERGED	narrowly oval	linear with a notched tip which can be expanded (looking like a spanner)		
Fruit	TERRESTRIAL	D 1.6–1.8 mm; stalkless; grey-brown when ripe; WING W 0.12–0.25 mm	STYLE recurved	D to 1.5 mm; black when ripe; WING W 0.1 mm	STYLE recurved + appressed	stalk L to 10 mm
	FLOATING		STYLE erect			stalkless
	SUBMERGED	absent			

SUBMERGED LF

RIPE FR brownish

SUBMERGED LF

RIPE FR black

PLANT

COMMON

PEDUNCULATE

J F M A M J J A S O N D   J F M A M J J A S O N D

77 Campanulaceae | **Bellflower** family   `3 spp. | 22 spp. B&I`

Fls distinctive; bell-shaped; blue; PETAL-LOBES spreading, in some species; OVARY inferior; obvious as a swelling below the flower.

Ⓟ **Harebell**
Campanula rotundifolia

FL

STEM-LVS

H to 40 cm. **Form** shortly spreading; stem erect. **Fls** L 15–20 mm; pale blue; **bell-shaped**; drooping; in few-flowered open branch infl. **Lvs** STEM linear; BASAL (may be absent) ± round to broadly triangular; long-stalked. **Hab** dry grassland, both acid and calcareous, cliff slopes, sand dunes, mountains.

BASAL LF

Ⓟ **Trailing Bellflower**
Campanula poscharskyana

PETAL-TUBE lobed; split > ½ way to base

L to 50 cm. **Form** trailing. **Fls** D 20–25 mm; greyish-blue; **lobes widely spreading**. **Lvs** broadly oval; deeply toothed. **Hab** walls and rocks, usually near gardens. **SS Adria Bellflower** *C. portenschlagiana* (N/I) [PETAL-TUBE lobed < 1/2 way to base].

LVS toothed oval

Ⓑ **Sheep's-bit**
Jasione montana

PETAL-TUBE lobed; split > ½ way to base

H to 30 cm. **Form** spreading; with basal rosettes from which arise erect flowering shoots. **Fls** D 5–7 mm; blue-violet; in **compact, rounded head** (D to 30 mm). **Lvs** L to 50 mm; narrowly oblong; with wavy margins. **Hab** grassland on acid soils, cliff slopes, sand dunes, walls. **SS Devil's-bit Scabious** (*p. 220*) [PETAL-TUBE split < ½ way to base; STAMENS well-separated].

LVS linear

78 Menyanthaceae | **Bogbean** family

1 sp. | 2 spp. B&I

Unmistakable marginal plant. **Fls** PETALS white; fringed. **Lvs** large; trifoliate.

P Bogbean *Menyanthes trifoliata*

L to 150 cm. **Form** creeping/floating with erect flowering spikes and leaves. **Fls** D 15–20 mm; white inside; pink outside; PETALS 5; fused at the base; 3-lobed; each lobe with long white fringes on upper face; flowers on spikes (L to 30 cm). **Lvs** trifoliate; LFLTS oval, L to 70 mm. **Hab** acidic ponds, pools in bogs and fens; spreading onto the muddy margins.

FL distinctive fringed petals

MONOCOTS (see p. 4)

79 Araceae | **Lords-and-Ladies + Duckweed** family

7 spp. | 13 spp. B&I

Two closely related (based on molecular studies) but completely different-looking and individually distinctive groups – the terrestrial **lords-and-ladies** and the aquatic **duckweeds**.

● **Lords-and-Ladies** | **Form** erect; H to 25 cm; spreading underground. **Fls** tiny; tightly packed into a cylindrical **spadix** that is partially enfolded by a sheath-like yellow-green **spathe**. **Lvs** long-stalked; triangular with backward-pointing basal lobes. **Fr** a spike of berries; green ripening red.

2 spp. | 2 spp. B&I

SPATHE

SPADIX

IDENTIFY BY: ► spadix length and colour ► leaf midrib colour

P Lords-and-Ladies *Arum maculatum* ✕

Spadix dull purple (a few yellow); generally ≤ ½ spathe L. **Spathe** usually unmarked but can have dark spots. **Lvs** appear in early spring; **often with blackish patches**; MIDRIB dark green, as leaf. **Fr** spike L 30–80 mm. **Hab** woods, hedges.

FR spike
L < 80 mm

LVS midrib as dark as leaf

P Italian Lords-and-Ladies *Arum italicum* ✕

Spadix yellow; generally ≤ ⅓ spathe L. **Spathe** always unmarked; tip droops on many. **Lvs** appear in early winter; only rarely black-spotted; MIDRIB + VEINS **paler than leaf** (ssp. *neglectum*), often very **distinctly whitish** (ssp. *italicum*). **Fr** spike L 100–150 mm. **Hab** gardens, woods, scrub, cliff slopes and stone walls.

FR spike
L > 100 mm

LVS pale midrib and veins

● Duckweeds |

5 spp. | 6 spp. B&I

Form small green fronds; on or under water, or stranded on mud; with or without roots.
Fls very reduced but rarely produced (and never seen unless specifically searched for!). **Hab** ponds, ditches, canals and still or slow-moving areas of streams and rivers. Autumn and winter specimens often lack the more distinctive features and may not be readily identifiable.

GREATER and COMMON DUCKWEEDS

Duckweeds aggregate; and more than one species may be found growing in the same patch.

IDENTIFY BY: ► leaf shape and size ► leaf venation (best viewed from below against light)

DUCKWEED IDENTIFICATION

Duckweeds		℗ Ivy-leaved *Lemna trisulca*	℗ Least *Lemna minuta*	℗ Common *Lemna minor*	℗ Fat *Lemna gibba*	℗ Greater *Spirodela polyrhiza*
Frond	L×W	to 15 mm × 5 mm	0·8–4·0 mm × 0·5–2·5 mm	2–5 mm × ± 3 mm	3·0–6·0 mm × ± 4 mm	3–10 mm × up to 8 mm
	COLOUR	pale green	dull green	green	green	bright green
	VEINS	usually 3	1 (obscure)	3 or 5; if 5, the outer 2 arising from a different point to the inner 3	4–5; radiating from one point	7–16
Roots		1 at most				7–16
Habit		submerged	at surface			
Habitat		ponds, ditches, canals and still/slow-moving areas of streams and rivers				
					also brackish	

IVY-LEAVED	LEAST	COMMON	FAT	GREATER
JFMAMJJASOND	JFMAMJJASOND	JFMAMJJASOND	JFMAMJJASOND	JFMAMJJASOND

highly distinctive leaf shape; leaves; often intermeshed and aggregated

UNDERSIDE
single vein, indistinct at best

UNDERSIDE
never inflated; 3(5) veins; 40–60 air spaces (W < 0·3 mm) visible as a net-like pattern

SIDE VIEW
usually inflated, but can be flat; 4–5 veins; 10–20 large air spaces (W ≥ 0·3 mm) visible as a net-like pattern

large size and multiple roots are distinctive

80 Alismataceae | **Water-plantain** family

1 sp. | 8 spp. B&I

Form rooted marginal. **Infl** open; whorled. **Fls** PETALS 3; SEPALS 3, green; STAMENS 6 or more; STIGMAS 1.

SS Other **water-plantains** (N/i) [leaf shape, flower and fruit details].

P Water-plantain *Alisma plantago-aquatica*

IN FRUIT

H to 100 cm. **Fls** opening in the afternoon; D 7–12 mm; **mauve to white, with yellow basal blotch**; in whorled, branched infl. **Form** erect, rooted in mud. **Lvs** AERIAL broadly oval, base rounded, long-stalked; SUBMERGED linear. **Fr** seeds in a single whorl, forming a ring. **Hab** shallow water and margins of ponds, ditches, canals and slow-moving rivers.

FR

SEED

LVS oval with rounded base, long stalked

81 Hydrocharitaceae | **Waterweed** family

2 spp. | 9 spp. B&I

Form diverse; aquatic. **Fls** typically PETALS 3 + SEPALS 3 (petals and sepals very similar in some). The two species covered in this book are introduced North American waterweeds with whorled leaves.

● **Waterweeds** | **Form** rooted, submerged aquatics of ponds, lakes, canals and slow-moving rivers. **Lvs** stalkless – LOWER opposite, UPPER in whorls of 3–5. **Fls** D±4 mm; white to reddish; solitary in axils; ♀s arising from a tubular sheath on long stalks such that flowers float on the surface.

CANADIAN

IDENTIFY BY: ► leaf apex details

SS Curly Waterweed *Lagarosiphon major* (N/i); garden escape – most like Nuttall's Waterweed – [even more strongly recurved leaves, arranged spirally around stem]. **Other leafy submerged aquatics** (see p.29).

n whorls

P Canadian Waterweed *Elodea canadensis* AQ

L to 3 m. **Fls** only ♀ recorded. **Lvs** L to 17 mm; not strongly recurved; W 0·8–2·3 mm near the apex; LEAF APEX **blunt to slightly pointed.**

FLS tiny

LVS shorter and wider than in Nuttall's; apex blunt

P Nuttall's Waterweed *Elodea nuttallii* AQ

L to 3 m. **Fls** not recorded. **Lvs** L to 35 mm; **recurved or twisted**; W 0·2–0·7 mm near the apex; LEAF APEX **sharply pointed.**

LVS slightly more recurved/twisted than in Canadian

LVS longer and narrower than in Canadian; apex pointed

82 Juncaginaceae | **Arrowgrass** family `2 spp. | 2 spp. B&I`

Form erect; plantain-like (p. 174); can be patch-forming. **H** to 60 cm. **Infl** unbranched spike of numerous stalked flowers. **Fls** tiny; TEPALS 6, sepal-like; STAMENS 6; ANTHERS stalkless. **Lvs** mostly basal; linear; sheathing; papery ligule at leaf/blade junction. **Aroma** spicy-sweet, coriander-like when crushed.

SS Both **arrowgrasses** are somewhat similar. Sea Plantain (p. 174) is superficially similar (can grow near Sea Arrowgrass); [FLS 4 papery sepals and 4 green or brown petals; scentless].

Ⓟ **Marsh Arrowgrass** *Triglochin palustris*

Stigmas long-fringed. **Infl** elongating in fruit. **Lvs** usually deeply furrowed near base; SHEATH LIGULE L < W (usually). **Aroma** strong. **Fr** L 7–10 mm; narrow; splitting when ripe. **Hab** fresh marsh, upper saltmarsh (rare); salted road verges (infrequent).

FR long; >7 mm

LIGULE W > L

Ⓟ **Sea Arrowgrass** *Triglochin maritima*

Stigmas not long-fringed. **Infl** not elongating in fruit. **Lvs** not furrowed; usually flat on upperside; SHEATH LIGULE L > W (usually). **Aroma** slight. **Fr** L 3–5 mm; egg-shaped; not splitting when ripe. **Hab** saltmarsh, salt-sprayed grassland; salted road verges (rare).

FR short; <5 mm

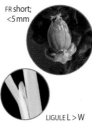

LIGULE L > W

83 Potamogetonaceae | **Pondweed** family `5 spp. | 24 spp. B&I`

Form floating or submerged aquatics. **Infl** above or below water surface; a spike of inconspicuous flowers arising from the base of the leaf or leaf-stalk. **Lvs** either grass-like or oval and stalked; usually with papery stipules and a flexible 'hinge' at the top of the stalk near the leaf-base.

INFL

FL

IDENTIFY BY: ▶ leaf shape and venation ▶ stipule details

SS Several other species and hybrids which require detailed examination to differentiate; Water-lilies (p. 67) and other **aquatic plants with floating leaves** such as **Amphibious Bistort** (p. 143).

Pondweeds with no surface-leaves; submerged leaves only

p. 234

Pondweeds with both surface and submerged leaves

p. 235

Pondweeds with no surface-leaves; submerged leaves only

LEAVES **narrow (<4 mm)**

℗ **Small Pondweed** *Potamogeton berchtoldii*

L to 60 cm. **Form** stem slightly compressed. **Infl** L to 10 mm; densely cylindrical. **Lvs** linear (W to 4 mm), **tip rounded** (can have hair-like point); 3 longitudinal veins; **midrib bordered with translucent patches; small nodules at base. Stipules** L 5–15 mm;

free from leaf. **Fr** L 1·8–3·0 mm. **Hab** ponds, lakes, canals, slow-moving rivers, ditches.

℗ **Fennel Pondweed** *Stuckenia pectinata*

L to 220 cm. **Form** well-branched. **Infl** L to 50 mm; densely cylindrical. **Lvs** linear (W to 4 mm), **tip pointed**; 3–5 longitudinal veins (lateral veins can be faint). **Stipules** fused to base of leaf, forming a sheath with a projecting membrane (L 5–15 mm) at tip. **Fr** L 3·3–4·7 mm. **Hab** ponds,

lakes, canals, slow-moving rivers; tolerant of brackish and nutrient-enriched water.

LF small nodules at the base

STIPULES **form a sheath**

LVS relatively flaccid

LF translucent patches bordering midrib

LF-TIP rounded

LF-TIP pointed

LVS relatively rigid

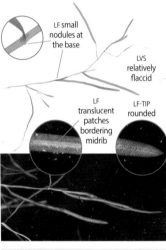

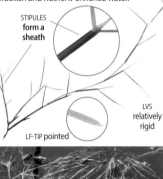

LEAVES **wide (>4 mm)**

℗ **Curled Pondweed** *Potamogeton crispus*

L to 150 cm. **Form** stem compressed. **Infl** emergent; L to 20 mm; loosely cylindrical. **Lvs** narrowly oval; usually with **toothed, wavy margins**, especially on those near surface; L to 95 mm × W to 12 mm; 3–5 longitudinal veins; stalk grooved. **Stipules** L 4–17 mm. **Fr** L 4–6 mm with **long, curved beak. Hab** ponds, lakes, canals, slow-moving rivers.

FR distinctive long beak

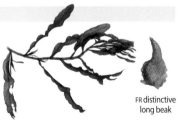

LF margins wavy

Pondweeds with both surface and submerged leaves

℗ Broad-leaved Pondweed
Potamogeton natans

L to 200 cm, can be longer in deep water. **Infl** emergent; L to 80 mm; densely cylindrical. **Lvs** SURFACE broadly oval, L 40–100 mm × W to 45 mm, **discoloured 'joint' at blade/stalk junction**; SUBMERGED linear, W to 3·5 mm, apex rounded. **Stipules** L 40–170 mm. **Fr** L 4–5 mm. **Hab** ponds, lakes, canals, slow-moving rivers.

℗ Bog Pondweed
Potamogeton polygonifolius

L to 50 cm. **Infl** emergent; L to 60 mm; densely cylindrical. **Lvs** SURFACE elliptical to oval, L 25–75 mm × W 10–65 mm, **no joint at blade/stalk junction**; SUBMERGED narrowly to broadly oval, L to > 160 mm × W to 25 mm. **Stipules** L 10–50 mm. **Fr** 1·9–2·6 mm. **Hab** pools and ponds in acid bogs.

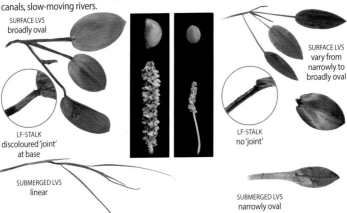

SURFACE LVS
broadly oval

SURFACE LVS
vary from
narrowly to
broadly oval

LF-STALK
discoloured 'joint'
at base

LF-STALK
no 'joint'

SUBMERGED LVS
linear

SUBMERGED LVS
narrowly oval

---

84 Nartheciaceae | **Bog Asphodel** family

`1 sp. B&I`

Form lily-like but classified in its own family based on findings of molecular research. **Fls** TEPALS 6, petal-like; STAMENS 2. **Lvs** mostly basal; held ± vertically; flat; UPP + UND faces both smooth.

℗ Bog Asphodel *Narthecium ossifragum*

H to 45 cm. **Fls** D 15–18 mm; **golden-yellow**; stalked; in spike of up to 20; filaments very hairy, hairs orange; anthers orange. **Form** erect. **Lvs** mostly basal; linear; curved; W 5 mm × L up to 30 cm. **Fr** capsule; **noticeably orange in late summer**; splits along centre of each cell; seeds finely pointed at each end. **Hab** bogs, damp heaths, moors.

85 Dioscoreaceae | **Black Bryony** family

1 sp. B&I

Form climbing and twining, tuberous plants. **Lvs** heart-shaped; net-veined. **Fls** inconspicuous; ♂ + ♀ on separate plants; TEPALS 6; sepal-like; in a spike-like inflorescence arising at the base of the leaf-stalk. **Fr** showy berries; red when ripe. The only representative in B&I of a largely tropical family.

ⓟ **Black Bryony** *Tamus communis* Ⓒ ✕

J F M A M J J A S O N D

L to 20 m; **H** to 5 m.
Form tuberous, twining climber. **Fls** D 3–6 mm; yellow-green; ♂ stalked; ♀ stalkless. **Lvs** glossy green, heart-shaped; L to 15 cm × W to 10 cm. **Fr** berry, bright red when ripe, to 13 mm across. **Hab** hedges, scrub, woodland edges.

LVS heart-shaped

NOTE: wholly unrelated to White Bryony (p.114); their only common feature being their climbing habit

---

86 Orchidaceae | **Orchid** family

15 spp. | 53 spp. B&I

COMMON TWAYBLADE

Although variable, the unique structure of an orchid flower makes it readily recognizable as such: **Fls** OVARY inferior (obviously so); TEPALS 6, usually petal-like, the lowest modified into an enlarged lip; ANTHERS 2, stalkless; POLLEN-MASSES 2, sticky.

IDENTIFY TO GROUP/SPECIES BY:	▶ lip shape and markings ▶ flower/inflorescence form

OVARY — LIP

Orchid group and species identification

LIP **1-lobed**

LIP **narrow**

BUTTERFLY-ORCHIDS
opposite

LIP **broad – 'petal'-like**

HELLEBORINES
opposite

LIP **broad – 'insect'-like**

'INSECT' ORCHIDS
opposite

LIP **2-lobed**

Common Twayblade
p.238

LIP **3-lobed**

Early-purple Orchid
p.238

SPUR **upcurved**

SPUR **broad, downcurved**

MARSH-ORCHIDS
SPOTTED-ORCHIDS
pp.238–239

SPUR **narrow, long**

Pyramidal Orchid
FRAGRANT-ORCHIDS
p.240

LIP 1-lobed

● **Butterfly-orchids** | **Form** erect, tuberous. **Fls** with long slender spur and conspicuous yellow pollen-masses; night-scented; TEPALS whitish, laterals spreading, upper incurved; LIP narrowly oblong; entire. **Lvs** 2–3 basal, with a few smaller ones up the stem; unspotted.

ⓟ Greater Butterfly-orchid
Platanthera chlorantha

H to 60 cm. **Fls** D 18–23 mm; white, often greenish; LIP L 10–16 mm. POLLEN-MASSES L 3–4 mm, **divergent**. **Spur** L 24–37 mm; usually downcurved. **Hab** grassland, open woodland; mostly on calcareous soils.

J F M A M J J A S O N D

ⓟ Lesser Butterfly-orchid
Platanthera bifolia

H to 50 cm. **Fls** D 11–18 mm; white, less greenish; LIP L 6–10 mm. POLLEN-MASSES L 2 mm, ± **parallel**. **Spur** L 14–20 mm; slightly downcurved. **Hab** grassland, open woodland; calcareous and acid soils.

J F M A M J J A S O N D

POLLEN-MASSES diverge top to bottom

POLLEN-MASSES ± parallel

Greater Butterfly-orchid is a generally larger species than Lesser.

● **Helleborines** | **Form** erect; spreading by underground runners. **Fls** LIP 'pinched'.

ⓟ Broad-leaved Helleborine
Epipactis helleborine

J F M A M J J A S O N D

H to 80 cm. **Spike** L to 30 cm; up to 50 fls. **Fls** TEPALS: OUTER greenish; INNER pink; LIP base W > L, pink/purplish with 2 brownish swellings. **Lvs** stem only; lower lvs W > L (usually). **Hab** woods, scrub, sand dunes. **SS** other helleborines (N/I) [leaf + lower lip details].

LIP pink/purplish; 'pinched'
LVS usually broad; arranged spirally

● **'Insect' orchids** | **Form** erect; rosette-forming. **Fls** LIP resembles an insect abdomen.

ⓟ Bee Orchid *Ophrys apifera*

J F M A M J J A S O N D

H to 45 cm. **Spike** loose, up to 7 fls. **Fls** TEPALS: OUTER pink; INNER upper 2 greenish; LIP convex, velvety, brown, cream-edged W-shape marking. **Hab** dry grassland, lawns and open scrub, especially on calcareous soils. **SS** other 'insect' orchids (N/I) [lower lip details].

LIP brown, velvety with a W-shaped marking

LIP 2-lobed

ⓅCommon Twayblade *Neottia ovata*

J F M A M J J A S O N D

H to 60 cm. **Infl** loose spike; L to 25 cm; up to 50 fls. **Fls** TEPALS **yellow-green**, upper 5 form a hood; LIP **deeply divided. Lvs** single opposite pair on lower stem; broadly oval; 3–5 prominent veins. **Hab** woodland, grassland, moorland, dune-slacks. **SS** Lesser Twayblade *N. cordata* (N/I) [much smaller].

LVS broad; opposite

LIP 3-lobed; SPUR upcurved; BRACTS inconspicuous

SPUR

ⓅEarly-purple Orchid *Orchis mascula*

J F M A M J J A S O N D

H to 50 cm. **Infl** loose spike; L to 15 cm. **Fls** TEPALS largely purple; outer lateral pair spreading; upper 3 form a hood (unmarked); LIP 3-lobed – side lobes shorter and bent back, middle lobe notched at apex; lip-base whitish and spotted purple; SEP plain. **Spur** purple.

Lvs BASAL LVS narrowly oval, W to 30 mm, in rosette; STEM-LVS sheathing, spirally arranged, usually with purplish-black blotches; BRACTS inconspicuous. **Hab** woodland, hedgebanks, calcareous grassland. **SS** marsh-orchids; **Green-winged Orchid** *Anacamptis morio* (N/I) [HOOD with green line markings].

LIP often notched and 'frilly'

LF-SPOTS more rounded than in marsh-orchids

LIP 3-lobed; SPUR broad, downcurved; BRACTS conspicuous

SPUR

● **Marsh- and Spotted-orchids** | **Form** erect; tuberous. **Infl** conical to cylindrical spike. **Fls** with broad downcurved spur; BRACTS green; TEPALS white to pink, purple and dark red; outer lateral pair spreading; the upper 3 forming a hood. **Lvs** spotted or unmarked.

SS All *Dactylorhiza* orchids are similar and highly variable. They are best identified by the shape and markings of the lower lip. They can be found growing together in various species combinations. All species readily hybridize between themselves over multiple generations, producing individuals that very closely resemble one of the parent species as well as plants that show a full range of intermediate features.

IDENTIFY BY: ► spike shape
► lower lip shape and markings

BRACTS conspicuous

LF-SPOTS more elongated than in Early-purple Orchid

Dactylorhiza orchids are often found in mixed-species colonies such as this group of Common Spotted- and Southern Marsh-orchids and their hybrid.

ⓟ Common Spotted-orchid *Dactylorhiza fuchsii*

H to 60 cm. **Lip** LOBES divided ± ½ depth; **central lobe longer** and ≥ ½ width of lateral lobes; white or pale pink; MARKINGS variable, purple spots, streaks and loops. **Lvs** W to 40 mm; spots transversely stretched. **Hab** meadows, marshes, open damp woodland, usually on calcareous soils.

J F M A M J J A S O N D

CENTRAL LOBE **longer** + **> ½ width of laterals**

MARKS typically more streaky and loopy

ⓟ Heath Spotted-orchid *Dactylorhiza maculata*

H to 50 cm. **Lip** LOBES divided < ½ depth; **central lobe shorter** and < ½ width of lateral lobes; white or pale pink; MARKINGS variable, purple spots, streaks and loops. **Lvs** W to 20 mm; spots ± rounded. **Hab** bogs and marshes, often peaty and acidic.

J F M A M J J A S O N D

CENTRAL LOBE **shorter** + **< ½ width** of laterals

MARKS typically more spotty; less loopy

Hab the three common marsh-orchids are all found in fens, marshes, damp meadows and dune-slacks; with Northern and Southern also found on brownfield.

ⓟ Northern Marsh-orchid	ⓟ Southern Marsh-orchid	ⓟ Early Marsh-orchid
Dactylorhiza purpurella	*Dactylorhiza praetermissa*	*Dactylorhiza incarnata*

H to 30 cm. **Lvs** typically umarked; marks (if present) usually small pale spots on upperside.

J F M A M J J A S O N D

H to 70 cm. **Lvs** typically unmarked; dark marks (if present) usually as rings on upperside.

J F M A M J J A S O N D

H to 40 cm. **Lvs** typically spotted on both sides (can be plain); leaf-tip markedly hooded.

J F M A M J J A S O N D

deep pink to reddish-purple

LIP diamond-shaped; **typically very shallowly lobed**

LAT LOBES ± flat

MARKS dashes and/or loops

pale to mid-pink

LIP rounded; **typically shallowly lobed**

LAT LOBES ± flat

MARKS dashes; loops rare

deep pink to reddish-purple

LIP highly variable; **appearing narrow**

LAT LOBES bent back

MARKS **usually 2 distinct loops**

SPIKE ± cylindrical; L to 6 cm

SPIKE dense, conical; L to 10 cm

SPIKE cylindrical; L to 12 cm

LIP 3-lobed; SPUR narrow, long

SPUR

ⓟ Pyramidal Orchid *Anacamptis pyramidalis*

J F M A M J J A S O N D

H to 50 cm. **Fls** unscented; deep pink, a few paler, even white. **Infl** dense 'pyramid' to cylindrical; 3 upper tepals form hood. **Lip** L 6–9 mm, deeply 3-lobed; SPUR long, **very slender**; L 14 mm. **Lvs** narrow, unspotted. **Hab** dry calcareous grassland, sand dunes.

INFL cylindrical

2 bumps at base of lip

INFL 'pyramid'

● **Fragrant-orchids** | **Form** erect; tuberous. **Fls** with long slender downcurved spur; aromatic; TEPALS pink to lilac; inner upper 2 incurved; outer laterals spreading. **Lvs** narrow.

HEATH CHALK MARSH

SS All three species very similar and can occur together (especially **Chalk** and **Marsh**).

IDENTIFY BY: ▶ scent *supplemented by:*
▶ lip shape ▶ outer tepals ▶ spike shape

J F M A M J J A S O N D   J F M A M J J A S O N D   J F M A M J J A S O N D

Fragrant-orchids		**ⓟ Heath** *Gymnadenia borealis*	**ⓟ Chalk** *Gymnadenia conopsea*	**ⓟ Marsh** *Gymnadenia densiflora*
Aroma		clove-like, sweet	sweet, not clove-like	clove-like, spicy
Flower	WIDTH	D 8–10 mm	D 10–11 mm	D 11–13 mm
	LIP	**L > W,** lobes shallow	**L ± W,** lobes obvious	**W > L,** lobes obvious
	OUTER LATERAL TEPALS	usually bent downwards	point downwards	**blunt, horizontal**
	SPUR	L 11–14 mm	L 12–14 mm	L 14–16 mm
Spike		± conical; ± dense; **H** to 30 cm	± cylindrical; ± loose; **H** to 50 cm	± cylindrical; dense; **H** to 90 cm
Habitat		upland grassland, bogs, moors	dry, calcareous grassland	fens, damp calcareous grassland

HEATH   CHALK   MARSH

87 Iridaceae | Iris family

Form erect. **Fls** TEPALS 6; petal-like; showy; fused at the base; STAMENS 3; STYLE 1; 3-branched, repeatedly in some, the branches are flattened and petal-like. **Lvs** mostly basal; held ± vertically; flat; UPP + UND faces both smooth.

ⓅYellow Iris *Iris pseudacorus*

H to 150cm.
Form stem often branched.
Fls yellow, with golden patch on outer tepals; D 70–100mm.
Lvs W 10–30mm.
Hab ditches, ponds, river and lake margins, wet meadows.

J F M A M J J A S O N D

ⓅStinking Iris *Iris foetidissima*

H to 80cm.
Form little-branched, angled on one side; crushed stem smells of **meaty gravy**. **Fls** D 50–70mm; dull purplish (can be pale yellow). **Lvs** W 10–25mm; winter-green. **Fr** bright orange berry in part-opened capsules. **Hab** woodland, hedgebanks, gardens, coastal scrub and cliff slopes.

J F M A M J J A S O N D

ⓅMontbretia
Crocosmia × crocosmiiflora

H to 90cm. **Fls** L 25–40mm; orange; tube curved; ± L lobes; in sparsely branched, 1-sided spike, with zig-zag axis. **Lvs** W 5–20mm; < L fl-stem. **Hab** gardens, hedges, road verges, open woodland, coastal scrub. **SS** Other variants (N/I) [FLS yellow to crimson].

J F M A M J J A S O N D

FR distinctive

FL-STALK zig-zags

FLS purplish or yellow

88 Amaryllidaceae | Daffodil + Onion family

Form bulbous, erect perennial. **Fls** distinctive, borne singly or in an umbel; TEPALS 6; petal-like; fused or free; STAMENS 6; BRACTS 1–2; papery; forming a sheath (spathe) at the base. **Lvs** linear; flat (some can have a keel) or be half-cylindrical in cross-section.

ⓅDaffodil *Narcissus pseudonarcissus*

H to 40cm. **Fls** L 20–40mm; solitary; TEPALS pale yellow; with a darker yellow (usually) **funnel-shaped centre** (corona), L ± tepal L. **Lvs** linear, flat, W 5–15mm wide. **Hab** woods, grassland, damp river valleys. **SS** many other *Narcissus* spp., hybrids and cultivars often escape from cultivation or are planted in the wild [CORONA shape + COLOUR (white to orange)].

J F M A M J J A S O N D

INSET: **Poet's Narcissus** *Narcissus poeticus* – one of the many garden species and cultivars that can be found naturalized.

241

ⓟ Snowdrop *Galanthus nivalis*

H to 20 cm. **Fls** solitary; nodding; TEPALS outer: semi-spreading; inner: **with green patch and notched apex**. **Spathe** 2 fused bracts, separated at tip. **Lvs** linear; flat; weakly keeled; L×W 150 × 10 mm. **Hab** woods, damp grassland, streamsides, hedgebanks. **SS** many other snowdrop spp., hybrids and cultivars escape from gardens or are planted in the wild [LF + TEPAL details].

J F M A M J J A S O N D

ⓟ Wild Onion *Allium vineale*

H to 100 cm. **Fls globular infl of flowers** (var. *capsuliferum* [rare]) or **purplish reproductive organs** (bulbils) (var. *compactum*) or both (var. *vineale*); TEPALS (if present) L 3–5 mm; pinky-green. **Spathe** 1 papery bract. **Lvs** hollow, half-cyclindrical; AROMA onion when crushed. **Hab** rough grassland, riverbanks, clay sea walls, sand dunes.

J F M A M J J A S O N D

var. *compactum*   var. *vineale*

ⓟ Ramsons *Allium ursinum*

H to 45 cm. **Form** patch-forming. **Fls flat-topped umbel** of 6–20 starry white flowers. **Spathe** 2 papery bracts. **Lvs** flat; AROMA garlic when crushed. **Hab** woodland, especially damp; hedgebanks.

J F M A M J J A S O N D

ⓟ Three-cornered Garlic *Allium triquetrum*

H to 45 cm. **Form** stem **sharply triangular** in section. **Fls** umbel of 3–15 drooping, long-stalked white fls; TEPALS green mid-line. **Lvs** flat; linear; keeled; AROMA garlic when crushed. **Hab** hedgebanks, woodland edges, brownfield, cultivated ground. **SS Some other *Allium* spp.** (N/I) [TEPAL + LF details].

J F M A M J J A S O N D

STEM sharply triangular in section

89 Asparagaceae | Bluebell [Asparagus] family   `4 spp. | 27 spp. B&I`

Form bulbous with basal leaves only. **Fls** TEPALS 6; petal-like; usually blue; STAMENS 6.

● **Bluebells** | **Infl** spike. **Fls** stalked; bell-shaped (L 14–20 mm); usually blue (can be pink or white) with a toothed mouth. **Lvs** linear; glossy green with hooded tip.

ANTHERS yellow; TEETH not strongly recurved

HYBRID

SS The two bluebells hybridize and the fully fertile hybrids exhibit a continuous range of intermediate features. Hybrids often occur in the absence of one or both parents and can be more common than the native Bluebell. Some hybrids can look almost identical to either of the parent species so it may be necessary to examine a range of individuals, particularly in habitats near houses, to establish if hybrids are present.

℗ Bluebell
Hyacinthoides non-scripta

H to 50 cm. **Infl** drooping, 1-sided. **Fls** TEETH **strongly recurved**; ANTHERS **yellow**; AROMA very sweet. **Lvs** W to 20 mm. **Hab** woods, hedgebanks, clifftop grassland, under Bracken.

℗ Hybrid Bluebell
H. × massartiana

Form mixed/ intermediate in all features. **Hab** common in gardens; spreading in woods that border urban areas.

℗ Spanish Bluebell
Hyacinthoides hispanica

H to 40 cm. **Infl** erect; not 1-sided. **Fls** TEETH not recurved; ANTHERS blue; AROMA slight. **Lvs** W to 35 mm. **Hab** woods, hedgebanks, gardens.

ANTHERS yellow blue

OUTER 3 STAMENS are fused to the tepal-tube

for >3/4 of their length / for <3/4 of their length

TEETH strongly recurved

TEETH ± straight

FLS paler than Bluebell

BLUEBELL   HYBRID   SPANISH

LF W [max]

● **Grape-hyacinths** | **Infl** egg-shaped to conical spike. **Fls** ± globular; usually blue; contracted into a toothed mouth.

℗ Garden Grape-hyacinth *Muscari armeniacum*

H to 35 cm. **Fls** topmost sterile, lower fertile (L 3.5–5.5 mm); **typically all the same blue** (sterile fls can be darker or paler in some cultivars). FERTILE FLS long, tepal-tube L > 2× tepal-lobe L; teeth white. **Lvs** linear; basal; underside rounded. **Hab** gardens, hedgebanks, road verges, brownfield. **SS** other *Muscari* spp. (1 native; 1 garden escape (N/i)) [flower shape and colour].

FLS lighten with age

243

90 Typhaceae | **Reedmace** [Bulrush] family

Form emergent. **Fls** unmistakable; either unisexual globes or densely cylindrical spikes, with ♂ + ♀ in separate portions of the spike. **Lvs** long; linear.

● **Reedmace** | **Form** erect plants of lakes, ponds, ditches and slow-moving rivers. **Lvs** linear; twisted in some. **Fls** unmistakable **dense cylindrical spikes of tiny individual flowers** in narrow, yellow ♂ and wider deep brown ♀ sections. ♀ sections erupt in late winter, releasing small seeds within a mesh of cottonwool-like fibres.

ⓟ **Greater Reedmace** *Typha latifolia*

H to 3 m. **Infl** flowering spikes **usually exceed leaves** (slightly lower at most); ♂ and ♀ sections **touching** (or separated by <25mm). **Lvs** W 10–24mm wide; often twisted.

J F M A M J J A S O N D

ⓟ **Lesser Reedmace** *Typha angustifolia*

H to 3 m. **Infl** flowering spikes **usually much lower than leaf-tips**; ♂ and ♀ sections normally **separated by 30–80mm**. **Lvs** W 3–6mm. **Hab** more tolerant of peaty and brackish water than Greater Reedmace.

J F M A M J J A S O N D

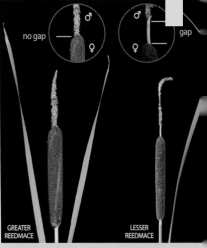

no gap ♂ ♀

gap ♂ ♀

GREATER REEDMACE

LESSER REEDMACE

● **Bur-reeds** | **Form** erect plants of ponds, lakes, canals, slow-moving rivers and ditches. **Fl-hds** globular, bur-like; smaller ♂s (W to 10mm) above larger ♀s (W to 10–20mm).

ⓟ **Branched Bur-reed**
Sparganium erectum

H to 150 cm. **Infl** branched; TEPALS **dark-tipped**. **Lvs** W 10–15mm; linear; keeled; floating and submerged leaves (if developed) also keeled.

J F M A M J J A S O N D

ⓟ **Unbranched Bur-reed**
Sparganium emersum

H to 60 cm. **Infl** unbranched; TEPALS **green**. **Lvs** W 3–12mm; linear; keeled; floating leaves ribbon-like, not keeled. **Hab** more restricted to open waterbodies than Branched Bur-reed.

J F M A M J J A S O N D

BRANCHED BUR-REED

UNBRANCHED BUR-REED

TEPALS tip dark

TEPALS tip green

♂

♀

♂

♀

RUSHES, SEDGES AND GRASSES

A group of monocots, comprising three families which have, to some extent, a similar structure and hence the whole group is often regarded as confusing.

Structurally, they are all somewhat leafy plants, with parallel-veined leaves that are typically linear. While there are exceptions, their flowering stems are usually long and thin with an inflorescence that is variable in appearance, from dense to open, branched to spike-like. The flowers, at first glance, may appear similar and all share the traits of being brownish, yellowish or greenish (*i.e.* not colourful).

However, a closer inspection, particularly of the flowering parts, should reveal reasonably easy differences that enable family recognition. From there, examination of the features relevant to each group should present a relatively clear pathway to a confident identification.

FLS bisexual; **6 tepals** – open in flower, closing in fruit but still 6 obvious parts

♂ ANTHER
♀ STIGMA
♀ FRUIT
♀ FRUIT
1 2 3 4 5 6 TEPALS
6 5 4 TEPALS

NOTE: 1–3 hidden behind fruit

Rush family
91 Juncaceae
below

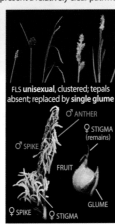

FLS **unisexual**, clustered; tepals absent; replaced by **single glume**

♂ ANTHER
♂ SPIKE
♀ STIGMA (remains)
FRUIT
GLUME
♀ SPIKE
♀ STIGMA

Sedge family
92 Cyperaceae
p. 250

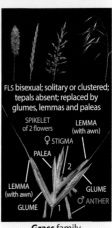

FLS bisexual; solitary or clustered; tepals absent; replaced by glumes, lemmas and paleas

SPIKELET of 2 flowers
LEMMA (with awn)
♀ STIGMA
PALEA
2
LEMMA (with awn)
GLUME
GLUME
1
♂ ANTHER

Grass family
93 Poaceae
p. 266

---

91 Juncaceae | **Rush +Woodrush** family

13 spp. | 36 spp. B&I

Distinguished from all similar grassy and sedgy families by possessing more 'traditional'-looking flowers. **Fls** SYMMETRY radial; TEPALS 6; STAMENS 6; STIGMAS 3. **Fr** in fruit, rushes can look more sedge-like, but the 6 tepals are still apparent. Many are characteristic of wetland habitats.

IDENTIFY TO GROUP BY:
► inflorescence shape and branching
► flower details

SS Club- and Bog-rushes (sedges – *p. 253*) [glumes; no tepals].

Rush groups identification

INFLORESCENCE **apparently lateral**; BRACTS **tubular**	INFLORESCENCE **terminal**; BRACTS **flat and leafy**		
	LVS **hollow**	LVS **not hollow** (flat or channelled)	
		LEAVES **not hairy**	LEAVES **hairy**
			WOOD-RUSHES
p. 248	*p. 247*	*p. 246*	*p. 249*

RUSHES | Inflorescence terminal; bracts flat and leafy

Short annual plants (typically < 10 cm tall); leaves narrow; channelled (not hollow or hairy)

Ⓐ Toad Rush *Juncus bufonius*

J F M A M J J A S O N D

H to 10 cm, rarely larger.
Fls TEPALS pointed, pale,
midrib green, no dark lines;
INFL flowers on branches in each
fork, along branch and at tip.
Form erect to spreading, widely
branched. **Lvs** W <1·5 mm.
Hab muddy, damp areas..

TEPALS
all pointed;
no dark lines

SS Two similar-looking, but scarcer, species that can grow with Toad Rush can be separated
by tepal details: **Frog Rush** *J. ranarius* (N/I) [TEPALS unmarked; inner blunt with short sharp
tip]; **Leafy Rush** *J. foliosus* (N/I) [TEPALS with dark line either side of midrib].

Tall perennial plants (typically > 10 cm tall); leaves flat or channelled (not hollow or hairy)

Ⓟ Heath Rush *Juncus squarrosus*

J F M A M J J A S O N D

H to 35 cm. **Form** tufted. **Fls**
TEPALS dark brown; singly
(rarely paired) on stalks (L >
3 mm) in spreading, branched
infl (branches angled down
after flowering). **Lvs** all basal;
W 3–4 mm; **channelled**. **Hab**
woods, hedges, moors.

Ⓟ Slender Rush *Juncus tenuis*

J F M A M J J A S O N D

H to 40 cm. **Form** densely
tufted. **Fls** TEPALS **yellow-
green**, pointed; 2 or 3 on
tips of branched infl. LOWEST
BRACTS **longer than infl**. **Lvs** all
basal; W ±1 mm; flat; yellowish-
green. **Hab** bare patches on
tracks, woodland rides.

Its tufted form, dark tepals and channelled
leaves make Heath Rush distinctive.

TEPALS dark

LEAVES channelled

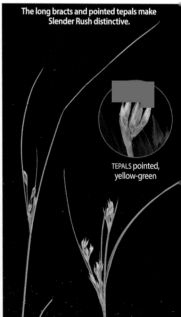

The long bracts and pointed tepals make
Slender Rush distinctive.

TEPALS pointed,
yellow-green

Leaves hollow with crosswise septa; bracts not extending beyond inflorescence tip

Three species with widely branched inflorescences. **Hab** marshes, bogs, moors and other damp habitats.

Septa are strengthening 'walls' within a stem – these can be felt by gently squeezing the stem using fingertips and moving them along the stem.

stem x-section

SEPTA

℗ Bulbous Rush
Juncus bulbosus

H to 30 cm. **Form** erect to spreading; (floating in shallow water); stem-base can be swollen. **Fls** TEPALS L 1·5–3·5 mm; green to dark brown. **Lvs** SEPTA indistinct. **Fr** matt, light brown or shiny, dark brown **blunt capsule**.

℗ Jointed Rush
Juncus articulatus

H to 80 cm. **Form** erect to spreading. **Fls** TEPALS L 2·5–3·5 mm; dark brown to black; outer pointed with **straight tips**; inner usually with pale edges. **Lvs** curved; flattened; SEPTA indistinct. **Fr** blackish, shiny capsule; **egg-shaped with a point**.

℗ Sharp-flowered Rush
Juncus acutiflorus

H to 100 cm. **Form** erect. **Fls** TEPALS L 1·5–2·7 mm; mid- to dark brown; long-pointed; **outer with curved tips**. **Lvs** straight; rounded; SEPTA very pronounced. **Fr** brown capsule; tapered to a point.

J F M A M J J A S O N D

FORM typically tufted and short

J F M A M J J A S O N D

FL-STALK usually relatively short

J F M A M J J A S O N D

FL-STALK usually relatively long

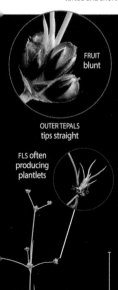

FRUIT blunt

OUTER TEPALS tips straight

FLS often producing plantlets

H 1/25

FRUIT with point

OUTER TEPALS tips **straight**

LVS flattened; SEPTA indistinct

H 1/25

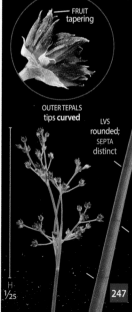

FRUIT tapering

OUTER TEPALS tips **curved**

LVS rounded; SEPTA distinct

H 1/25

247

RUSHES | Inflorescence apparently lateral, the 'stem' continuing above inflorescence as a tubular bract; **leaves basal only, much reduced.**

Three densely tufted, somewhat similar, species that can be found growing in close proximity. **Hab** marshes, bogs and other wet places (*e.g.* damp woodland).

IDENTIFY BY: ► stem texture and colour
► pith inside stem

HARD   SOFT-   COMPACT

J F M A M J J A S O N D   J F M A M J J A S O N D   J F M A M J J A S O N D

Stem	Hard	Soft-	Compact
COLOUR	blue-green	green; **glossy**	green or grey-green; **dull**
TEXTURE	**12–28 strong ridges**	>30 fine ridges = **smooth**	12–30 ridges
PITH (inside stem)	**interrupted**	not interrupted	

Stem-ridges are most apparent just below the inflorescence and easily felt by rolling the stem between finger and thumb.

SOFT-RUSH

The compact tufts of 'lateral flowering' rushes are a familiar sight in many damp habitats.

SS The coastal **Sea Rush** *J. maritimus* (N/I) [straw-coloured tepals] – most like Hard Rush.

ⓟ Hard Rush
Juncus inflexus

H to 120 cm. **Fls** in heads on erect or spreading stalks; TEPALS dark brown.

ⓟ Soft-rush
Juncus effusus

H to 140 cm. **Fls** in heads on erect or spreading stalks (infl of var. *subglomeratus* compact); TEPALS pale brown.

ⓟ Compact Rush
Juncus conglomeratus

H to 100 cm. **Fls** typically in very compact head; **rarely stalked**; TEPALS dark red-brown.

STEM **strongly ridged;** PITH interrupted

TEPALS dark brown

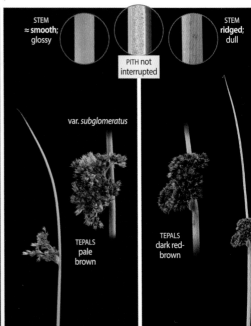

STEM ≈ **smooth; glossy**

STEM **ridged; dull**

PITH not interrupted

var. *subglomeratus*

TEPALS pale brown

TEPALS dark red-brown

WOOD-RUSHES | Leaves grass-like with long, spidery white hairs, at least at the base

 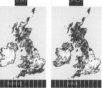

J F M A M J J A S O N D   J F M A M J J A S O N D   J F M A M J J A S O N D   J F M A M J J A S O N D

IDENTIFY BY: ▶ inflorescence shape and branching ▶ flower details

SS All wood-rushes are similar though **Field** and **Heath Wood-rushes** are the most readily confused. Other rushes have hairless leaves.

Ⓟ Hairy Wood-rush
Luzula pilosa

H to 35 cm. **Form** tufted. **Fls** TEPALS dark brown; singly (rarely paired) on stalks (L > 3 mm) in spreading, variably branched, infl (branches reflex after flowering). **Lvs** W 3–4 mm. **Hab** woods, hedges, moors.

Ⓟ Great Wood-rush
Luzula sylvatica

H to 80 cm. **Form** erect; densely tufted; tussock-forming; shortly spreading. **Fls** groups of 3–5 in loose, widely branched infl. **Lvs** W 8–12 mm; glossy. **Hab** woods, streamsides, moors.

Ⓟ Field Wood-rush
Luzula campestris

H to 35 cm. **Form** tufted, patch-forming. **Fls** TEPALS chestnut brown; ANTHER L 3–4× filament L; STIGMA L > 1 mm; INFL 1 stalkless and 3–6 stalked flower clusters, each with 3–12 FLS. **Lvs** W 2–4 mm. **Hab** grassland, lawns.

Ⓟ Heath Wood-rush
Luzula multiflora

H to 40 cm. **Form** tufted. **Fls** TEPALS red-brown; ANTHER L ≤2× filament L; STIGMA L <0·9 mm; INFL up to 10 stalked flower clusters, each with 8–16 FLS. **Lvs** W 3–6 mm. **Hab** acid grassland, heathland, moors.

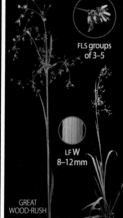

in flower

FLS typically single

FLS groups of 3–5

branches reflex after flowering

LF W 3–4 mm

LF W 8–12 mm

HAIRY WOOD-RUSH

GREAT WOOD-RUSH

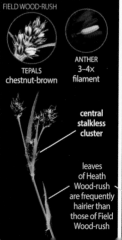

FIELD WOOD-RUSH

TEPALS chestnut-brown

ANTHER 3–4× filament

central stalkless cluster

leaves of Heath Wood-rush are frequently hairier than those of Field Wood-rush

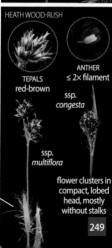

HEATH WOOD-RUSH

TEPALS red-brown

ANTHER ≤ 2× filament

ssp. *congesta*

ssp. *multiflora*

flower clusters in compact, lobed head, mostly without stalks

249

92 Cyperaceae | Sedge family

37 spp. | 109 spp. B&I

'Grass-like' with simplified flowers. **Form** STEM solid; triangular in cross-section in many species; lacking obvious nodes ('knees'). **Fls** much reduced; lacking tepals; one bract (glume) per flower. **Infl** typically separated, adjacent or contiguous spikes of single-sexed flowers but can be unisexual. **Fr** a seed that can be enclosed within a distinctively shaped sac (utricle), some with a prominent beak. **Lvs** have a sheathing base, and a papery ligule at the junction of sheath and blade.

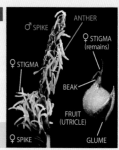

IDENTIFY TO GROUP BY: ► inflorescence form
► number and arrangement of ♂ and ♀ spikes

Sedge group identification

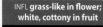

INFL **grass-like in flower; white, cottony in fruit**	INFL **terminal; small;** LVS **absent or much reduced**	INFL **apparently lateral;** LVS **present**

COTTONGRASSES *opposite*	DEERGRASSES; SPIKE-RUSHES *p.252*	BOG-RUSHES; CLUB-RUSHES *p.253*

Carex sedges

INFL **terminal;** SPIKE **1;** LVS **present**	SPIKE >**1;** ♂ + ♀ look similar	SPIKE >**1;** ♂ + ♀ look different	
		UTRICLES **hairy**	UTRICLES **not hairy**

p.254	*pp.255–257*	*p.258*	*pp.259–265* ► *p.254*

SS Some rushes are superficially similar: **Hard**, **Soft**- and **Compact Rushes** (*p.248*) [look like Bog- and club-rushes]; **Heath Rush** (*p.246*) [look like sedges with a compact spike]. All rushes can be told by their 6 tepals and absence of glumes.

♂ spike count is variable in these species – it is worth examining a range of flowering spikes to see if the majority have 1 or >1♂ spikes

● **Cottongrasses** | easily told from other members of the sedge family by the bristle-like flower segments that elongate in fruit into a conspicuous white cottony head. **Hab** marshes, bogs and other wet places (*e.g.* damp woodland).

IDENTIFY BY: ▶ inflorescence shape
▶ glume + flower-stalk details

COMMON COTTONGRASS

in flower   *in fruit*

SS Slender Cottongrass *E. gracilis* (N/I) and **Broad-leaved Cottongrass** *E. latifolium* (N/I) – most like Common Cottongrass – [SPIKELET-STALKS rough; GLUMES: **Broad-leaved** has narrow silver margins; **Slender** has several veins].

℗ Common Cottongrass
Eriophorum angustifolium

J F M A M J J A S O N D

H to 75 cm. **Form** creeping, patch-forming. **Infl** 3–7 **drooping spikelets** on smooth; **hairless** stalks; GLUMES 1-veined; broad silvery margins. **Lvs** W 2–6 mm; v-shaped in cross-section; tip triangular.

℗ Hare's-tail Cottongrass
Eriophorum vaginatum

J F M A M J J A S O N D

H to 60 cm. **Form** densely tufted. **Infl** single, erect, **terminal spikelet**; GLUMES 1-veined; base and tip silvery. **Lvs** W 0·5–1·0 mm; triangular in cross-section; tip triangular; UPPER LF-SHEATHS **inflated**.

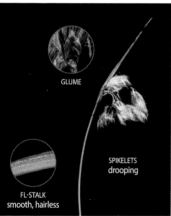

GLUME

SPIKELETS
drooping

FL-STALK
smooth, hairless

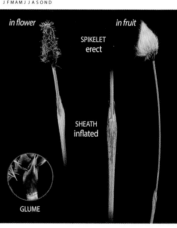

in flower   *in fruit*

SPIKELET
erect

SHEATH
inflated

GLUME

Areas of cottongrass in fruit are obvious, even at distance

SEDGES | Spike-rushes and deergrasses

● **Spike-rushes** | **Form** tufted. **Infl** 1 terminal spikelet. **Lvs** UPPERMOST LF-SHEATH (+ most or all others) **without a blade.**

IDENTIFY BY: ► flower + bract details ► leaf-sheath angle

COMMON SPIKE-RUSH

ⓟ Common Spike-rush
Eleocharis palustris

H to 75 cm. **Spklt** L 5–30mm; **20–70 fls**; STIGMAS **2**; GLUME ± ½ encircling spikelet at its base; L much < spikelet L. **Lvs** UPPERMOST LF-SHEATH apex **not angled. Hab** marshes, ditches, ponds; in or by water.

ⓟ Few-flowered Spike-rush
Eleocharis quinqueflora

H to 30 cm. **Spklt** L 4–10mm; **3–7 fls**; STIGMAS **3**; GLUME ± encircling spikelet at its base, and >½ spikelet L. **Lvs** UPPERMOST LF-SHEATH apex **obliquely angled. Hab** marshes, fens, dune-slacks.

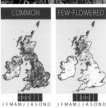

COMMON   FEW-FLOWERED

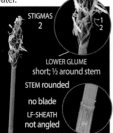

STIGMAS **2**

LOWER GLUME
short; ½ around stem

STEM rounded

no blade

LF-SHEATH
not angled

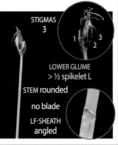

STIGMAS **3**

LOWER GLUME
> ½ spikelet L

STEM rounded

no blade

LF-SHEATH
angled

SS Other spike-rushes (N/I) [stigma count and details of stem, leaf-sheath, glumes and seeds].

J F M A M J J A S O N D   J F M A M J J A S O N D

● **Deergrasses** | **Form** tufted. **Infl** 1 terminal spikelet; L to 8mm. **Lvs** UPPERMOST LF-SHEATH **with short blade.**

IDENTIFY BY: ► bract + seed details ► leaf-sheath opening

COMMON DEERGRASS

ⓟ Northern Deergrass
Trichophorum cespitosum

H to 25 cm. **Spklt** with **3–10 fls**; never produces plantlets; GLUME **yellow-brown** midrib. **Lvs** UPPER LF-SHEATH opening ± **circular;** L ±1mm; LEAF-BLADE L < 3× L opening. **Fr** seed; dark brown (blackish when ripe); usually shiny. **Hab** bogs, wet moors, heathland (usually in wetter areas than Common).

ⓟ Common Deergrass
Trichophorum germanicum

H to 40 cm. **Spklt** with **8–20 fls**; may produce plantlets; GLUME **green** midrib. **Lvs** UPPER LF-SHEATH opening **oval;** L 2mm × W 1mm; LEAF-BLADE L > 5× L opening. **Fr** seed; brown when ripe (may have grey coating). **Hab** bogs, wet moors, heathland.

NORTHERN   COMMON

J F M A M J J A S O N D   J F M A M J J A S O N D

Form Both densely tufted but **Northern** typically less dense and with thinner stems. Hybrids may outnumber parents in mixed populations; – best recognized by their 'headless' stalks in July–August.

flowering completed

leaf-blade

GLUME MIDRIB
yellow-brown

LF-SHEATH
± circular;
L ± 1mm

in flower

NB leaf-blade hidden

GLUME MIDRIB
green

LF-SHEATH
oval;
L 2mm

SEDGES | Club-rushes and bog-rushes

● **Infl** clustered and apparently lateral as the BRACT is stem-like and exceeds the height of the clustered inflorescence. **Club-rushes** are of two broad types: those of the genus *Isolepis* are small, slender annuals; others are tall, robust perennials, usually growing out of water. **Bog-rushes** are somewhat similar but the extending bract is angled and not so stem-like.

IDENTIFY BY: ► size ► bract details ► stem details

Ⓐ Bristle Club-rush *Isolepis setacea*

J F M A M J J A S O N D

H to 10 cm. **Form** densely tufted, stem wiry. **Infl** 1–4 unstalked spikelets; **apparently lateral** as LOWEST BRACT stem-like; tip > infl; GLUME red-brown, midrib green. **Lvs** LF-SHEATHS basal (can have short blades). **Hab** open wet areas in fens, marshes, wet-heaths; pond margins. **SS** Slender Club-rush *I. cernua* (N/I) [bract < infl].

GLUME

clusters of 1–4 spikelets

Ⓑ Black Bog-rush *Schoenus nigricans*

J F M A M J J A S O N D

H to 75 cm. **Form** densely tufted, tussock-forming. **Infl** terminal head of 5–10 spikelets (each of 1–4 flowers); LOWEST BRACT leaf-like; dark brown; tip > infl. **Spklt** blackish-brown; flattened. **Lvs** narrow; W 1–2 mm; basal; inrolled; crescent-shaped in cross-section. **Hab** bogs, wet heaths, fens, dune-slacks, upper saltmarshes.

Ⓒ Common Club-rush *Schoenoplectus lacustris*

J F M A M J J A S O N D

H to 300 cm. **Form** patch-forming; STEM round, smooth, W up to 15 mm, bright green (mud on stem can affect appearance). **Infl** terminal; loose cluster of stalked heads of spikelets; LOWEST BRACT L ± infl. **Spklt** egg-shaped; red-brown; unstalked. **Lvs** narrow (SUBMERGED wider). **Hab** lakes, slow-moving rivers; sometimes in deep water. **SS** Grey Club-rush *S. tabernaemontani* (N/I) [STEM blue-green]; **Sea Club-rush** *Bolboschoenus maritimus* (N/I) [STEM sharply 3-angled] – both can be abundant by the coast.

BLACK BOG-RUSH

SPIKELET

black heads in tussocks distinctive, even at distance

COMMON CLUB-RUSH

submerged strap-like leaves

SPIKELET

● **Carex sedges** | 'grass'-like plants with distinctive inflorescences, consisting of: **Infl** 1-flowered spikelets in spikes in various arrangements; each flower has a single glume; ♂ and ♀ flowers can be intermixed, or in separate spikes. **How flowers are arranged is a good primary differentiator of *Carex* sedges. Stems** rounded or, more typically, 3-angled – the angles range from blunt to sharp. Angularity (depicted as cross-section icons on these pages) can be difficult to assess but is of some help as a supporting identification feature.

28 spp. | 75 spp. B&I

IDENTIFY TO GROUP (and some species) BY: ► inflorescence arrangement

IDENTIFY TO SPECIES BY: ► ♀ glume shape and colour ► ripe fruit (utricle) details

Carex sedge groups identification

SPIKE **1**; terminal	SPIKE **>1**; ♂+♀ similar	SPIKE **>1**; ♂ + ♀ look different			
♂ ♀	♂ ♀	UTRICLES **hairy**			HAIRY
		UTRICLES **not hairy**			p. 258
		♀ SPIKES	SPIKE COUNT	OTHER	page
		± cylindrical	1–2♂ above 3–5♀	♀ spikes **all drooping**	259
			1–3♂ above 1–3♀	PLANT **yellowish-green**	260
		L < 2×W	1♂ above 1–3♀		261
		L >2×W	1♂ above 1–4♀	LVS green	262
				LVS **at least** UND **grey-green**	263
		± cylindrical; widely spaced down stem	usually 1♂ above 1–4♀	♀s **lowest stalked; all others unstalked**	264
			> 1♂ above up to 5♀	♀s **all short-stalked**	264–5
below	*opposite*				

Carex sedges | **1 spike; terminal**

SS 4 other species in this group; told by spike form, glume and fruit details.

Ⓟ **Dioecious Sedge** *Carex dioica*

H to 30 cm. **Form** erect, creeping; ♂ + ♀ **on separate plants. Spike** L 5–20 mm; ♂ long + slender; ♀ shorter + wider; ♀ GLUME **broadly oval, tip blunt, purplish-brown, midrib green; edges pale. Lvs** very narrow, W <1 mm. **Utr** L 2·3–3·5 mm with beak (L 0·5–1·3 mm). **Hab** bogs, fens.

J F M A M J J A S O N D

STEM rounded

Ⓟ **Flea Sedge** *Carex pulicaris*

H to 30 cm. **Form** tufted, often densely. **Spike** ♂ above ♀s; ♂ GLUMES **narrowly oval, purple-brown;** ♀ GLUME broader, red-brown, edges pale. **Lvs** very narrow, W <1 mm. **Utr** L 3·5–6·0 mm with beak (L 0·2–0·5 mm). **Hab** fens, base-rich flushes, damp grassland, mountain rock ledges.

J F M A M J J A S O N D

STEM rounded

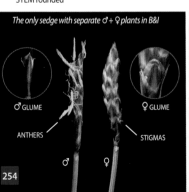

The only sedge with separate ♂ + ♀ plants in B&I

♂ GLUME

♀ GLUME

ANTHERS

STIGMAS

♂

♀

in flower

in fruit

♂

♂ GLUME

♀

♀ GLUME

♀

When ripe, ♀ glumes fall and fruits point downwards and seem to 'jump off' when touched – hence the English name.

Carex sedges | more than 1 spike; ♂ and ♀ look similar   1/2

SS 11 other species in this group; told by spike form, glume and fruit details.

Ⓟ Remote Sedge
Carex remota

H to 60 cm.
Form tufted. **Infl** L to 200 mm; 3–10 well-spaced spikes – **each with long bract**; upper fls ♀; ♀ GLUME oval, pale, midrib green; BRACTS long, leaf-like, some exceeding the whole infl. **Lvs** W 1·5–2·0 mm. **Utr** green; L 2·5–3·5 mm with beak (L 0·5–0·8 mm). **Hab** damp woodland.

STEM 3-angled

STIGMAS
ANTHERS
LOWEST SPIKE
♀ fls **above** ♂

Ⓟ Grey Sedge
Carex divulsa

H to 80 cm.
Form densely tufted. **Infl** L 50–150 mm; **5–8 spikes; mostly well-separated**; UPPER SPIKES all ♀; LOWER SPIKES ♂ above ♀; ♀ GLUME sharply pointed, pale brown, midrib green, edges pale; BRACT L ± spike. **Lvs** W 2–3 mm; dark green; L ± stem L. **Utr** yellow-brown turning dull black; L 3–4 mm; narrowed into indistinct beak. **Hab** rough grassland, hedgebanks.
SS ssp. *leersii*; scarce (calcareous soils) [LVS yellower; INFL more contracted].

STEM bluntly 3-angled

Ⓟ White Sedge
Carex canescens

H to 80 cm.
Form loosely tufted. **Infl** L 30–50 mm; 4–8 congested or separated **egg-shaped spikes**; upper fls ♀; ♀ GLUME whitish, midrib green; BRACTS short, brownish, not leaf-like. **Lvs** W 2–3 mm. **Utr** distinctly pale greenish; L 2–3 mm with beak (L 0·5–0·7 mm). **Hab** bogs, wet heaths and mountains.

STEM 3-angled

STIGMAS
ANTHERS
LOWEST SPIKE
♀ fls **above** ♂

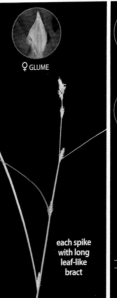

♀ GLUME

each spike with long leaf-like bract

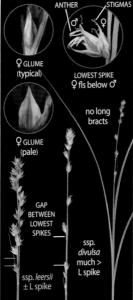

ANTHER   STIGMAS
♂   ♀

♀ GLUME (typical)

LOWEST SPIKE
♀ fls below ♂

♀ GLUME (pale)

no long bracts

GAP BETWEEN LOWEST SPIKES

ssp. *divulsa* much > L spike

ssp. *leersii* ± L spike

UTRICLE   ♀ GLUME

in fruit

no long bracts

***Carex* sedges** | >1 spike; ♂ and ♀ look similar

ⓟ Greater Tussock-sedge
Carex paniculata

H to 150 cm.
Form densely
tufted; **forms
large tussocks.
Infl** L 100–
150 mm;
numerous
spikes; upper
fls ♂; ♀ GLUME
narrowly triangular, orange-
brown, tip sharp, margin
broad + translucent. **Lvs**
W 5–7 mm; rough; L ± stem L.
Utr dark brown; rounded on
both faces; ribbed; strongly
winged in upper half; L
3–4 mm with notched beak
(L 1·0–1·5 mm). **Hab** wet
woodland, marshes, fens.

▼ STEM sharply 3-angled

ⓟ False Fox-sedge
Carex otrubae

H to 100 cm.
Form tufted.
Infl L 30–
70 mm; usually
greenish-
brown; **very
congested;**
hard to discern
individual
spikes; GLUME with long point,
pale red to orange-brown,
midrib green; LOWEST BRACT
> **L** spike. **Lvs** W 4–10 mm;
LIGULE long, pointed, L > W.
Utr green turning brown;
shiny; L 4·5–6·0 mm with
gradually tapering beak (L
1·0–1·5 mm). **Hab** pond and
ditch edges, damp grassland.

▼ STEM sharply 3-angled

ⓟ Spiked Sedge
Carex spicata

H to 80 cm.
Form densely
tufted. **Infl** L
20–30 mm;
3–10 spikes; at
least the **lower
ones separated
from the rest;**
GLUME red-
brown, midrib green. **Lvs**
W 2–4 mm; LIGULE long,
pointed. **Utr** red-brown; one
side flat; the other rounded;
thickened and corky at the
base; L 4–5 mm with beak
(L 1–2 mm). **Hab** rough
grassland. **SS** Prickly Sedge *C.
muricata* (N/I) [LIGULE blunter;
FRUIT more globular].

▼ STEM sharply 3-angled

♀ GLUME

UTRICLE

the large tussocks
are distinctive

LIGULE

♀ GLUME

UTRICLE
rounded on
one side; flat
on the other

♀ GLUME

LIGULE

UTRICLE
rounded on
one side; flat
on the other

glumes
pointed,
giving an
overall spiky
appearance

P Star Sedge
Carex echinata

H to 40 cm. **Form** densely tufted. **Infl** L to 30 mm; 2–5 ± globular spikes; separated; upper fls ♀. ♀ GLUME broadly oval; purplish-brown; green midrib; pale edges; blunt tip. **Lvs** W 1–2·5 mm; dark green; L ± stem L. **Utr** L 2·8–4·0 mm; **spreading star-shaped in fruit**. **Hab** bogs and marshes.

P Brown Sedge
Carex disticha

H to 100 cm. **Form** patch-forming. **Infl** L 30–70 mm; 15–30 **spikes**; **congested**; upper + lower fls ♀; middle ♂; ♀ GLUME pale red-brown, midrib brown; LOWEST BRACT usually bristle-like. **Lvs** W 2–4 mm; LIGULE blunt. **Utr** red-brown; wings narrow, toothed; L 4·5–5·0 mm + beak (L 1·0–1·5 mm). **Hab** fens, damp grassland, dune-slacks. **SS** Sand Sedge *C. arenaria* (N/I) [smaller; LIGULE sharp; upper spikes all ♂].

P Oval Sedge
Carex leporina

H to 60 cm. **Form** densely tufted. **Infl** L 20–40 mm; 2–9 ± **egg-shaped spikes** – all with ♀ fls above ♂ (lowest spike can be all ♀); ♀ GLUME reddish-brown, midrib green, margin pale; LOWEST BRACT bristle-like. **Lvs** W 1–3 mm; dark green; L ± stem L. **Utr** red-brown; distinctly winged in upper half; L 3·8–5·0 mm + beak (L 1·0–1·5 mm). **Hab** acid grassland, heaths.

▽ STEM bluntly 3-angled

▽ STEM sharply 3-angled

▽ STEM sharply 3-angled

UTRICLES spread to form a star shape when ripe

in fruit

UTRICLE

♀ GLUME

♂ fls 'sandwiched' between ♀s

♂

♀

in fruit

UTRICLE

♀ GLUME

♀

♀ fls at top of spike above ♂

♂

lowest spike can be all ♀

♀ GLUME

257

***Carex* sedges | ≥1 spike; ♂ and ♀ look obviously different**

UTRICLES hairy

SS 9 other species in this group; told by spike form, glume and fruit details. **Glaucous Sedge** (*p. 263*) fruits have hair-like projections and can look hairy.

Ⓟ Hairy Sedge
Carex hirta

H to 70 cm.
Form tufted.
Infl 2–3
♂ spikes
above 2–3
♀ s; the lowest
near base of
stalk; compact;
♀ GLUME pale
green, midrib paler, projecting
as a hairy point; LOWEST
BRACT usually < L infl; sheaths
hairy. **Lvs** W 2–5 mm; **hairy,
(at least on sheath)**; L ± stem
L. Utr hairy; ribbed; green;
L 4·5–7·0 mm with notched
beak (L 1·0–1·5 mm; notch
> 0·5 mm). **Hab** grassland,
usually damp.

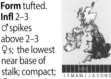

 STEM bluntly 3-angled

Ⓟ Spring-sedge
Carex caryophyllea

H to 30 cm.
Form loosely
to densely
tufted;
creeping. **Infl** 1
♂ spike above
2–3 clustered
♀ s; ♀ GLUME
broad oval
with projecting point, brown,
midrib green; LOWEST BRACT
leafy or bristle-like, sheathing
stem, < L its spike, sheath
L 3–5 mm. **Lvs** W 2–3 mm;
mostly basal; shiny green; L
± stem L. **Utr shortly hairy
all over**; L 2–3 mm with short
beak (L < 0·3 mm). **Hab** short
grassland.

  STEM rounded to
triangular

Ⓟ Pill Sedge
Carex pilulifera

H to 40 cm.
Form densely
tufted; FL-
STALK often
spreading/
sprawling.
Infl 1 ♂ spike
above 3–4 ♀ s;
congested.
♀ GLUME with pointed tip, red-
brown, midrib green, edges
pale. LOWEST BRACT leafy or
bristle-like, barely forming a
sheath. **Lvs** W 2 mm; LIGULE
blunt. **Utr hairy towards
beak**; green; L 2·0–3·5 mm
with beak (L 0·3–0·5 mm). **Hab**
heaths, moors, dry grassland.

 STEM sharply 3-angled

in fruit

in fruit

in flower ♂

♀ GLUME

LEAF +
LF-SHEATH
hairy

UTRICLE

UTRICLE ♀ GLUME

STEM
slightly
arching
at most

in fruit

♀ GLUME
MIDRIB green
typically extends
wider than in
Spring-sedge

UTRICLE

STEM **often spreading, in a long arch**

UTRICLES not hairy

SS 32 other species in this group; told by spike form, glume and fruit details. To add further confusion, some of these species hybridize, with their offspring having intermediate characters. Consequently, there will be individual plants that may not be identifiable in the field and which require detailed, even microscopic, examination and measurement.

The following 14 species are part of the largest, and most confusing, group of sedges which can need careful attention to a number of detailed features for confident identification. The following order is based on the number of male spikes; **but beware that spike count is only a guide, as it can vary within a population.**

INFLORESCENCE All ♀ spikes noticeably drooping; 1–2 ♂ spikes above 3–5 ♀ spikes

℗ Pendulous Sedge *Carex pendula*

H to 200 cm. **Form** densely tufted. **Infl** drooping; 1–2 ♂ spikes above 4–5 long (L to 160 mm) ♀s on short stalks; ♀ GLUME narrowly oval with pointed tip, red-brown, midrib pale. **Lvs** bright green above, blue-green below; W 15–20 mm; edges rough; LIGULE long (L to 80 mm), pointed. **Utr** 3-sided; L 3.0–3.5 mm with beak (L 0.3–0.6 mm). **Hab** woodland, hedgerows, stream- and pond-sides on clay.

J F M A M J J A S O N D

 STEM 3 rounded to to sharp angles

℗ Wood-sedge *Carex sylvatica*

H to 70 cm. **Form** tufted. **Infl** drooping; 1 ♂ spike above 3–5 well-separated, 'untidy' ♀s on long stalks; ♂ can be with ♀s; ♀ GLUME **pale green, midrib green**, pointed tip. **Lvs** W 3–6 mm, floppy; LIGULE **short (2–4 mm)**. **Utr** egg-shaped, yellowish-green; L 3–5 mm with long beak (L 1.0–2.5 mm). **Hab** woods, especially along rides. **SS** Thin-spiked Wood-sedge *C. strigosa* (N/I) [INFL erect; fl-stalks very short; UTR short beak].

J F M A M J J A S O N D

STEM 3 rounded angles (almost round)

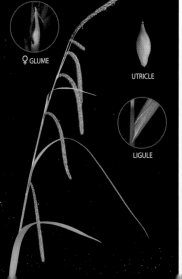

♀ GLUME

UTRICLE

LIGULE

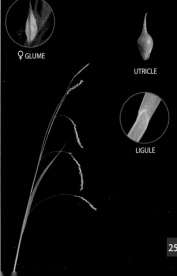

♀ GLUME

UTRICLE

LIGULE

Carex sedges | >1 spike; ♂ and ♀ look **obviously different**

UTRICLES not hairy

INFLORESCENCE ♀ spikes **egg-shaped**; **L < 2×W**; yellowish-green; 1–3 ♂ spikes above 1–3 ♀ spikes

SS Two other **yellow-sedges** (which all hybridize) have overlapping characters (N/I) [best separated by utricle details]; **Long-bracted Sedge** *C. extensa* (N/I) is superficially similar [BRACTS very long; LVS grey-green].

The yellow-green foliage of the yellow-sedges is fairly conspicuous in the field.

Ⓟ Common Yellow-sedge
Carex demissa

H to 50 cm. **Form** densely tufted; STEM usually at an angle, not erect. **Infl** ♂ spike usually on **straight stalk**; lowest ♀ usually **well-separated** from those above; STIGMAS 3. **Lvs** W 2–5 mm; L ≈ stem; LIGULE **rounded; notched**; W much > L. **Utr** yellow-green; L 3–4 mm with **straight beak** (L 0·8–1·3 mm). **Hab** fens, bogs, flushes.

 J F M A M J J A S O N D

 STEM bluntly 3-angled

Ⓟ Long-stalked Yellow-sedge
Carex lepidocarpa

H to 75 cm. **Form** loosely tufted. **Infl** ♂ spike usually on **angled** (10–20°) **stalk**; ♀s usually **slightly separated**; STIGMAS 3. **Lvs** W 2–4 mm; L ½ of stem; LIGULE **rounded; W much > L**. **Utr** yellow-green; L 3·0–5·5 mm with **sharply recurved beak** (L 1·5–2·0 mm). **Hab** fens, marshes, upland flushes.

J F M A M J J A S O N D

 STEM 3-sided

♀ GLUME L 3·5 mm; oval with slightly pointed tip; pale, edged brown with green midrib

♀ GLUME L 3·5 mm; oval with slightly pointed tip; pale, edged brown with green midrib

♂ SPIKE usually straight

UTRICLE beak **straight**

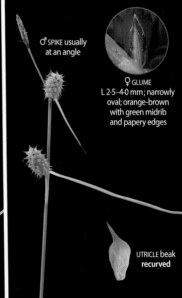

♂ SPIKE usually at an angle

♀ GLUME L 2·5–4·0 mm; narrowly oval; orange-brown with green midrib and papery edges

UTRICLE beak **recurved**

2/4

2/4

INFLORESCENCE ♀ spikes **long egg-shaped; L >2×W;**
1 ♂ spike above **1–3** ♀ spikes

P Pale Sedge
Carex pallescens

H to 60 cm. **Form** densely
tufted. **Infl** 1 ♂ spike above
2–3 ♀ (♂ can be nestled with
♀s); ♀ GLUME oval, tapered
to a sharp point, very pale
brown or translucent, broad
midrib green; LOWEST BRACT
crimped at base, L much >
infl. **Lvs** W 2–5 mm; LOWER LF-
SHEATHS hairy. **Utr** egg-shaped, shiny green;
L 2·5–3·5 mm, no beak. **Hab** damp grassland,
woodland rides.

P Tawny Sedge
Carex hostiana

H to 65 cm. **Form** loosely
tufted. **Infl** 1 ♂ spike above
1–3 well-separated cylindrical
to egg-shaped ♀s (L
10–20 mm) on short stalks;
♀ GLUME **broadly oval,**
pointed, dark brown, midrib
green, broad edges silvery;
LOWEST BRACT L < infl L.
Lvs W 2–5 mm with solid long triangular tip.
Utr yellowish with several prominent ribs;
L 2·5–3·5 mm with forked beak (L 0·8–1·2 mm;
notch > 0·5 mm). **Hab** marshes, fens, bogs,
wet grassland.

▼ STEM sharply 3-angled

▼ STEM bluntly 3-angled

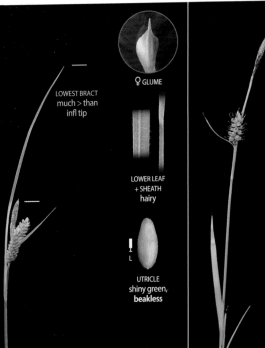

LOWEST BRACT
much > than
infl tip

♀ GLUME

LOWER LEAF
+ SHEATH
hairy

UTRICLE
shiny green,
beakless

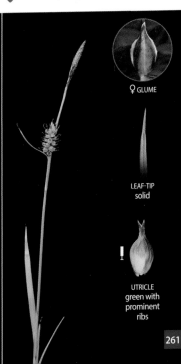

♀ GLUME

LEAF-TIP
solid

UTRICLE
green with
prominent
ribs

261

Carex sedges | >1 spike; ♂ and ♀ look obviously different

UTRICLES not hairy

INFLORESCENCE ♀ spikes ± **cylindrical; L >2×W; widely spaced down stem; 1** ♂ above **1–4** ♀

ⓟ Smooth-stalked Sedge *Carex laevigata*

H to 120 cm. **Form** densely tufted; FL-STALK often spreading/sprawling. **Infl** 1–2 ♂ spikes above 2–4 widely spaced ♀s (L >30 mm) – at least the lowest drooping on long stalk; ♀ GLUME oval to narrowly oval, **tapering to long point**; pale brown, midrib green; **edge not pale**. LOWEST BRACT L < infl L. **Lvs** W 5–12 mm; LIGULE long (L 7–15 mm), pointed. **Utr** green with **small reddish dots**; L 4–6 mm, tapering into forked beak (L 1–2 mm). **Hab** damp woods, especially on clay.

 STEM 3 rounded angles

ⓟ Green-ribbed Sedge *Carex binervis*

H to 100 cm. **Form** densely tufted; FL-STALK often spreading/sprawling. **Infl** 1 ♂ spike above 2–4 widely spaced ♀s (L15–45 mm) – lowest can be drooping on longish stalk; ♀ GLUME broadly oval with pointed tip, purplish brown, midrib pale or green, edge pale. LOWEST BRACT L < infl L. **Lvs** W 2–6 mm; tip tapered; LIGULE short (L 1–2 mm). **Utr** green, brown or purplish with **2 prominent green ribs**; L 3·5–4·5 mm with beak (L 1·0–1·5 mm). **Hab** damp heaths, moors, mountains.

 STEM 3 rounded angles (almost round)

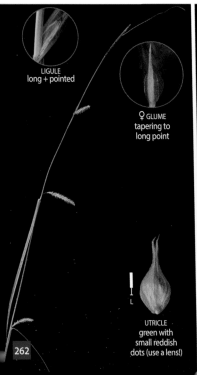

LIGULE
long + pointed

♀ GLUME
tapering to
long point

UTRICLE
green with
small reddish
dots (use a lens!)

LIGULE
short

♀ GLUME
oblong with
pointed tip

♀ SPIKE
compact and
dense

LEAF-TIP
tapered

UTRICLE
green with
2 prominent
green ribs

3/4

3/4

INFLORESCENCE ♀ spikes ± cylindrical; **L >2× W**; widely spaced down stem; **1 ♂ above 1–4 ♀**; LVS **distinctly grey-green, at least on underside**

Ⓟ Glaucous Sedge *Carex flacca*

J F M A M J J A S O N D

H to 60 cm. **Form** loosely tufted. **Infl** 1–3 ♂ spikes above 1–6 separated, ± cylindrical ♀s (L 15–50 mm) – **at least the lowest drooping on long stalk**; ♀ GLUME oval, with blunt tip, purple-black, midrib + edges pale; LOWEST BRACT **L ± infl L**. **Lvs** dull green above; grey-green below; W 2–6 mm; with channelled tip; LIGULE rounded (L 2–3 mm). **Utr** green (turning black); with tiny hair-like projections; L 2·5–3·5 mm with tiny beak (L 0·8–1·2 mm). **Hab** grassland, especially on base-rich sites.

STEM 3 rounded angles

Ⓟ Carnation Sedge *Carex panicea*

H to 60 cm. **Form** loosely tufted. **Infl** 1 ♂ spike above 1–3 well-separated, cylindrical, **few-flowered ♀s on short stalks**; ♀ GLUME broadly oval with pointed tip; purplish-brown; midrib pale; edges chaffy; LOWEST BRACT **L < infl L**. **Lvs** grey-green above and below; W 2–4 mm; with solid, 3-angled tip; LIGULE short (L 2 mm). **Utr** olive-green to brownish; L 3–4 mm with beak (L 0·5 mm). **Hab** damp neutral to acid grassland, heaths, bogs.

J F M A M J J A S O N D

STEM 3 rounded angles

LIGULE short

LOWEST BRACT > infl tip

♀ GLUME oval with blunt tip

♀ SPIKE more (and smaller) utricles than Carnation Sedge

LEAF-TIP channelled

UTRICLE tiny beak; minute projections (use a lens!)

upper | LEAF | under

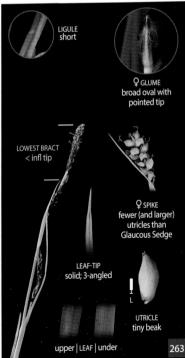

LIGULE short

LOWEST BRACT < infl tip

♀ GLUME broad oval with pointed tip

♀ SPIKE fewer (and larger) utricles than Glaucous Sedge

LEAF-TIP solid; 3-angled

UTRICLE tiny beak

upper | LEAF | under

Carex sedges | >1 spike; ♂ and ♀ look **obviously different**

UTRICLES not hairy

INFL. normally 1 unstalked ♂ spike above 1–4 ♀; all except lowest unstalked

INFLORESCENCE > 1 ♂ spikes above **up to 5 ♀;**

SS Slender Tufted-sedge *C. acuta* (N/I) [LOWEST BRACT tip > infl top] and **Tufted-sedge** *C. elata* (N/I) [LOWEST BRACT tip much lower (±½ of infl); in dense tussocks].

SS Bladder Sedge *C. vesicaria* (N/I) [UTR tapers into beak]; **Cyperus Sedge** *C. pseudocyperus* (N/I) [BRACTS longer; ♀ spikes more pendulous].

Ⓟ Common Sedge *Carex nigra*

H to 70 cm. **Form** tufted.
Infl 1–2 ♂ spikes above 1–4 ♀s – all unstalked except for lowermost; STIGMAS 2; ♀ GLUME **broadly oval, with blunt tip, black, midrib green, edge pale;** LOWEST BRACT L ± infl. Lvs greyish-green; W 2–4 mm; LOWER LF-SHEATHS hairy. **Utr** wider than glume; faintly ribbed; green with variable black; L 2·5–3·5 mm with **short beak** (L < 0·6 mm). **Hab** fens, marshes, bogs.

J F M A M J J A S O N D

▽ STEM bluntly 3-angled

Ⓟ Bottle Sedge *Carex rostrata*

H to 100 cm. **Form** patch-forming. **Infl** 2–4 ♂ spikes above 2–5 ♀s on short stalks; STIGMAS 3; ♀ GLUME **narrowly diamond-shaped, brown, midrib pale;** LOWEST BRACT L usually > infl. **Lvs** blue-green above, shiny dark green below; W 2–7 mm; margins can be inrolled; LIGULE L 2–3 mm, rounded. **Utr** yellow-green; wider than glume; L 3·5–6·5 mm, abruptly contracted into forked beak (L 1·0–1·5 mm). **Hab** ponds, lakes, reedbeds.

J F M A M J J A S O N D

▽ STEM 3 rounded angles

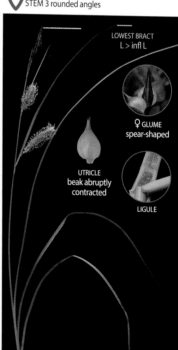

LOWEST BRACT L ± infl L

♀ GLUME + FRUIT (wider than glumes)

LIGULE

UTRICLE beak tiny

Two other sedges in B&I (**Carnation** and **Glaucous Sedge** (both *p.263*) have utricles with minute beaks and appear 'beakless'.

LOWEST BRACT L > infl L

♀ GLUME spear-shaped

UTRICLE beak abruptly contracted

LIGULE

4/4

4/4

♀ spikes all short-stalked; some at least clustered around base of lower ♂ spikes

The loosely tufted, patch-forming **pond-sedges** of ponds, ditches and marshes have the lowest bract exceeding the inflorescence. In both species, the upper ♀ spikes may also be ♂ at their top.

SS Pond-sedges often grow together, along with **Slender Tufted-sedge** *C. acuta* (N/I) [LVS much narrower; STIGMAS 2]. All these species hybridize giving rise to intermediate plants.

<table>
<tr><td>

℗ Lesser Pond-sedge *Carex acutiformis*

H to 150 cm. **Infl** 2–3 ♂ spikes above 3–4 short-stalked ♀s; STIGMAS 3 (rarely 2); ♂ GLUME tip **rounded**; ♀ GLUME purple-brown, sharp-pointed. **Lvs** W 7–20 mm; LIGULE L 5–15 mm, **pointed**. **Utr** L 3·5–5·0 mm with shallowly notched beak (L ±0·5 mm).

♀ GLUMES
Greater broader;
Lesser narrower

▼ STEM sharply 3-angled

</td><td>

℗ Greater Pond-sedge *Carex riparia*

H to 150 cm. **Infl** 3–6 ♀ spikes above 1–5 short-stalked ♀s (L 30–100 × W 10–12 mm) – lowest can be long-stalked and drooping; STIGMAS 3; ♂ GLUME tip **pointed**; ♀ GLUME purple-brown, point usually long. **Lvs** W 6–15 mm; LIGULE L 5–10 mm, **rounded**. **Utr** L 5–8 mm with deeply notched beak (L 1–2 mm).

▼ STEM sharply 3-angled

</td></tr>
</table>

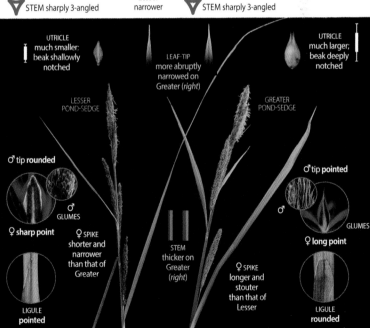

UTRICLE
much smaller: beak shallowly notched

LEAF-TIP more abruptly narrowed on Greater (*right*)

UTRICLE
much larger; beak deeply notched

LESSER POND-SEDGE

GREATER POND-SEDGE

♂ tip **rounded**

♂ GLUMES

♀ **sharp point**

♀ SPIKE shorter and narrower than that of Greater

STEM thicker on Greater (*right*)

LIGULE **pointed**

♂ tip **pointed**

♂ GLUMES

♀ **long point**

♀ SPIKE longer and stouter than that of Lesser

LIGULE **rounded**

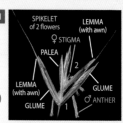

93 Poaceae | Grass family · 67 spp. | 183 spp. B&I

With similarly highly simplified flowers (and an associated botanical jargon all of their own - see *below*), grasses can be distinguished from sedges by: **Form** stem rounded or flattened; hollow between the nodes ('knees') in many. **Fls** usually bisexual. **Lvs** comprising an enfolding sheath and a free blade, with a distinctive, usually papery, ligule at the junction and, in some species, claw-like appendages (auricles) that clasp the stem at the top of the sheath.

SPIKELET of 2 flowers — LEMMA (with awn) — ♀ STIGMA — PALEA — 2 — LEMMA (with awn) — GLUME — ♂ ANTHER — GLUME — 1

How to identify grasses

Grasses, whether encountered as individual plants or as a diverse range of species in a meadow, can appear daunting to identify. The following pathway provides a sound platform from where to start. There is no getting away from the fact that to arrive at a confident identification for some species one needs to have a good grasp of some small details and also be prepared to examine and, in a few cases, measure them. In these situations a hand lens, or the ability to take a macro photograph, is almost essential.

Grass identification pathway

1 Using the galleries opposite; identify the **shape and structure of the inflorescence**

2 Go to the page number indicated and identify **the type of spikelet and/or flower**

3 Examine any indicated detailed primary features (*e.g.* **glume, ligule, awn, leaf**)

4 Check any supporting subjective characteristics (*e.g.* **annual or perennial, habitat**)

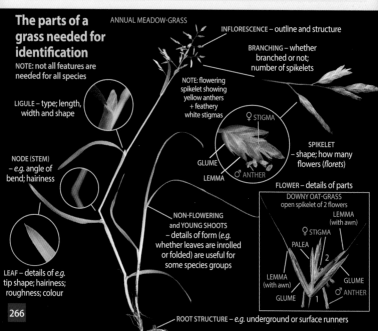

The parts of a grass needed for identification

NOTE: not all features are needed for all species

ANNUAL MEADOW-GRASS

INFLORESCENCE – outline and structure

BRANCHING – whether branched or not; number of spikelets

NOTE: flowering spikelet showing yellow anthers + feathery white stigmas

LIGULE – type; length, width and shape

♀ STIGMA

SPIKELET – shape; how many flowers (*florets*)

NODE (STEM) – *e.g.* angle of bend; hairiness

GLUME

LEMMA

♂ ANTHER

FLOWER – details of parts

NON-FLOWERING and YOUNG SHOOTS – details of form (*e.g.* whether leaves are inrolled or folded) are useful for some species groups

DOWNY OAT-GRASS open spikelet of 2 flowers

LEMMA (with awn)

♀ STIGMA

PALEA

2

LEMMA (with awn)

GLUME

♂ ANTHER

GLUME

1

LEAF – details of *e.g.* tip shape; hairiness; roughness; colour

ROOT STRUCTURE – *e.g.* underground or surface runners

IDENTIFY BY: ▶ inflorescence outline ▶ presence/absence of flower-bearing branches ▶ length + arrangement of these branches ▶ number + arrangement of spikelets/flowers

BEWARE: The form of a grass inflorescence is rather subjective and therefore hard to pigeonhole. The inflorescence shapes shown in the grasses introductory keys are best regarded as examples along a continuum. In addition, some grasses change with age (*e.g.* from contracted or upright when young to open or drooping when mature). Such changes are allowed for in the following guide.

1 ▶*p.268*

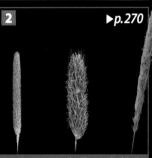

INFLORESCENCE **an unbranched spike with stalkless or short-stalked spikelets**

2 ▶*p.270*

INFLORESCENCE **spike-like; ± cylindrical to narrowly egg-shaped with short branches ± concealed by dense crowded clusters of ± unstalked spikelets/flowers**

3 ▶*p.272*

INFLORESCENCE **somewhat spike-like; narrow to egg-shaped; spikelet-stalks short or inconspicuous**

BEWARE: includes some open, branched species with contracted inflorescences due to *e.g.* age

4 ▶*p.274*

INFLORESCENCE **open; very obvious branches and/or spikelet-stalks**

from
1
p. 267

INFLORESCENCE **an unbranched spike with stalkless or short-stalked spikelets**

SPKLTS **1-sided arrangement**	SPKLTS **alternating; in two opposite rows**			
in flower	SPIKELETS **narrow side faces stem**	SPIKELETS **broad side faces stem**		SPIKELETS **in distinctive group of 3**
highly distinctive post-flowering		GLUMES **narrow**	GLUMES **broad**	2 3 1
Mat-grass *below*	**RYE-GRASSES** *below*	**COUCHES** *opposite*	**Bread Wheat** *p. 270*	**BARLEYS** *p. 270*

SPKLTS **1-sided arrangement**

Ⓟ Mat-grass *Nardus stricta*

H to 50 cm.
Form densely tufted/mat-forming with persistent, fibrous sheath-remains. **Infl** L to 9·5 cm.
Spklt L 2·0–3·5 mm; FLS 1; GLUMES **shorter than spklt** (upper often absent); LEMMAS to 9 mm + awn to 3 mm. **Lvs** bristle-like. **Hab** acid heaths and moors. **SS** Sheep's-fescue (see p. 285).

SPKLTS **in two opposite rows; narrow side faces stem**

**Ⓟ Rye-grasses |
Form** tufted.
Infl stiff unbranched spike; SPKLTS flattened with their **narrower side facing the stem.**
Lvs dull above; shiny below.

in flower

PERENNIAL   ITALIAN

J F M A M J J A S O N D   J F M A M J J A S O N D

Ⓟ Perennial Rye-grass
Lolium perenne

H to 90 cm. **Infl** L to 30 cm.
Spklt L to 20 mm, up to 14 FLS; LEMMAS typically awnless.
Lvs W to 6 mm; AURICLES **small, hooked. Hab** grassland (especially if 'improved' or damaged).

Ⓐ Italian Rye-grass
Lolium multiflorum

H to 100 cm. **Infl** L to 35 cm.
Spklt L to 25 mm, up to 25 FLS; LEMMAS typically with awn (L to 15 mm). **Lvs** W to 11 mm; AURICLES **long, overlapping. Hab** 'improved' and re-sown grassland, field margins, brownfield.

SPIKELET **single-flowered**

'1-sidedness' obvious on old stem

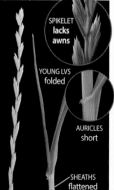

SPIKELET **lacks awns**

YOUNG LVS **folded**

AURICLES **short**

SHEATHS **flattened**

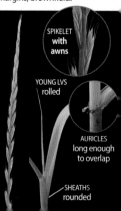

SPIKELET **with awns**

YOUNG LVS **rolled**

AURICLES **long enough to overlap**

SHEATHS **rounded**

SPKLTS **in two opposite rows; broad side faces stem**

● **Couches and Lyme-grass** |
Form tufted; some patch-forming.
Infl stiff, unbranched spike; SPKLTS
stalkless, flattened, alternating
along the stem with their **broader
side facing the stem** (couches
spklts single; **Lyme-grass** spklts in
pairs); LIGULE short (L to 1 mm).

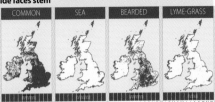

COMMON | SEA | BEARDED | LYME-GRASS

J F M A M J J A S O N D   J F M A M J J A S O N D   J F M A M J J A S O N D   J F M A M J J A S O N D

SS Common Couch and Sea Couch hybridize freely; producing sterile, intermediate plants.

ⓟ Common Couch
Elymus repens

H to 120 cm. **Form**
patch-forming; long
underground runners.
Infl L to 15 cm. **Spklt** L
to 20 mm, up to 8 FLS;
LEMMAS L to 13 mm
with stiff awn (L to
7 mm) – may be absent.
Lvs W to 13 mm; dull
green; TASTE faint
herby; AURICLES short,
curved. **Hab** arable
fields, gardens, rough
grassland.

SPIKELET
almost
parallel-sided

LVS **dull green**, flat;
STEM hairless

AURICLES
short

ⓟ Sea Couch
Elymus athericus

H to 120 cm. **Form**
patch-forming; long
underground runners.
Infl L to 20 cm. **Spklt** L
to 20 mm; up to 12 FLS;
LEMMAS L to 13 mm
with stiff awn (L to
6 mm) – may be absent.
Lvs W to 6 mm; bluish-
green; AROMA/TASTE
faint, sweet, herby;
AURICLES short, curved.
Hab maritime grassland,
upper saltmarshes, sea
walls.

SPIKELET
narrowly
diamond-
shaped

LVS **bluish-green**, often
inrolled; STEM hairless

AURICLES
short

ⓟ Bearded Couch
Elymus caninus

H to 120 cm. **Form**
loosely tufted; runners
absent. **Infl** L to 20 cm.
Spklt L to 20 mm, up
to 6 FLS; LEMMAS L to
13 mm; typically with
long, wavy awn (L
to 20 mm). **Lvs** W to
13 mm; pale to bright
green. **Hab** woods, river
banks, road verges.

SPIKELET
with wavy
awns

STEM **hairy, especially
below the nodes**

ⓟ Lyme-grass
Leymus arenarius

H to 200 cm. **Form**
densely tufted; long
underground runners.
Infl L to 35 cm. **Spklt**
L to 30 mm; **in pairs**;
FLS up to 6; LEMMAS
L to 25 mm; typically
awnless. **Lvs** W to
20 mm; grey-green;
rigid; rolled when dry.
Hab sand dunes. **SS** in
same habitat as **Marram**
(*p. 271*).

SPIKELETS
in pairs

GLUMES **as long
as spikelets**

INFLORESCENCE **an unbranched spike with stalkless or short-stalked spikelets** 2/2

SPKLTS **in two opposite rows along the stem; not flattened**

SPKLTS **stalkless; 1 at each node** SPKLTS **short-stalked; in groups of 3 at each node**

Ⓐ Bread Wheat
Triticum aestivum

H to 150 cm.
Form erect.
Infl L to 18 cm.
Spklt with up
to 9 FLS. **Lvs**
W to 15 mm;
can be hairy;
LIGULE L to
3 mm.
Hab cultivated; farmyards,
roadsides, disturbed areas.

Ⓐ Two-rowed Barley
Hordeum distichon

H to 90 cm.
Form erect.
Infl L to
12 cm (excl.
awns). **Spklt**
in groups
of 3; each
with 1 fl; only
middle spklt
fertile. **Lvs** W 10–15 mm;
hairless; LIGULE L to 1 mm.
Hab cultivated; roadsides,
brownfield and other
disturbed areas.

Ⓐ Wall Barley
Hordeum murinum

H to 60 cm.
Form erect or
spreading.
Infl L to 12 cm
(excl. awns).
Spklt all
fertile. **Lvs** W
to 8 mm, hairy,
lower sheaths
inflated; LIGULE L to 1 mm.
Hab walls, brownfield,
disturbed areas. **SS Meadow
Barley** *H. secalinum* (N/I)
[GLUMES margins not hairy].

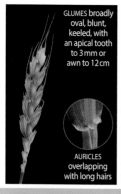

GLUMES broadly
oval, blunt,
keeled, with
an apical tooth
to 3 mm or
awn to 12 cm

AURICLES
overlapping
with long hairs

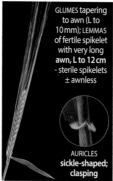

GLUMES tapering
to awn (L to
10 mm); LEMMAS
of fertile spikelet
with very long
awn, L to 12 cm
- sterile spikelets
± awnless

AURICLES
**sickle-shaped;
clasping**

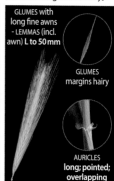

GLUMES with
long fine awns
– LEMMAS (incl.
awn) **L to 50 mm**

GLUMES
margins hairy

AURICLES
**long; pointed;
overlapping**

---

from 2 *p. 267* INFLORESCENCE **spike-like; ± cylindrical to narrowly egg-shaped with short branches ± concealed by dense crowded clusters of ± unstalked spikelets/flowers**

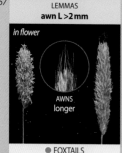

LEMMAS
awn L >2 mm

in flower

AWNS
longer

● FOXTAILS
opposite

LEMMAS
awn L <2 mm

in flower

AWNS
shorter

● CAT'S-TAILS
p. 272

LEMMAS **with tiny awn at most;**
HABITAT **sand dunes**

CENTRAL SPKLT
GLUMES
margins hairy

● **Marram**
opposite

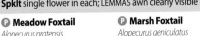

● **Foxtails** | **Infl** densely cylindrical with pointed or tapering tip (*right*). **Spklt** single flower in each; LEMMAS awn clearly visible beyond spikelet.

℗ Meadow Foxtail
Alopecurus pratensis

H to 120 cm. **Form** erect, tufted. **Infl** L×W to 12 cm ×12 mm, TIP pointed. **Spklt** L to 6 mm; GLUMES **conspicuously hairy**, fused above base; LEMMAS with angled awn, L > spklt by 3–5 mm. **Lvs** LIGULE L to 2·5 mm, blunt. **Hab** meadows, rough grassland.

℗ Marsh Foxtail
Alopecurus geniculatus

H to 50 cm. **Form** spreading; STEM **strongly bent at lower node, often rooting**. **Infl** L×W to 70 ×7 mm, TIP pointed. **Spklt** L to 4 mm; GLUMES fused at base only, hairs appressed; LEMMAS with slightly angled awn, L > spklt by 1·5–3·0 mm. **Lvs** LIGULE L to 5 mm, rounded. **Hab** wet meadows, water margins.

Ⓐ Black-grass
Alopecurus myosuroides

H to 90 cm. **Form** erect, tufted. **Infl** L×W to 12 cm ×6 mm, TIP tapering. **Spklt** L to 7 mm; GLUMES enclose spklt; LEMMAS with angled awn, **L > spklt by 4–8 mm**. **Lvs** LIGULE L to 5 mm, blunt. **Hab** arable fields, brownfield and other disturbed areas.

ANTHERS can be purplish

AWN 3–5 mm showing

in flower

GLUMES lower ¼–½ fused

in flower

SHEATHS ± cylindrical

LOWER NODE ± straight

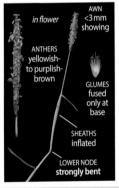

in flower

AWN <3 mm showing

ANTHERS yellowish- to purplish-brown

GLUMES fused only at base

SHEATHS inflated

LOWER NODE strongly bent

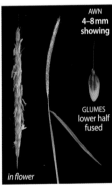

AWN 4–8 mm showing

GLUMES lower half fused

in flower

LEMMAS **with tiny awn at most;** HABITAT **sand dunes**

℗ Marram | *Ammophila arenaria*

H to 120 cm. **Infl** L to 22 cm; straw-coloured; ± cylindrical tapering to a point. **Spklt** L to 14 mm; almost enclosed by glumes; 1 FL. **Form** tussock-forming. **Lvs** L to 100 cm; **inrolled**; W to 6 mm when opened out; stiff, sharp-pointed; upper surface deeply ridged; greyish-green; LIGULE L to 30 mm, long and pointed. **Hab** sand dunes.

LIGULE very long and pointed

INFLORESCENCE **spike-like; ± cylindrical to narrowly egg-shaped**

2/2

LEMMAS **with awn L <2mm**

● **Cat's-tails** | **Form** erect, tufted. **Infl** tightly **cylindrical** with blunt tip. **Spklt** FL 1; GLUMES truncate; with small awns; LEMMAS awnless, blunt.

SS Timothy and Smaller Cat's-tail are very difficult to separate confidently.

❶ Timothy
Phleum pratense

H 70–150cm at flowering. **Infl** L to 150mm; W to 10mm. **Spklt** L 4·0–5·5mm. **Lvs** W to 10mm, rough at tip; LIGULE L to 6mm, blunt. **Hab** grassland, brownfield.

SPIKELET: GLUMES with awn L 1–2mm

LIGULE **rounded**

J F M A M J J A S O N D

❶ Smaller Cat's-tail
Phleum bertolonii

H to 70cm at flowering. **Infl** L to 80mm; W to 7·5mm. **Spklt** L < 4mm. **Lvs** W to 6mm, rough at tip; LIGULE L to 4mm, pointed. **Hab** grassland.

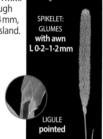

SPIKELET: GLUMES with awn L 0·2–1·2mm

LIGULE **pointed**

J F M A M J J A S O N D

from **3**

p.267

INFLORESCENCE **somewhat spike-like; narrow to egg-shaped; spikelet-stalks short or inconspicuous**

BEWARE: This section is without doubt the most subjective and therefore open to alternative interpretations. **If your grass cannot be found in this section it is most likely to be a plant with a contracted inflorescence.** Look at the spikelet shape of your specimen and check Gallery **4** (*p.274*). In an attempt to minimise this some species and types feature here as well as in Gallery **4**.

INFL **narrowly egg-shaped;** SPIKELETS **short-stalked**

not to scale

a

SPIKELETS 2 FLS

LEMMAS bent awns

see also **4v** *p.275*

● Early Hair-grass *p.280*

b

SPIKELETS 4–6 FLS

LIGULE **fringe of hairs**

● Heath-grass *p.277*

c

SPIKELETS 3 FLS

● Sweet Vernal-grass *p.277*

d

SPIKELETS 2–3 FLS

GLUMES silvery edges

● Crested Hair-grass *p.277*

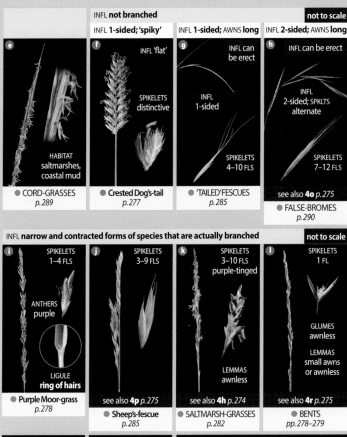

INFL **not branched**

not to scale

INFL 1-sided; 'spiky'   INFL 1-sided; AWNS **long**   INFL 2-sided; AWNS **long**

e

HABITAT
saltmarshes,
coastal mud

● CORD-GRASSES
p. 289

f

INFL 'flat'

SPIKELETS
distinctive

● Crested Dog's-tail
p. 277

g

INFL can
be erect

INFL
1-sided

SPIKELETS
4–10 FLS

● 'TAILED' FESCUES
p. 285

h

INFL can be erect

INFL
2-sided; SPKLTS
alternate

SPIKELETS
7–12 FLS

see also **4o** *p. 275*

● FALSE-BROMES
p. 290

INFL **narrow and contracted forms of species that are actually branched**   **not to scale**

i

SPIKELETS
1–4 FLS

ANTHERS
purple

LIGULE
ring of hairs

● Purple Moor-grass
p. 278

j

SPIKELETS
3–9 FLS

see also **4p** *p. 275*

● Sheep's-fescue
p. 285

k

SPIKELETS
3–10 FLS
purple-tinged

LEMMAS
awnless

see also **4h** *p. 274*

● SALTMARSH-GRASSES
p. 282

l

SPIKELETS
1 FL

GLUMES
awnless

LEMMAS
small awns
or awnless

see also **4r** *p. 275*

● BENTS
pp. 278–279

m

SPIKELETS
2 FLS

LEMMA OF
UPPER FL
long awn

in flower

● False Oat-grass
p. 288

n

SPIKELETS up
to 20 FLS;
± cylindrical

see also **4j** *p. 274*

● SWEET-GRASSES
pp. 286–287

The importance of spikelets

Although many groups of grasses are structurally distinctive in flower, others are less so but an examination of the spikelet form should enable an identification to group.

'BROME'   FESCUE   MEADOW-GRASS

BENT 'awnless'   SOFT-GRASS   BENT 'awned'   HAIR-GRASS

273

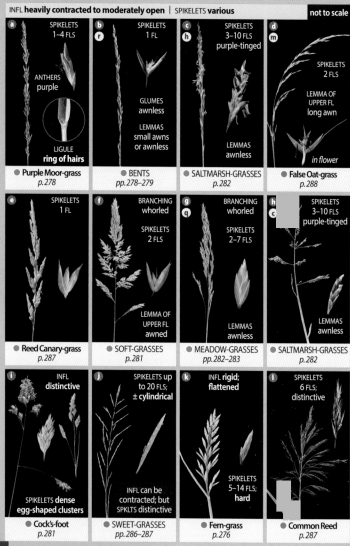

from **4** p. 267

INFLORESCENCE **open; very obvious branches and/or spikelet-stalks**

IDENTIFY TO GENUS/SPECIES BY: ▶ inflorescence 'openness' ▶ spikelet + flower details

NOTE: outlines range from heavily contracted and dense, to broad and open, depending on species and/or plant maturity – some commonly encountered examples are shown here, with some duplicated (denoted by paler letter codes) to show variation. **Spikelet details will confirm the group.**

INFL heavily contracted to moderately open | SPIKELETS **various** | **not to scale**

a SPIKELETS 1–4 FLS
ANTHERS purple
LIGULE **ring of hairs**
● Purple Moor-grass *p.278*

b **r** SPIKELETS 1 FL
GLUMES awnless
LEMMAS small awns or awnless
● BENTS *pp.278–279*

c **h** SPIKELETS 3–10 FLS purple-tinged
LEMMAS awnless
● SALTMARSH-GRASSES *p.282*

d **m** SPIKELETS 2 FLS
LEMMA OF UPPER FL long awn
in flower
● False Oat-grass *p.288*

e SPIKELETS 1 FL
● Reed Canary-grass *p.287*

f BRANCHING whorled
SPIKELETS 2 FLS
LEMMA OF UPPER FL awned
● SOFT-GRASSES *p.281*

g **q** BRANCHING whorled
SPIKELETS 2–7 FLS
LEMMAS awnless
● MEADOW-GRASSES *pp.282–283*

h **c** SPIKELETS 3–10 FLS purple-tinged
LEMMAS awnless
● SALTMARSH-GRASSES *p.282*

i INFL **distinctive**
SPIKELETS **dense egg-shaped clusters**
● Cock's-foot *p.281*

j SPIKELETS up to 20 FLS; ± cylindrical
INFL can be contracted; but SPKLTS distinctive
● SWEET-GRASSES *pp.286–287*

k INFL **rigid; flattened**
SPIKELETS 5–14 FLS; **hard**
● Fern-grass *p.276*

l SPIKELETS 6 FLS; distinctive
● Common Reed *p.287*

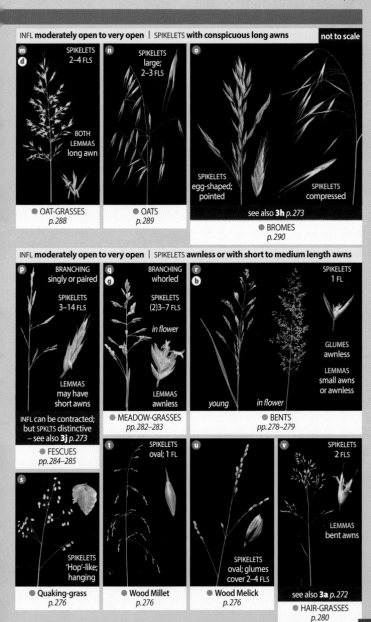

INFL **moderately open to very open** | SPIKELETS **with conspicuous long awns** `not to scale`

m
d
SPIKELETS
2–4 FLS

BOTH
LEMMAS
long awn

● OAT-GRASSES
p.288

n
SPIKELETS
large;
2–3 FLS

● OATS
p.289

o
SPIKELETS
egg-shaped;
pointed

SPIKELETS
compressed

see also **3h** p.273

● BROMES
p.290

INFL **moderately open to very open** | SPIKELETS **awnless or with short to medium length awns**

p
BRANCHING
singly or paired

SPIKELETS
3–14 FLS

LEMMAS
may have
short awns

INFL can be contracted;
but SPKLTS distinctive
– see also **3j** p.273

● FESCUES
pp.284–285

q
g
BRANCHING
whorled

SPIKELETS
(2)3–7 FLS

in flower

LEMMAS
awnless

● MEADOW-GRASSES
pp.282–283

r
b
SPIKELETS
1 FL

GLUMES
awnless

LEMMAS
small awns
or awnless

young   in flower

● BENTS
pp.278–279

s

SPIKELETS
'Hop'-like;
hanging

● Quaking-grass
p.276

t
SPIKELETS
oval; 1 FL

● Wood Millet
p.276

u
SPIKELETS
oval; glumes
cover 2–4 FLS

● Wood Melick
p.276

v
SPIKELETS
2 FLS

LEMMAS
bent awns

see also **3a** p.272

● HAIR-GRASSES
p.280

275

INFLORESCENCE **distinctive**

FORM **hairless annual;** INFL **twice-divided (fern-like); rigid; flattened**

A **Fern-grass** *Catapodium rigidum*

H to 30 cm. **Form** stiff; tufted or solitary; winter annual. **Infl** L to 80 mm; scarcely branched. **Spklt** L to 8 mm; short-stalked; 5–9 FLS; GLUMES L to 2 mm, LEMMAS L to 3 mm; unawned. **Lvs** incurved when young; LIGULE L to 3 mm, blunt. **Hab** walls, bare habitats on thin soils, especially coastal. **SS** Sea Fern-grass *C. marinum* (INSET) [SPKLTS mostly stalkless; MAIN INFL-AXIS stout; not angled].

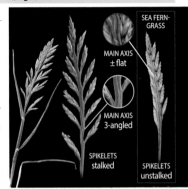

SEA FERN-GRASS

MAIN AXIS ± flat

MAIN AXIS 3-angled

SPIKELETS stalked

SPIKELETS unstalked

SPIKELETS **'hop'-like**

P **Quaking-grass** *Briza media*

H to 80 cm. **Form** erect. **Infl** L to 18 cm; conical; very open; **Spklt** L 4–7 mm; nodding; 4–12 FLS. **Lvs** W to 4 mm; LIGULE L to 2 mm, blunt. **Hab** grassland, especially calcareous. **SS** Lesser Quaking-grass *B. minor* (N/I) [SPKLT L ≥ 5 mm], Greater Quaking-grass *B. maxima* (N/I) [SPKLT L 10–25 mm].

SPIKELETS **egg-shaped**

P **Wood Melick** *Melica uniflora*

H to 60 cm. **Form** delicate, loosely patch-forming. **Infl** L to 20 cm; very open; few branches bearing a few spikelets. **Spklt** egg-shaped; L to 7·5 mm, usually 1 fertile FL; GLUMES enclosing spikelet. **Lvs** soft; L×W to 20 cm × 7 mm; tapering to fine point; LIGULE L to 2 mm, flat-topped. **Hab** woods, shady banks.

P **Wood Millet** *Milium effusum*

H to 120 cm. **Form** elegant, loosely tufted, erect. **Infl** L to 40 cm; very open, with widely spaced whorls of branches. **Spklt** L to 4 mm; 1 FL; GLUMES enclosing spikelet. **Lvs** L×W to 30 cm × 15 mm; LIGULE L to 10 mm, blunt. **Hab** woodland, often damp. **SS** Tufted Hair-grass (p. 280); meadow-grasses (p. 282).

SPIKELETS **flattened;** L<3×W; GLUMES **bulging;** LEMMAS **overlapping**

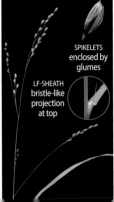

SPIKELETS enclosed by glumes

LF-SHEATH bristle-like projection at top

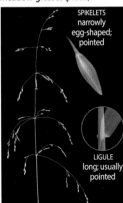

SPIKELETS narrowly egg-shaped; pointed

LIGULE long; usually pointed

INFLORESCENCE **1-sided; 'spiky'**

Ⓟ Crested Dog's-tail *Cynosurus cristatus*

H to 80 cm. **Form** compact tufts. **Infl** L to 10 cm; narrow; 1-sided; prickly looking. **Spklt** L to 6 mm; on very short stalks; 1–6 FLS; **fertile fls diamond-shaped; hidden by fan-shaped sterile ones. Lvs** L×W to 15 cm × 4 mm; basal sheaths yellowish; LIGULE L to 1 mm. **Hab** grassland.

in flower

sterile — sterile

SPIKELETS fertile diamond-shaped; sterile larger – a fan of up to 18 narrow, pointed bracts

---

INFLORESCENCE **narrowly egg-shaped;** SPIKELETS **short-stalked**

Ⓟ Heath-grass *Danthonia decumbens*

H to 60 cm. **Infl** L to 6 cm, contracted, with 3–12 spikelets; **Spklt** L to 10 mm; 4–6 FLS; GLUMES equal, blunt. **Lvs** L×W to 25 cm × 4 mm; stiff; usually folded; UPP dull; UND shiny; TIP hooded; SHEATHS flattened; LIGULE L to 0·5 mm. **Hab** heaths, moors, acid grassland.

SS Blue Moor-grass *Sesleria caerulea* (N/I) [LIGULE papery, fringed with hairs].

SPIKELET plump and shiny

LIGULE **dense fringe of hairs, with long collar-whiskers**

Ⓟ Sweet Vernal-grass *Anthoxanthum odoratum*

H to 100 cm. **Aroma** vanilla-like; taste of new-mown hay from leaves and stem when damaged. **Infl** L to 12 cm, in a loose spike. **Spklt** L to 10 mm; 3 FLS (lower 2 sterile); GLUMES very unequal, pointed, persist after seeds shed. **Lvs** L×W to 12 cm × 5 mm; LIGULE L to 5 mm, blunt, can be ragged. **Hab** pastures, heaths, open woodland.

in flower

ANTHERS large; L to 4·5 mm

SPIKELETS

LF-SHEATH **with bearded 'shoulders' at apex, beside the ligule**

Ⓟ Crested Hair-grass *Koeleria macrantha*

H to 60 cm. **Form** tufted or patch-forming, with many non-flowering shoots. **Infl** L to 10 cm; compact, lobed, or interrupted; silvery or purplish. **Spklt** L to 6 mm; 1–3 FLS; GLUMES pointed, persist after seeds shed. **Lvs** L×W to 25 cm × 4 mm; greyish-green; stiff; upperside deeply ribbed. **Hab** dry calcareous and sandy grassland.

GLUMES + LEMMAS with thin shiny margins; lemmas may have a short awn from near tip

INFLORESCENCE **variable: dense to open**

FORM **tussock-forming;** LIGULE **ring of hairs**

❶ Purple Moor-grass
Molinia caerulea

H to 120 cm. **Form** raised tussocks. **Infl** L to 40 cm; erect; variable – dense to open; usually purplish. **Spklt** L to 9 mm; 2–4 FLS; GLUMES equal, narrow; LEMMAS blunt, narrow. **Lvs** L×W to 45 cm × 7 mm; soft; grey-green. LIGULE L to 0·6 mm. **Hab** damp or wet heaths, moors.

SPIKELET

INFL variable; dense to open

LIGULE **dense fringe of short hairs**

● **Bent-grasses** | **Infl** delicate; branches in whorls. **Spklt** 1 FL; surrounded by glumes; lemmas may have awns, but variable between species and individual plants.

SS Bents have some recognition-based notoriety. However, examination of the ligules of non-flowering (vegetative) shoots alongside other features should enable identification.

IDENTIFY BY: ▶ **vegetative shoot (VS) ligule shape** ▶ **root type** ▶ **plant size + leaf width**

VEGETATIVE SHOOT (VS) LIGULES **papery; pointed**

❶ Velvet Bent *Agrostis canina*

H to 60 cm. **Form** spreading with surface runners; tufts of shoots at nodes; loosely turf-forming. **Infl** L to 18 cm; OUTLINE conical to egg-shaped. **Spklt** L to 3 mm; LEMMAS usually with bent awn (L to 3·5 mm). **Lvs** narrow, flat; W to 3 mm; VS LIGULE L to 4 mm; pointed, becoming ragged. **Hab** grassland on heavy, damp or acid soils.

❶ Brown Bent *Agrostis vinealis*

H to 60 cm. **Form** creeping; slender underground runners form a fine turf. **Infl** L to 16 cm; OUTLINE densely contracted to loosely conical. **Spklt** L to 3 mm; LEMMAS usually with bent awn (L to 4 mm). **Lvs** variable – inrolled and bristle-like (W to 0·8 mm), or flat (W to 3 mm); VS LIGULE L to 5 mm; pointed, becoming ragged. **Hab** acid grassland, especially on heaths and uplands.

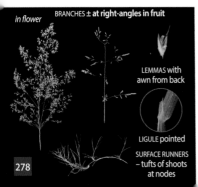

in flower BRANCHES ± **at right-angles in fruit**

LEMMAS with awn from back

LIGULE pointed

SURFACE RUNNERS – tufts of shoots at nodes

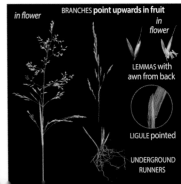

in flower BRANCHES **point upwards in fruit** *in flower*

LEMMAS with awn from back

LIGULE pointed

UNDERGROUND RUNNERS

Bents at a glance	VEGETATIVE SHOOT LIGULE	INFLORESCENCE OUTLINE			SPREAD	FORM
		PRE-FLOWER	IN FLOWER	POST-FLOWER		
Velvet Bent	flat-topped, blunt or rounded	contracted	open	open	surface	tufted; loose turf
Brown Bent		contracted	open	contracted	underground	creeping; fine turf
Creeping Bent	pointed; becoming ragged	contracted	open	contracted	surface	creeping; turf
Common Bent		contracted	open	open	surface or underground	loosely tufted; turf
Black Bent		open	open	open		larger than others

VEGETATIVE SHOOT (VS) LIGULES **papery; flat-topped, blunt or rounded** (not pointed)

🅿 Creeping Bent
Agrostis stolonifera

H to 40 cm. **Form** loosely tufted, turf-forming; creeping. **Infl** L to 30 cm; narrowly cylindrical to conical. **Spklt** L to 3 mm; LEMMAS usually awnless. **Lvs** broad, flat; W to 8 mm; **VS LIGULE** long; L to 6 mm; blunt, becoming ragged. **Hab** grassland (damp), pond margins, disturbed sites.

🅿 Common Bent
Agrostis capillaris

H to 50 cm. **Form** loosely tufted, turf-forming. **Infl** L to 20 cm; egg-shaped to conical. **Spklt** L to 3·5 mm; LEMMAS usually awnless (occasionally short awn from the back). **Lvs** W to 4 mm, flat; **VS LIGULE short; L** to 2 mm; blunt, becoming ragged. **Hab** grasslands, brownfield.

🅿 Black Bent
Agrostis gigantea

H to 150 cm. **Form** erect, loosely tufted. **Infl** L to 25 cm; egg-shaped to conical. **Spklt** L to 3 mm; LEMMAS usually awnless. **Lvs** broad, flat; W to 7 mm; **VS LIGULE** long; L to 6 mm; blunt, becoming ragged. **Hab** arable margins, roadsides, brownfield. **SS** Tufted Hair-grass (*p. 280*).

BRANCHES **with spikelets near the base**

LIGULE L to 6 mm

point upwards in fruit

INFL **narrowly cylindrical to conical**

SPIKELET awn (if present) **arises from apical ⅓ of lemma**

LVS W to 8 mm

BRANCHES ± **bare of spikelets near the base**

LIGULE L to 2 mm

± at right-angles in fruit

INFL egg-shaped to conical

in flower

SPIKELET awn (if present) **arises from basal ⅓ of lemma**

LVS W to 4 mm

± at right-angles in fruit

LIGULE L to 6 mm

SPIKELET awn (if present) **arises from apical ⅓ of lemma**

FORM **erect; not turf-forming; much larger than other bents**

279

INFLORESCENCE **open; very obvious branches and/or spikelet-stalks**

SPIKELETS **typically with short to medium-length awns**

● **Hair-grasses** | **Infl** open, well-branched (NOTE: Early Hair-grass compact). **Spklt** FLS 2; GLUMES ± enclosing spikelet; LEMMAS with ± bent awn from back.

Ⓐ Silver Hair-grass
Aira caryophyllea

H to 25 cm. **Form** delicate; tufted; **twinkles diffusely at a distance. Infl** L to 120 mm; **distinctly branched. Spklt** L 2·0–3·5 mm. **Lvs** bristle-like; W to 0·5 mm; LIGULE L to 5 mm, toothed. **Hab** dry thin, often acid, soils.

Ⓟ Wavy Hair-grass
Avenella flexuosa

H to 80 cm. **Form** tufted; can form extensive tufty swards. **Infl** L to 150 mm, well-branched. **Spklt** L to 6 mm; GLUMES shiny. **Lvs** bristle-like; L×W to 20 cm×0·8 mm; shiny; bright green; LIGULE L to 3·5 mm, bluntly pointed. **Hab** heaths, moors.

Ⓟ Tufted Hair-grass
Deschampsia cespitosa

H to 200 cm. **Form** usually large tussocks. **Infl** L to 50 cm; branches in whorls. **Spklt** L to 7 mm. **Lvs** L×W to 60 cm×5 mm; rough; LIGULE L to 15 mm, long. **Hab** damp pastures and woods, moor, grassland. **SS** Black Bent (*p.279*); **Wood Millet** (*p.276*).

BRANCHES **not crinkled**

AWN bent

SPIKELET **shorter** than its stalk

LIGULE **toothed**

BRANCHES **crinkled in many**

AWN **weakly bent**

SPIKELET

LIGULE **blunt point**

SPIKELET

LIGULE **long; narrow; pointed**

Ⓐ Early Hair-grass *Aira praecox*

H to 15 cm. **Form** delicate, tufted, **glistening at a distance. Infl** L to 50 mm, **tightly compact,** branches hard to see. **Spklt** L 2·5–3·5 mm; 2 FLS; GLUMES ± enclosing spikelet, ± equal; LEMMAS notched at tip, short tuft of hairs at base, AWN bent. **Lvs** W to 0·5 mm; SHEATH open, smooth; LIGULE L to 2 mm, bluntly pointed. **Hab** dry thin soils, often acid. **SS** reminiscent of young **Annual Meadow-grass** (*p.282*).

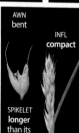

AWN bent

INFL **compact**

SPIKELET **longer** than its stalk

Early Hair-grass looks unlike the other hair-grasses

SPIKELETS **typically with short to medium-length awns**

● **Soft-grasses** | **Form** erect. **Infl** contracted then spreading in flower; branches in whorls. **Spklt** FLS 2; GLUMES oval; hairy; enclosing 2 LEMMAS – the upper of which is awned. **Lvs** flat; soft; grey-green with pinkish sheaths.

℗ **Yorkshire-fog** *Holcus lanatus*

H to 100 cm. **Form** tufted, softly hairy; runners absent. **Infl** L to 20 cm; open or compact; erect or drooping; whitish to pinkish. **Spklt** L to 6 mm; usually 2 FLS; upper ♀ only in some; GLUMES equal. **Lvs** L×W to 20 cm × 10 mm; soft. LIGULE L to 4 mm; blunt or rounded. **Hab** meadows, pastures, open woodland.

J F M A M J J A S O N D

℗ **Creeping Soft-grass** *Holcus mollis*

H to 110 cm. **Form** mat-forming, creeping by underground runners. **Infl** L to 20 cm, usually open; pale straw-coloured; **Spklt** L to 6 mm; usually 2 FLS; upper ♀ only; GLUMES unequal. **Lvs** L×W to 20 cm × 12 mm; hairless to softly hairy; LIGULE L to 5 mm; blunt; ragged in many. **Hab** open woodland, shady hedgebanks, occasionally infertile grassland.

J F M A M J J A S O N D

UPPER **LEMMA** with **awn < glume** (L to 2 mm); recurved ('fish-hook') when dry

GLUMES equal

BASAL LF-SHEATH striped purplish – **'striped pyjamas'**

STEM **hairy**; STEM NODES **hairy**

UPPER **LEMMA** with slightly bent awn > glume (L to 5 mm)

GLUMES unequal

STEM NODES are visible at a distance as tiny white horizontal dashes in the sward – **'hairy knees'.**

STEM **hairless**; STEM NODES **hairy**

INFL **reminiscent of a chicken's foot when open;** SPIKELETS **dense egg-shaped clusters**

℗ **Cock's-foot** *Dactylis glomerata*

H to 140 cm. **Form** tufted, in small tussocks with many flattened non-flowering shoots. **Infl** L to 25 cm; **1-sided; clumped**; contracted unless flowering. **Spklt** L to 9 mm, stalkless, flattened, with up to 5 florets; GLUMES L to 6·5 mm, pointed; LEMMAS L to 7 mm with rigid awn (L to 1·5 mm). **Lvs** L×W to 45 cm × 14 mm; sticky to touch; boat-shaped tip; LIGULE L to 12 mm, bluntly pointed. **Hab** meadows, pastures, rough grassland.

J F M A M J J A S O N D

in flower

SPIKELETS flattened

INFL lower branches bare at base

INFLORESCENCE **open; very obvious branches and/or spikelet-stalks**

SPIKELETS **awless**

● **Saltmarsh-grasses** | **Form** salt-tolerant; erect; tufted; patch-forming.
Infl branches stiff. **Spklt** single flower in each; GLUMES unequal; LEMMAS
awnless, rounded; 5-veined. **Lvs** grey-green; SHEATH flattened. **Hab** saltmarshes,
coastal habitats, salted road verges.

IDENTIFY BY: ► inflorescence shape ► spikelet details

℗ **Common Saltmarsh-grass**
Puccinellia maritima

H to 80 cm. **Form** tufted; with
surface runners; forming
extensive turf. **Infl** L to 25 cm;
contracted to open; BRANCHES
2–3 at each node, can be
clustered towards top **Spklt** L
7–13 mm; 3–10 FLS, GLUMES L
to 4 mm; LEMMAS L 2·8–5·0 mm;
ANTHERS L 2 mm. **Lvs** L×W to
10 cm×3 mm; LIGULE L to 3 mm, blunt.

℗ **Reflexed Saltmarsh-grass**
Puccinellia distans

H to 60 cm. **Form** tufted,
patch-forming; **runners
absent. Infl** L to 25 cm; open;
BRANCHES 4–6 at each node,
lower branches become
strongly bent back with age.
Spklt L to 4–7 mm; 2–9 FLS;
GLUMES L to 2 mm; LEMMAS L
<2·5 mm; ANTHERS L <1 mm.
Lvs L×W to 10 cm×3 mm; LIGULE L to 2 mm.

BRANCHES **pointing out or upwards;**
bunched towards the top in many

SPIKELETS typically
larger than in
Reflexed SM-G

ANTHERS
L 2 mm

LVS parallel-sided;
grey-green;
fleshy; rough on
upper surface

LOWER BRANCHES bare in basal half;
becoming **strongly bent back with age**

SPIKELETS typically
smaller and more
purplish than in
Common SM-G

ANTHERS
L 0·5–0·8 mm

LVS tapering;
grey-green;
stiff; ribbed on
upper surface

● **Meadow-grasses** | **Form** ranges from solitary to tufted.
Spklt egg-shaped with tip ± pointed; FLS typically ≥3; LEMMAS awnless.
Lvs TIP **'boat'-shaped**; SHEATH flattened to some degree.

LF-TIP **'boat'-shaped**

IDENTIFY BY: ► branch architecture ► ligule + leaf + glume details

---

Ⓐ **Annual Meadow-grass** *Poa annua*

H to 30 cm. **Form** tufted. **Infl** L
to 12 cm; BRANCHES spreading;
1–2 at each node. **Spklt** L
3–10 mm, 3–10 FLS. **Lvs** L×W
to 15 cm×10 mm; can have
transverse crinkles. LIGULE
L 2–5 mm; rounded. **Hab**
arable, gardens, paving cracks,
brownfield. **SS** Early Meadow-
grass *P. infirma* (INSET) [anther L
<0·5 mm].

BRANCHES
1–2 at each node

ANNUAL   EARLY

GLUMES
L to 3 mm;
± equal

FORM
tufted

LIGULE
rounded;
L 2–5 mm

Smooth Meadow-grass *Poa pratensis*

H to 60 cm. **Form** tufted, with obvious underground runners. **Infl** L to 20 cm; BRANCHES whorls of 2–5; **Spklt** L 4–6 mm; 3–5 FLS; GLUMES L to 4 mm, **unequal**; LEMMAS L to 4 mm. **Lvs** L×W to 30 cm × 6 mm; SHEATH smooth; only slightly flattened; LIGULE L 1–3 mm. **Hab** meadows, pastures.

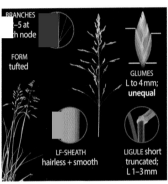

BBRANCHES –5 at h node

FORM tufted

GLUMES L to 4 mm; **unequal**

LF-SHEATH hairless + smooth

LIGULE short truncated; L 1–3 mm

Rough Meadow-grass *Poa trivialis*

H to 100 cm. **Form** solitary or loosely tufted; can have leafy runners. **Infl** L to 20 cm, open (can be contracted); BRANCHES clustered together. **Spklt** L 3–4 mm; 3–4 FLS; GLUMES L to 3·5 mm, ± equal; LEMMAS L to 3·5 mm. **Lvs** L×W to 20 cm × 6 mm; SHEATH **rough**; shiny, light green; LIGULE **long, pointed**, L 4–10 mm. **Hab** lowland grassland, marshes.

BRANCHES 3–7 at each node

FORM solitary or loosely tufted

GLUMES L to 3·5 mm; ± equal

LF-SHEATH hairless + rough (rub to feel)

LIGULE **long, pointed**; L 4–10 mm

Spreading Meadow-grass *Poa humilis*

H to 20 cm. **Form** solitary or tufted; extensive underground runners. **Infl** L to 11 cm; BRANCHES sparse. **Spklt** compressed; L to 7 mm; 3–4 FLS; GLUMES L to 4 mm, ± equal; LEMMAS L to 5 mm. **Lvs** L×W to 18 cm × 4 mm; dark bluish-green; LIGULE L 1–2 mm. **Hab** meadows, roadsides, dunes.

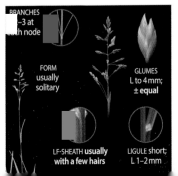

BBRANCHES –3 at each node

FORM usually solitary

GLUMES L to 4 mm; ± equal

LF-SHEATH **usually with a few hairs**

LIGULE short; L 1–2 mm

Wood Meadow-grass *Poa nemoralis*

H to 90 cm. **Form** slender, loosely tufted. **Infl** L to 20 cm; BRANCHES whorls of 2–5. **Spklt** elongate; compressed; L 3–6 mm; 3–5 FLS; GLUMES L to 3·5 mm; LEMMAS L to 3·6 mm. **Lvs** narrow; L×W to 18 cm × 2·5 mm; SHEATH rough, slightly flattened; LIGULE tiny; L <1 mm. **Hab** woods, hedgebanks.

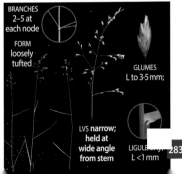

BRANCHES 2–5 at each node

FORM loosely tufted

GLUMES L to 3·5 mm;

LVS **narrow; held at wide angle from stem**

LIGULE tiny; L <1 mm

● **Perennial fescues** (*Festuca* and *Schedonorus*) | **Form** tufted. **Infl** open (contracted when young); branches usually in pairs; never in whorls. **Spklt** each with 3 or more flowers; LEMMAS pointed; AWNS arising from the tip in some species/ individuals. *Schedonorus* fescues are quite substantial plants with broad leaves and auricles that are large and clasping unlike the small rounded auricles of *Festuca*.

Festuca auricle

ⓟ Meadow Fescue
Schedonorus pratensis

H to 120 cm.
Form loosely tufted. **Infl** L to 35 cm; **shorter branches with 1–3 spklts**. **Spklt**
L to 20 mm;
5–14 FLS, overlapping; GLUMES L to 4 mm; LEMMAS L to 7 mm; **typically awnless** but some with small awn (L to 1·2 mm). **Lvs** W to 8 mm; LIGULE L to 1 mm. **Hab** grassland on rich soil, water meadows, road verges.

ⓟ Tall Fescue
Schedonorus arundinaceus

H to 200 cm.
Form densely tufted, tussock-forming. **Infl** L to 50 cm; **all branches with 4–18 spklts**. **Spklt** L to 18 mm; 3–8 FLS, overlapping; GLUMES L to 6 mm; LEMMAS L to 9 mm; **typically awnless** but some with small awn (L to 4 mm). **Lvs** W to 12 mm; LIGULE L to 2 mm. **Hab** grassland on heavy soil, water meadows, dry calcareous pastures.

ⓟ Giant Fescue
Schedonorus giganteus

H to 150 cm.
Form densely tufted, tussock-forming. **Infl** L to 50 cm; shorter branches with 2–16 spklts. **Spklt**
L to 20 mm; 3–10 FLS; overlapping at first but spaced at maturity; GLUMES L to 7 mm; LEMMAS L to 9 mm with **fine, wavy awn** (L 10–18 mm). **Lvs** hairless; W to 18 mm; LIGULE L to 2·5 mm. **Hab** damp, shady woodland rides, ditches, verges. **SS** Wood Brome (*p.290*) [LVS hairy].

SPIKELET 5–14 fls; very short awn at most

INFL **shorter branches with 1–3 spikelets**

AURICLES narrow, spreading

SPIKELET 3–8 fls; short awn at most

INFL **all branches with 4–18 spikelets**

AURICLES narrow, spreading, creamy-white, **fringed with hairs** (may rub off!)

SPIKELET **with wavy awns**

INFL **shorter branches with 2–16 spikelets**

AURICLES **large, pincer-like, hairless**   NODES **dark purple**

P Red Fescue *Festuca rubra*

H to 100 cm. **Form** loosely tufted. **Infl** L to 170 mm; not spreading widely. **Spklt** L to 14 mm; 4–10 FLS; GLUMES L to 9 mm; pointed; LEMMAS L to 10 mm with awn (L to 4·5 mm). **Lvs** L to 50 cm; BASAL bristle-like, STEM flat (W 0·5–3·0 mm); shiny green, LIGULE tiny papery rim. **Hab** wide range of grasslands/man-made habitats.

LF-SHEATH
closed – 'zipped like a sleeping bag'

SPIKELET

new shoots arise from
outside lowest lf-sheath

P Sheep's-fescue *Festuca ovina*

H to 60 cm. **Form** densely tufted. **Infl** L to 80 mm; open or contracted. **Spklt** L to 7 mm; 4–8 FLS; GLUMES L to 4 mm, pointed; LEMMAS L to 5 mm with awn (L to 1·5 mm). **Lvs** L to 25 cm; bristle-like, tightly rolled/folded; dull or bluish-green; LIGULE tiny papery rim. **Hab** grassland on well-drained soils.

LF-SHEATH
open – 'rolled like a duvet'

SPIKELET

new shoots arise from
inside lowest lf-sheath

● Annual fescues (Vulpia) | **Form** tufted, slender. **Infl** 1-sided. **Spklt** GLUMES markedly unequal; shorter than spikelet; LEMMAS narrow into awn. **Lvs** minutely hairy on upper surface.

A Squirreltail Fescue *Vulpia bromoides*

H to 60 cm. **Infl** L to 100 mm, 1-sided; many slightly nodding. **Spklt** L 8–12 mm; 5–10 FLS; GLUMES finely pointed; LEMMAS L 5·5–7·0 mm with awn (L to 14 mm). **Lvs** L to 14 cm; LIGULE L to 0·5 mm. **Hab** dry grassland, heaths, dunes, walls, paths.

A Rat's-tail Fescue *Vulpia myuros*

H to 70 cm. **Infl** L to 30 cm; 1-sided; **markedly nodding**. **Spklt** L 6·5–12 mm, 4–7 FLS; GLUMES finely pointed, LEMMAS L 5–7 mm with awn (L to 15 mm). **Lvs** L to 15 cm; LIGULE L to 1 mm. **Hab** walls, pavements, embankments, brownfield.

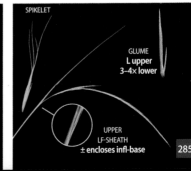

SPIKELET

GLUME
**L upper
≤ 2× lower**

UPPER
LF-SHEATH
well below infl-base

SPIKELET

GLUME
**L upper
3–4× lower**

UPPER
LF-SHEATH
± encloses infl-base

INFLORESCENCE **open; very obvious branches and/or spikelet-stalks**

● **Sweet-grasses** | **Spklt** distinctive; ± cylindrical. **Lvs** ribbon-like leaves. **Hab** generally found in wetter and aquatic margin habitats.

SS Reed Sweet-grass is distinctive, with smaller spikelets (L < 6×W; others L ± 10×W). The other 3 spp. need careful differentiation, and hybrids occur.

IDENTIFY BY: ▶ **inflorescence form + spikelet size** ▶ **lemmas/ligule/anther details**

SPIKELETS **cylindrical; L to 25mm; 7–16 FLS**

Ⓟ **Small Sweet-grass**
Glyceria declinata

H to 50cm.
Form weakly erect or prostrate, loosely tufted.
Infl L to 38cm; open; slender; 1-sided; 1–3 branches at each node. **Spklt** LEMMAS L to 5mm; tip with 3 sharp teeth.
Lvs L×W to 18 cm × 10 mm; greyish-green; rough; SHEATH rough, can be hairy near blade; LIGULE L 4–9mm, bluntly pointed. **Hab** water margins, wet meadows.

Ⓟ **Floating Sweet-grass**
Glyceria fluitans

H to 100cm.
Form weakly erect or floating; in loose mats.
Infl L to 50cm; open; slender; stiffly flexible; 1–2 branches at each node. **Spklt** LEMMAS L to 7mm; tip bluntly pointed.
Lvs L×W to 25 cm × 10 mm; green; tapering to pointed tip; SHEATH **smooth**, hairless; LIGULE L 5–15mm, bluntly pointed. **Hab** water margins, mud, shallow water.

Ⓟ **Plicate Sweet-grass**
Glyceria notata

H to 75cm.
Form weakly erect or prostrate; in loose mats.
Infl L to 40cm, open; 2–5 branches at each node.
Spklt LEMMAS L to 5mm, tip lobed or bluntly 3-toothed.
Lvs L×W to 30 cm × 10 mm; yellowish-green; hairy; SHEATH **rough**, can be hairy near blade; LIGULE L 2–8mm, broadly rounded. **Hab** water margins, mud, shallow water.

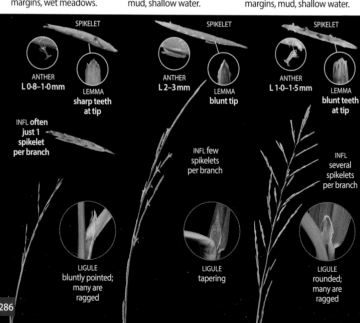

SPIKELET

ANTHER
L 0·8–1·0 mm

LEMMA
sharp teeth
at tip

INFL **often
just 1
spikelet
per branch**

LIGULE
bluntly pointed;
many are
ragged

SPIKELET

ANTHER
L 2–3 mm

LEMMA
blunt tip

INFL few
spikelets
per branch

LIGULE
tapering

SPIKELET

ANTHER
L 1·0–1·5 mm

LEMMA
blunt teeth
at tip

INFL
several
spikelets
per branch

LIGULE
rounded;
many are
ragged

SPIKELETS **cylindrical; L to 12 mm; 4–10** FLS

P **Reed Sweet-grass**
Glyceria maxima

H to 250 cm. **Form** erect, robust, forming dense stands. **Infl** L to 45 cm; open; branches numerous and clustered, spreading. **Spklt** L to 12 mm; 4–10 FLS; LEMMAS L to 4 mm, tip rounded; PALEA shallow notch at tip. **Lvs** L×W to 60 cm × 20 mm; hooded, abruptly contracted to pointed tip; SHEATH rough, hairless; LIGULE L to 6 mm. **Hab** water margins, ditches, wet meadows.

● **Reed-like grasses** | There are several species that have a similar form and structure to **Common Reed** but, if there is any doubt, they can be differentiated easily using flower and ligule characteristics as shown below.

	LIGULE	FLS	AWNS
Common Reed	ring of hairs	2–6	no
Canary-grasses *Phalaris*	papery	1 + 2 sterile (tiny)	no
Small-reeds *Calamagrostis* (N/I)		1	short

P **Common Reed**
Phragmites australis

H to 350 cm. **Form** in dense stands. **Infl** L to 40 cm; purplish-brown; erect to nodding. **Spklt** L 10–16 mm; 2–6 FLS. **Lvs** L×W to 60 cm × 50 mm; stiff; grey-green; LIGULE dense fringe of short (L to 1 mm) hairs. **Hab** swamps, water margins, shallow water, including brackish.

P **Reed Canary-grass**
Phalaris arundinacea

H to 200 cm. **Form** robust, patch-forming. **Infl** L to 28 cm; branched into 'clumps'. **Spklt** L 5–6 mm; FLS 1 fertile + 2 tiny, hairy sterile lemmas; GLUMES L×W to 35 cm × 18 mm; rough, hairless; LIGULE L 2–16 mm, often ragged. **Hab** margins of waterbodies, marshes.

INFL many spikelets per branch

SPIKELET

LEMMA
tip rounded

LIGULE
'curly bracket'
point in middle

LIGULE **fringe**
of short hairs

LIGULE long;
papery

SPIKELET

INFL spikelets
clustered, on
very short stalks

SPIKELET

INFLORESCENCE **open; very obvious branches and/or spikelet-stalks**

SPIKELETS **with obvious awns; 1–3** FLS

● **Oat-grasses** | **Form** erect, tufted (loosely in some). **Infl** open, typically drooping. **Spklt** large and somewhat glossy; LEMMAS typically with conspicuous angled awn.

IDENTIFY BY: ► **flower and awn count** ► **branch count** ► **details of awns**

ⓟ False Oat-grass
Arrhenatherum elatius

H to 180 cm. **Infl** L to 30 cm; well-branched. **Spklt** L to 11 mm; 1–2 FLS; lower ♂, upper bisexual; GLUMES long, papery; LEMMAS lower (♂) awned (L to 20 mm). **Lvs** W to 10 mm; tapering to fine point; dull green; sparsely hairy on upperside; **bitter aftertaste**; LIGULE L to 3 mm; can be ragged. **Hab** rough grassland, road verges, hedgebanks.

ⓟ Downy Oat-grass
Avenula pubescens

H to 100 cm. **Infl** L to 20 cm; lowest node 3–4 branches. **Spklt** L to 17 mm; 2–3 FLS; GLUMES L to 13 mm, unequal, narrow; LEMMAS L to 14 mm with angled awn (L to 20 mm). **Lvs** L×W to 30 cm × 6 mm; LVS + SHEATHS softly and densely hairy; LIGULE **L to 8 mm, pointed**. **Hab** lowland grassland, especially calcareous. **SS Meadow Oat-grass** *Helictochloa pratensis* (N/I) [SPKLTS 3–6 fls; LVS + SHEATHS hairless; LOWEST NODE 1–2 branches].

ⓟ Yellow Oat-grass
Trisetum flavescens

H to 60 cm. **Infl** L to 17 cm; many branches. **Spklt** L to 7·5 mm; **shiny golden-yellow** (introduced ssp. *purpurascens* tinged purple) usually 3 FLS; GLUMES long, papery; LEMMAS with long, angled awn (L to 9 mm) arising from the back. **Lvs** L×W to 15 cm × 4 mm; tapering to fine point; softly hairy on upper side; LIGULE L to 2 mm. **Hab** pastures, road verges, calcareous hill grassland.

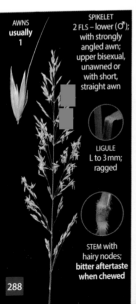

AWNS **usually 1**

SPIKELET 2 FLS – lower (♂); with strongly angled awn; upper bisexual, unawned or with short, straight awn

LIGULE L to 3 mm; ragged

STEM with hairy nodes; **bitter aftertaste** when chewed

AWNS **2–3**

SPIKELET 2–3 FLS – each with angled awn

LIGULE L to 8 mm; pointed

SPIKELET 3 FLS – each with angled awn

AWNS **usually 3**

LIGULE L to 2 mm; pointed

SPIKELETS **large; drooping; typically with long awns; 2–3** FLS

● **Oats** | **Form** erect, with all shoots flowering. **Infl** tightly cylindrical with blunt tip.
Spklt chunky; dangling; FLS 2–3; **GLUMES** long and papery; enclosing spikelet.

Ⓐ Oat *Avena sativa*

H to 130 cm. **Infl** L to 26 cm, open, with ascending or spreading branches. **Spklt** L to 27 mm; LEMMAS L to 21 mm, **awnless or lowest with ± straight awn** (L 25–40 mm). **Lvs** L×W to 40 cm×30 mm; flat; LIGULE L to 5 mm, blunt. **Hab** cultivated; disturbed ground, road verges.

J F M A M J J A S O N D

Ⓐ Wild-oat *Avena fatua*

H to 150 cm. **Infl** L to 40 cm, open, with drooping or spreading branches. **Spklt** L to 28 mm; LEMMAS L to 21 mm with 2–4 short teeth; AWN **bent**; L 25–40 mm. **Lvs** L×W to 45 cm×30 mm; flat; LIGULE L to 6 mm, blunt. **Hab** arable and other disturbed ground.

J F M A M J J A S O N D

SPIKELET
2–3 FLS;
LEMMAS
awnless, or
awn from
back of lowest
lemma only

LEMMAS **smooth;**
lacking beard of hairs;
awn (if present) angled
or straight, L to 40 mm

SPIKELET
2–3 FLS

GLUMES
remain after
seeds fall

LEMMAS
**beard of
brown silky
hairs on
lower part**

INFLORESCENCE **several not or barely spread stiff spikes coming from the stem apex**

● **Cord-grasses** | **Highly distinctive saltmarsh and mudflat grasses.**
Form erect, patch-forming. **Infl** long spikes with a bristle at the apex.

Ⓟ Common Cord-grass *Spartina anglica*

H to 130 cm. **Form** patch-forming with underground runners. **Infl** compound, 2–8 erect spikes; L to 23 cm. **Spklt** L to 20 mm; 1 FL; GLUMES unequal; elongate; downy; LEMMAS L to 17 mm.

J F M A M J J A S O N D

Lvs L×W to 50 cm×15 mm; stiff; inrolled when dry; LIGULE L to 3 mm. **Hab** intertidal mudflats, saltmarshes. **SS** Small Cord-grass *S. maritima* (N/I) [SPIKE-BRISTLE (L<20 mm), LIGULE + ANTHERS shorter].

INFL **each spike _____
ending in bristle**
(L to 50 mm)

FL anthers
and stigmas
conspicuous

LIGULE dense
fringe of
hairs

INFLORESCENCE **loosely branched (bromes) or alternately arranged spikelets (false-bromes)**

WOOD   SOFT-   FALSE   BARREN

● **Bromes and False Brome** | Form ± hairy. **Fls** variable, however each genus has a relatively distinctive spikelet.

IDENTIFY BY:
► **spikelet and awn shape/size**
► **leaf-sheath opening**

J F M A M J J A S O N D   J F M A M J J A S O N D   J F M A M J J A S O N D   J F M A M J J A S O N D

ⓟ **Wood Brome** *Bromopsis ramosa*

H to 190 cm. **Form** hairy; loosely tufted. **Infl** L to 50 cm; open; **branches flop to one side**. **Spklt** L to 45 mm; ± **spear-shaped; flattened**; 4–11 FLS; GLUMES L to 7 mm; LEMMAS L to 15 mm + AWN L to 11 mm (can be absent). **Lvs** L×W to 60 cm × 16 mm; LIGULE L to 6 mm; ragged. **Hab** open woodland, hedgebanks, in partial shade. **SS** Lesser Wood Brome *B. benekenii* (N/I) [INFL >2 branches at lowest node; scale below node hairless].

ⓐ **Soft-brome** *Bromus hordeaceus*

H to 120 cm. **Form** erect or spreading; loosely tufted; all shoots flowering. **Infl** L to 10 cm; compact, but can droop to one side at maturity. **Spklt** L to 25 mm; ± **egg-shaped, tapering to a point**; up to 12 FLS; LEMMAS hairy; L to 11 mm + AWN L to 10 mm. **Lvs** W to 7 mm; flat; hairy; LIGULE L to 2·5 mm; toothed; hairy. **Hab** dry grassland, sand dunes, cliff slopes, disturbed ground.

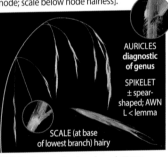

AURICLES **diagnostic of genus**

SPIKELET ± spear-shaped; AWN L < lemma

SCALE (at base of lowest branch) hairy

LIGULE toothed   LF-SHEATH hairy

INFL branches shorter than spikelets

SPIKELET ± egg-shaped; AWN L ≤ lemma

ⓟ **False Brome** *Brachypodium sylvaticum*

H to 90 cm. **Form** hairy; tufted. **Infl** L to 20 cm; drooping. **Spklt** L to 40 mm; **cylindrical**; 7–16 FLS; GLUMES L to 11 mm; LEMMAS L to 15 mm + AWN L 7–15 mm. **Lvs** L×W to 35 cm × 12 mm; LIGULE L to 6 mm; blunt. **Hab** open woodland, hedgebanks, in partial shade. **SS** Heath False-brome *B. pinnatum* (N/I) [AWN L 3–5 mm; LVS flat]; **Tor-grass** *B. rupestre* (N/I) [AWN L 2–4 mm; LVS inrolled].

ⓐ **Barren Brome** *Anisantha sterilis*

H to 125 cm. **Form** erect or spreading. **Infl** L to 25 cm; open ± nodding to one side; INFL-AXIS downy at most. **Spklt** L to 25 mm; **wedge-shaped when mature**; up to 12 FLS; LEMMAS L to 23 mm + AWN L to 35 mm. **Lvs** flat; floppy; LIGULE L to 4 mm; toothed. **Hab** arable margins, roadsides, brownfield. **SS** Great Brome *A. diandra* (N/I) [SPKLT L to 50 mm; INFL AXIS densely hairy].

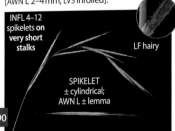

INFL 4–12 spikelets on **very short stalks**

LF hairy

SPIKELET ± cylindrical; AWN L ± lemma

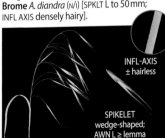

INFL-AXIS ± hairless

SPIKELET wedge-shaped; AWN L ≥ lemma

Gymnosperms Those included in this book are known collectively as conifers. Many are evergreen, though a few (larches) are deciduous. They all have needle-like or scale-like leaves. The 'flowers' are not actually as such, but are separate male and female structures – the males have a single (or grouped) cone-like sporangium containing spores that are motion-released (*e.g.* by wind) when mature; the females are typically hard (but can be soft) cones, or cone-like structures, with fused scales that part when mature to release the seeds. **Yew** is an exception as the 'cones' are fleshy.

IDENTIFY FAMILIES BY: ● growth form   ● leaf shape and arrangement
● female cone details   ● male cone details

Gymnosperms

Alders (*p. 109*) have cone-like fruits but are flowering trees

Pine
94 Pinaceae
▶ *pp. 292–294*

PINES
Form evergreen. **Lvs** long (L>7 cm) needle-like; in groups of 2, 3 or 5.
♀ **cones** large (L>5 cm); woody; scales fused to open.

FIRS and SPRUCES
Form evergreen. **Lvs** short (L<0·6–2·5 cm); needle-like.
♀ **cones** large (L>5 cm); woody; **hanging**; scales fused to open.

LARCHES
Form deciduous. **Lvs** short; (L<3 cm); needle-like; in clusters or singly. ♀ **cones** small (L<5 cm); woody; scales fused to open.

Juniper (incl. cypresses)
96 Cupressaceae
▶ *pp. 295–296*

Yew
95 Taxaceae
▶ *p. 295*

CYPRESSES
Form evergreen; TWIGS flattened. **Lvs** scale-like.
♀ **cones** small; woody with fused scales; open when mature.

Juniper
Form evergreen; TWIGS spreading. **Lvs** needle-like in **whorls of 3**. ♀ **cones** soft-scaled; green (young) to **black** (mature).

Yew
Form evergreen. **Lvs** narrow (2–3 mm wide); flat. **Fr** berry-like; flesh partially surrounding seed; green (young) to **red** (mature).

▶ Conifer botanical terms *p. 15*

94 Pinaceae | **Pine + Larch** family **9 spp. | 23 spp. B&I**
Form typically evergreen trees. **Lvs** linear (needles), spiralled
around stem, resin-scented. **Fr** seeds in ± woody cones.

● **Pines** | **Form** large evergreen trees with conical crowns.
Lvs needle-like; long (L > 7 cm), in groups of 2, 3 or 5 arising
from short, lateral woody spurs; BUDS resinous and sticky.
Cones similar across most species; ♂ CONES small, rounded;
yellow with pollen in elongate clusters near tip of shoots with
a tuft of needles above; ♀ CONES red to purple, soft at first,
growing and ripening into woody pine cones over 1–3 years.

IDENTIFY BY: ▶ needle length, colour and form   ▶ ripe ♀ cone details

♂ CONES clusters
near shoot tips;
pollen yellow

♀ CONES red
to purple;
soft when
young

SCOTS   BLACK   LODGEPOLE

'Flowers' of all the *Pinus* species are similar and of limited
use in identification.

J F M A M J J A S O N D   J F M A M J J A S O N D   J F M A M J J A S O N D

P Scots Pine *Pinus sylvestris*

H to 35 m. **Bark** upper trunk
and branches orange-red;
peeling. **Needles** L to 8 cm
(shorter in native plants); blue-
green; twisted; in pairs.
Ripe ♀ cones L to 75 mm
(shorter in native plants);
scales without a prickle.
Hab upland northern forests;
plantations and gardens
especially on sandy soils.

♀ CONE (RIPE)
no scale-prickle

BARK
lower, grey; upper,
orange-red   SCOTS   AUSTRIAN   CORSICAN

P Black Pine *Pinus nigra*

H to 45 m. **Bark** whole trunk
grey, often dark. **Needles**
light to dark green; **barely
twisted**; in pairs; Austrian (ssp.
nigra) L 8–12 cm; Corsican
(ssp. *laricio*) L 10–18 cm.
Ripe ♀ cones L to 90 mm;
scales usually without a
prickle though can be present.
Hab plantations and gardens.

♀ CONE (RIPE)
tiny scale-prickle
may be present

larger cones
than in
Scots Pine

BARK
all grey

P Lodgepole Pine
Pinus contorta

H to 30 m. **Bark** whole
trunk dark reddish-brown.
Needles L to 10 cm; bright
green; twisted. **Ripe ♀ cones**
L to 60 mm; **scales with a
prickle**. **Hab** plantations,
occasionally self-seeds.

♀ CONE (RIPE)
scale-prickle

BARK
all reddish-
brown   LODGEPOLE

SCOTS

NEEC

● **Firs** and **Spruces** | **Form** large evergreen trees. **Lvs** needle-like; short (< 0·6–2·5 cm). Ripe ♀ **cones** woody; hanging downwards.

Ⓟ Douglas Fir *Pseudotsuga menziesii*

H to 60m. **Twigs** sparsely hairy. **Needles** L to 35 mm; green above; two whitish stripes below; borne singly; stalkless; those on the underside of a branch spreading laterally; strongly resin-scented. **Ripe ♀ cones** L to 100mm long; drooping; **3-pointed bract** emerging from scales. **Hab** plantations and parks; self-seeds on occasion.

Ⓟ Western Hemlock-spruce *Tsuga heterophylla*

H to 50m. **Twigs** hairy. **Needles** L to 20mm; dark green above; two broad whitish stripes below; borne singly on a short green stalk. **Ripe ♀ cones** L to 25 mm, egg-shaped. **Hab** plantations, self-seeds.

♀ CONE (RIPE)
3-pointed
bract

♀ CONE (RIPE)

upper

NEEDLE

under

upper

NEEDLE

under

♂ FLS green-yellow to purple

Norway and **Sitka Spruces** are similar in structure and form. Needles are borne singly on a short woody peg. Cones hang downwards. **Hab** plantations and parks, self-seeds.

Ⓟ Norway Spruce *Picea abies*

H to 50m. **Needles** L to 25mm; **4-angled; glossy green above and below**; not twisted at base. **Ripe ♀ cones** L to 200mm.

Ⓟ Sitka Spruce *Picea sitchensis*

H to 60m. **Needles** L to 30mm; **flattened; dark green above; two broad whitish stripes below**; twisted at base. **Ripe ♀ cones** L to 100mm.

NEEDLE

x-section
upper

under

NEEDLE

x-section
upper

under

'Christmas-tree' when young (TOP); 'pine-tree' when mature (BOTTOM)

SS Many pine, fir and spruce species are planted for ornamental reasons. Identification of these is by needle and cone detail but needs specialist keys beyond the scope of this book.

● **Larches** | **Form** deciduous trees. **Lvs** short (L < 3 cm); needle-like; borne singly on leading (new) shoots; in rosettes on side shoots. **Ripe ♀ cones** small (L < 3 cm); woody; erect.

IDENTIFY BY: ► needle underside   ► ♂ cone colour
► ripe ♀ cone shape and form

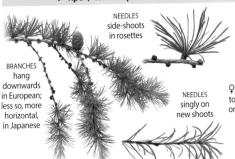

NEEDLES side-shoots in rosettes

BRANCHES hang downwards in European; less so, more horizontal, in Japanese

NEEDLES singly on new shoots

♀ CONE green to burgundy; on upperside of twig

♂ CONE yellowish; on underside of twig

SS The larches are similar; they also hybridize readily; the hybrid is often found planted in shelterbelts. **Hybrid Larch** *L.* × *marschlinsii* (N/I) [features intermediate and/or mixed – RIPE ♀ CONE is perhaps the most reliable differentiating feature].

Ⓟ European Larch *Larix decidua*

J F M A M J J A S O N D

H to 45m. **Bark** red-brown. **Needles** green to greyish-green with two inconspicuous pale green stripes below. **Ripe ♀ cones** L to 40mm; elongate egg-shaped; L to 1·5× W; cone-scales wavy, not curved out at tip. **Hab** plantations and parks; often self-seeds.

Ⓟ Japanese Larch *Larix kaempferi*

J F M A M J J A S O N D

H to 35m. **Bark** grey-brown, rough. **Needles** grey-green with two whitish stripes below. **Ripe ♀ cones** L to 35mm; erect; egg-shaped; cone-scales curved out at tip. **Hab** plantations; self-seeds on occasion.

♀ CONE (RIPE)
H 1·25–1·50 × W scales wavy at most, **never curved out**

HYBRID LARCH

Highly variable, but typically cone shape as **European** with scales curved out like in **Japanese**

NEEDLES typically less grey-green than Japanese; **inconspicuous pale stripes** on underside

♀ CONE (RIPE)
H 1·0–1·25 × W scales typically **curved out**

NEEDLES typically greyer-green than European; **conspicuous pale stripes** on underside

TWIGS
Those of **European Larch** (*left*) are grey brown; those of **Japanese Larch** (*right*) are typically more reddish-brown

294

95 Taxaceae | **Yew** family   `1 sp. B&I`

Form tree or large evergreen shrub; often with several trunks.
Lvs linear; flattened; borne either side of a branch and
± flattened into one plane. **Fr** a single seed, almost surrounded
(when mature) by red, fleshy upgrowth (aril).

Ⓟ **Yew** *Taxus baccata* ✘

H to 25m. **Bark** red-brown; flaking. **Needles**
L×W to 30mm × 2–3mm; **linear**; dark green
with two pale stripes on underside. **Fls** small
cone-like structures (L to 3mm); ♂ + ♀ on
separate plants; ♂ yellow with pollen; ♀ with
a droplet of liquid which acts as a pollen
receptor. **Fr** D to 9mm; red when ripe.
Hab woodland and scrub on limestone or
sandstone; **churchyards** (planted).

J F M A M J J A S O N D

♂ FL

♀ FL

FRUIT ripens from
green to red

NEEDLES
flattened

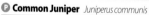

CONES on separate trees; ♂ on leaf upperside; ♀ on underside.

96 Cupressaceae | **Juniper** family   `4 spp. | 14 spp. B&I`

Form evergreen trees or shrubs. **Lvs** linear; spine-tipped or scale-like;
AROMA distinctive to species. **Fr** seeds in globular woody or berry-like cones.

● **Juniper** | **Form** evergreen shrub to small tree; erect to
spreading in a wide array of forms. **Lvs** needle-like; in whorls of 3;
borne on spreading twigs. **Fr** a soft-scaled cone; black when mature.

Ⓟ **Common Juniper** *Juniperus communis*

H to 7m. **Needles** L to 20mm; stiff; linear tipped
with a spine; **in whorls of three**; green with a
whitish band on upper side. **Fls** cones; ♂ + ♀ on
separate plants; ♂ small, rounded, yellow with
pollen; ♀ insignificant. **Ripe ♀ cones** fleshy;
more or less globular (D to 10mm), green,
ripening to **blue-black; aromatic (of gin)**
when crushed. **Hab** scrub, especially upland,
in grassland, heathland, moorland and open
woodland on calcareous and acid soils.

J F M A M J J A S O N D

CONE ripens from
green to black

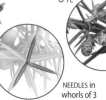

♂ FL

AROMA likened
to spicy apple
from needles
when rubbed

NEEDLES in
whorls of 3

♀ FL

295

● **Cypresses and Cedars** | **Form** evergreen trees; very variable in form and leaf colour (yellowish- to dark green; BARK red-brown; twig sprays flattened. **Lvs** scale-like. **Fr** a small woody cone with fused scales that open when mature.

IDENTIFY BY: ► leaf details ► ripe ♀ cone details
► aroma of crushed leaves

SS Other planted cypresses (N/I) [identification can be difficult and needs specialist keys beyond the scope of this book]; Tamarisk *Tamarix gallica* (N/I) [FLOWERS pink].

LEAVES **longer than cypresses**; L to 6 mm	LEAVES **shorter than Western Red-cedar; scale-like, pointed, appressed in opposite pairs; L to 3 mm**	
LF AROMA **sweet pineapple**	LF AROMA **parsley**	LF AROMA **slightly citrus**

℗ Western Red-cedar
Thuja plicata

H to 45m.
Branches not drooping at tip. **Lvs** glossy deep green; whitish 'butterfly' marks on underside.
♂ **cones** blackish. **Ripe** ♀ **cones** egg-shaped; D to 12mm; scales with a recurved spike at the apex. **Hab** plantations; self-seeds.

J F M A M J J A S O N D

℗ Lawson's Cypress
Cupressus lawsoniana

H to 40m.
Branches drooping at tip. **Lvs** obvious translucent gland above; waxy white patches below.
♂ **cones** reddish. **Ripe** ♀ **cones** globular; D to 10mm; scales with a conical spine. **Hab** gardens, plantations; self-seeds.

J F M A M J J A S O N D

℗ Leyland Cypress
Cupressus × leylandii

H to 36m.
Branches more erect. **Lvs** lacking translucent gland above; no waxy patches below.
♂ **cones** yellowish. **Ripe** ♀ **cones** globular; D to 20mm; scales with a near-conical spine. **Hab** gardens, plantations; does not usually set viable seed.

J F M A M J J A S O N D

BRANCHES **not drooping**

BRANCHES **drooping**

BRANCHES **more erect**

♂ CONE blackish

♂ CONE red

♂ CONE yellow

LVS upper under

LVS upper 'butterfly' marks under translucent gland

LVS upper whitish marks under no glands or marks

♀ CONE (RIPE)

♀ CONE (RIPE)

♀ CONE (RIPE)

Pteridophytes include a range of diverse-looking non-flowering plants, from the well-known ferns and horsetails, to the lesser-known adder's-tongues, clubmosses and spike-mosses. They all reproduce via spores, in sporangia that take a variety of forms – from prominent cones in the clubmosses and horsetails, to a spike in the adder's-tongues and small structures on the underside of the fronds in ferns.

IDENTIFY FAMILIES BY: ▶ growth form ▶ frond shape and arrangement
▶ sporangia details

Pteridophytes

Horsetails
97 Equisetaceae
▶ *pp. 298 – 299*

SPORANGIA IN CONE

FERTILE

STERILE

SS Mare's-tail
(*p. 180*)

Upright plants without branches or with very thin branches in whorls; some have separate fertile (March–April) and sterile stems; sporangia typically a conspicuous cone-like structure.

CLUBMOSSES and SPIKEMOSSES
98 Lycopodiaceae & **99** Selaginellaceae
▶ *p. 300*

Moss-like plants with 1-veined leaves and often branched stem; sporangia structures conspicuous and cone-like or unobtrusive, rounded and situated at the base of the leaves.

Adder's-tongues
100 Ophioglossaceae
▶ *p. 300*

Distinctive plants with two blades; a leaf-like sterile blade and a fertile spike.

SS Lords-and-Ladies (*p. 230*);
Greater Plantain (*p. 175*)

FERNS
8 FAMILIES **101** – **108**
▶ *pp. 301–308*

Fern stalks are rounded, or shallowly 'U'-shaped, and ungrooved

SPORANGIUM

SORI

SS Cow Parsley
(*p. 222*)

Plants with fronds in a variety of shapes, outlines and architecture; reproduction via spores that are contained within sori situated on the underside of the frond.

SS Leaves of a few flowering plants, particularly those of Cow Parsley (*above*) and some other members of the carrot family (*p. 222*) which have grooved 'V'-shaped stalks

▶ Parts of a fern *p. 14* | Fern identification *p. 301*

HORSETAILS

GREAT
HORSETAIL

97 Equisetaceae | **Horsetail** family   **5 spp. | 9 spp. B&I**

Form creeping perennials with jointed, ridged stem with or without whorls of branches at the nodes, and a sheath of fused leaves at each node. Some species have both fertile and sterile stems – in these, fertile stems (bearing a cone at the shoot tip) typically appear in March–April, before sterile ones.

SS Mare's-tail (*p. 180*), an aquatic plant [whorls of obvious leaves, rather than the angled branches of horsetails]; **other horsetails** (N/I) [told by stem-sheath and branch details].

IDENTIFY BY: MAR–APR ▶ **sheath-teeth** ▶ **general form**
APR onwards ▶ **branch architecture** ▶ **sheath-teeth**

WATER | MARSH

J FMAMJ J ASOND J FMAMJ J ASOND

FIELD | WOOD

J FMAMJ J ASOND J FMAMJ J ASOND

GREAT

J FMAMJ J ASOND

Horsetails with a single (sterile and fertile) stem

ⓟ Water Horsetail
Equisetum fluviatile

Stem L to 150 cm; branches typically absent although a few can be present; STEM-RIDGES 10–30, indistinct; SHEATH-TEETH black-tipped; CONES (if present) blunt. **Hab** shallow lakes, extending into surrounding marshy vegetation.

ⓟ Marsh Horsetail
Equisetum palustre

Stem L to 50 cm; usually with at least some branches although these can be absent; LOWEST BRANCH SEGMENT black-tipped, L < stem-sheath L; STEM-RIDGES 5–9; SHEATH-TEETH black with broad white edges; CONES (if present) pointed. **Hab** marshes, damp grassland.

Horsetail – key parts for identification

STEM-SHEATH + TEETH   STEM-RIDGES

STERILE

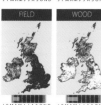

FERTILE

CONE

STERILE

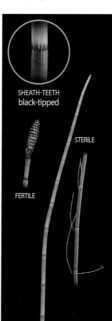

SHEATH-TEETH black-tipped

STERILE

FERTILE

BRANCH SEGMENT (LOWEST) **shorter** than sheath, **dark-tipped**

SHEATH-TEETH black with **broad pale edges**

STERILE

FERTILE

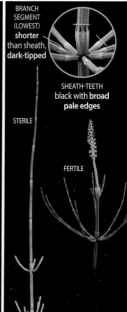

Unbranched fertile stem (Mar–Apr)

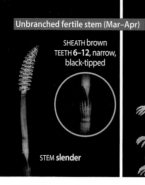

SHEATH brown
TEETH **6–12**, narrow,
black-tipped

STEM **slender**

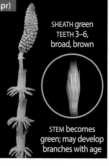

SHEATH green
TEETH 3–6,
broad, brown

STEM becomes
green; may develop
branches with age

SHEATH green to brown
TEETH **20–30**, narrow,
black-tipped

STEM **thick**

Ⓟ Field Horsetail
Equisetum arvense

Stem FERTILE L to 25 cm;
D 3–6 mm; white or pinkish;
branches absent; STERILE
L to 50 cm, green, with
unbranched or barely
branched branches; LOWEST
BRANCH SEGMENT **pale-
tipped, L ± to > stem-sheath
L**; STEM-RIDGES 6–20; SHEATH-
TEETH black-tipped. **Hab** dry
grassy and disturbed areas.

Ⓟ Wood Horsetail
Equisetum sylvaticum

Stem L to 60 cm; FERTILE
D 3–5 mm; white or cream,
unbranched, becoming
green, usually with
branches; STERILE green
with numerous **branched
branches that droop** at
the tip; STEM-RIDGES 10–18;
SHEATH green with 3–6
reddish-brown teeth. **Hab**
woodland and moorland.

Ⓟ Great Horsetail
Equisetum telmateia

Stem FERTILE L to 30 cm;
D 10–15 mm; pale brown;
unbranched; STERILE **L to
200 cm, D 10 mm, whitish**
with numerous green,
upward-pointing, arching or
lax unbranched branches;
STEM-RIDGES **20–40**; SHEATH
green with dark brown
teeth. **Hab** damp grassland
and woodland.

Sterile (green) stem (April onwards)

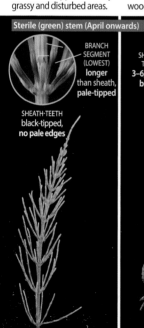

BRANCH
SEGMENT
(LOWEST)
longer
than sheath,
pale-tipped

SHEATH-TEETH
black-tipped,
no pale edges

SHEATH-
TEETH
3–6, broad,
brown

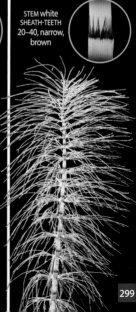

STEM white
SHEATH-TEETH
20–40, narrow,
brown

98 Lycopodiaceae | **Clubmoss** family

1 sp. | 7 spp. B&I

Form like large mosses with erect, often branched, stems and sporangia often in cone-like structures. **Lvs** entire or minutely toothed at most.

ⓟ Fir Clubmoss *Huperzia selago*

H to 15 cm. **Form** erect; stem fork into equal branches. **Lvs** L 4–8 mm; narrow; pointed; tips colourless. **Sporangia** cones not obvious; spore-bearing bodies rounded, yellowish; **solitary at base of upper leaves**; can be replaced by vegetative buds. **Hab** heaths, moors, mountains. **SS** other clubmosses (N/i) [obvious cones]; **Lesser Clubmoss** [LVS margins deeply toothed; tips green].

sporophytes

99 Selaginellaceae | **Spikemoss** family

1 sp. | 2 spp. B&I

Form moss-like; creeping stems with upright spore-producing shoots that bear poorly defined cones. **Lvs** toothed margins and small papery projection (ligule) at base.

ⓟ Lesser Clubmoss *Selaginella selaginoides*

L to 15 cm; **H** to 6 cm. **Form** erect; spreading; irregularly branched. **Lvs** L 1–3 mm, green points, margins deeply toothed. **Sporangia poorly defined terminal cone**; spore-bearing structures (♂ yellowish, ♀ green) at base of leaves. **Hab** montane grassland and rocks; dune-slacks. **SS** Fir Clubmoss; **Marsh Clubmoss** *Lycopodiella inundata* (N/i) [LVS untoothed; scarce]; **Krauss's Clubmoss** *S. kraussiana* (N/i) [greenhouse escape; creeping stem strongly flattened].

poorly defined cone

♂

100 Ophioglossaceae | **Adder's-tongue** family

1 sp. | 5 spp. B&I

Only distantly related to true ferns. **Form** shoots in two parts: a single photosynthetic blade and a separate spore-producing fertile spike.

ⓟ Adder's-tongue *Ophioglossum vulgatum*

Sterile blade L to 15 cm; broadly oval; pointed; ± succulent. **Fertile spike** L to 7 cm; slightly flattened; with 10–40 pairs of sporangia. **Hab** grassland (damper especially), dune-slacks, and heathland under Bracken (p. 308). **SS** rare adder's-tongues (N/i) [all much smaller].

SPORANGIA in pairs, on spike

Non-fern SS Lords-and-Ladies *Arum* spp. (p. 230) [SPADIX cylindrical]. **Greater Plantain** (p. 175) [LEAVES in a rosette] and **Common Twayblade** (p. 238) [LEAVES paired] have similar leaves, but both have prominent longitudinal veins.

FERN IDENTIFICATION

▶ Initial differentiation is by the architecture of the frond – *i.e.* how many times the frond is divided from the central stem.

For some individuals, and species, it can be hard to establish how divided the frond is – this situation is allowed for in the species accounts that follow.

Frond architecture

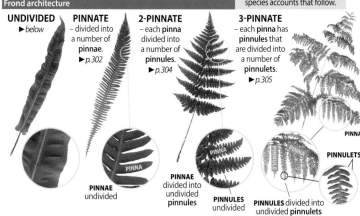

UNDIVIDED
▶ *below*

PINNATE – divided into a number of **pinnae**. ▶ *p.302*

2-PINNATE – each **pinna** divided into a number of **pinnules**. ▶ *p.304*

3-PINNATE – each **pinna** has **pinnules** that are divided into a number of **pinnulets**. ▶ *p.305*

PINNA

PINNULETS

PINNAE undivided

PINNA

PINNAE divided into undivided **pinnules**

PINNULES undivided

PINNULES divided into undivided **pinnulets**

▶ After that the shape and details of the sori, indusia and sporangia (*located on the frond underside – see p.14*) and details (such as shape and toothing on the margin) of the pinnae and pinnules are used for specific identification.

Sori shape examples

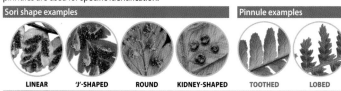

LINEAR   **'J'-SHAPED**   **ROUND**   **KIDNEY-SHAPED**

Pinnule examples

TOOTHED   **LOBED**

FERNS | FRONDS undivided

Hart's-tongue is a member the **Spleenwort** family (Aspleniaceae *p.303*), a family that has all levels of frond architecture from simple to 2–3-pinnate. Given that the key to identification here is frond complexity, splitting the family across several pages is the best way of enabling the correct identity to be reached.

ⓟ Hart's-tongue *Asplenium scolopendrium*

Frond L to 60 cm, **unlobed**, glossy, often with wavy margin; OUTLINE narrow; SORI linear, **very long**.
Hab shady woodland, banks, rocks and walls.

J F M A M J J A S O N D

SORI very long

FERNS | FRONDS once-divided (pinnate)

101 Polypodiaceae | **Polypody** family

2 spp. | 3 spp. B&I

● **Form** fronds borne singly along length of a runner; can be found growing on trees (*right*).
Fronds 12–30 ladder-like pinnae on each side of a midrib; SORI rounded; lacking an indusium. Polypodies are variable and structural features overlap; accurate identification requires microscopic examination of sporangia. The features here are thus only indicative of a species.

lower pinnae not tapering at frond base

ⓟ **Polypody** *Polypodium vulgare*

Frond L to 25 cm, L 3–6 × W; OUTLINE **lower part ± parallel-sided**; PINNAE margin ± untoothed; SORI ± circular. **Hab** rocks, walls, earth banks, trees (usually oaks); generally on acid soils. **SS** Southern Polypody *P. cambricum* (N/I) [typically fronds L <2 × W].

J F M A M J J A S O N D

± parallel-sided

SORI ± circular

PINNAE TIP ± blunt

SPORANGIA (fully formed but immature) annulus* **orange**

ⓟ **Intermediate Polypody**
Polypodium interjectum

Frond L to 40 cm, L 2–4 × W; OUTLINE **lower part less parallel-sided than Polypody**; PINNAE margin can be toothed; SORI ± oval. **Hab** as Polypody; often in shadier and more base-rich areas; often grows on Elder or Hawthorn.

J F M A M J J A S O N D

less parallel-sided

SORI ± oval

PINNAE TIP pointed to ± blunt

SPORANGIA (fully formed but immature) annulus* **green**

*an annulus is a line of water-filled cells which facilitate spore dispersion by contraction

SS Polypodies and Hard-fern could be confused but Hard-fern has tapering lower pinnae (see *insets*).

102 Blechnaceae | **Hard-fern** family

1 sp. | 2 spp. B&I

● **Form** patch-forming; loosely tufted; **Fronds** FERTILE ± erect, pinnae narrow and spaced, produced from centre of rosette; STERILE less erect and wider, more narrowly spaced pinnae compared to fertile fronds.

FERTILE

lower pinnae **tapering** toward frond base

STERILE

ⓟ **Hard-fern** *Blechnum spicant*

Frond L to 50 cm, **W** 4 cm, leathery; OUTLINE narrowly oval; PINNAE can be toothed; SORI **linear**, continuous either side of the pinna midrib. **Hab** woods, heaths, moors, usually on acid (often rather dry) soils. **SS** possibly polypodies.

J F M A M J J A S O N D

STERILE

SORI linear

FERTILE

103 Aspleniaceae | **Spleenwort** family **6 spp.** | 11 spp. B&I

● **Form** tufted with a variety of frond shapes, such that some could be mistaken for other ferns. However, spleenworts are distinctive in having linear to oblong sori which in some are hidden within a dense covering of scales.

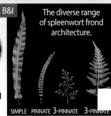

The diverse range of spleenwort frond architecture.

Spleenworts have **elongated sori**

SIMPLE   PINNATE   3-PINNATE   3-PINNATE

IDENTIFY BY: ► frond architecture
► midrib colour ► underside texture

Other spleenworts | FRONDS UNDIVIDED **Hart's-tongue** (p. 301); FRONDS IRREGULAR (REGARDED AS 2-PINNATE)
Wall-rue (p. 304); FRONDS 2-PINNATE Black Spleenwort (p. 304) – NOTE: may look 3-pinnate

---

ⓟ **Rustyback** *Asplenium ceterach*

Frond L to 15 cm;
OUTLINE narrowly oval;
PINNAE underside **densely clothed with scales**, at first whitish becoming red-brown;
SORI hidden among the dense scales. **Hab** walls, clefts in rock faces.

underside

SORI

ⓟ **Maidenhair Spleenwort**
Asplenium trichomanes

Frond L to 20 cm;
OUTLINE ± linear;
MIDRIB **blackish**;
SORI linear/oblong.
Hab rocks and walls.

MIDRIB blackish

SORI

ⓟ **Green Spleenwort** *Asplenium viride*

Frond L to 15 cm;
OUTLINE ± linear;
MIDRIB **green**;
SORI linear/oblong.
Hab base-rich rock crevices, occasionally on walls.

MIDRIB green

SORI

SORI **linear to oblong**

ⓟ Wall-rue *Asplenium ruta-muraria*

Frond L to 10 cm; irregularly pinnate to 2-pinnate; STALK green; PINNULES **diamond-shaped. Hab** walls, rocks.

ⓟ Black Spleenwort
Asplenium adiantum-nigrum

Frond L to 50 cm, stalk comprising half the length; OUTLINE triangular; STALK dark brown to black at base. **Hab** rocks, walls. **SS Brittle Bladder-fern** *Cystopteris fragilis* (N/I) [STALK all green; SORI rounded].

SORI

may look 3-pinnate

STALK **black at base**

SPORANGIA **at frond apex;** PINNULES **broad**

104 Osmundaceae | **Royal Fern**

● Fronds large; the outer sterile, the inner fertile, with terminal pinnae replaced by sporangia. 1 sp. B&I

ⓟ Royal Fern *Osmunda regalis*

Form clump-forming. **Frond** L to 3 m; PINNAE 5–15 pairs; PINNULES **broad**, flat L to 80 mm × W 15 mm. **Hab** damp woodland, heaths and fens, wet rocks, sea cliff slopes; sometimes naturalized from gardens.

sporangia like **upright bunches of little 'grapes'**, golden-brown, on 2-pinnate structures at the apex of the inner fronds

---

SORI **crescent or 'J'-shaped;** INDUSIA **edge ragged**

105 Athyriaceae | **Lady-fern** family

● Fronds SORI crescent-shaped; otherwise resembles other ferns with 2-pinnate fronds. 1 sp. | 2 spp. B&I

ⓟ Lady-fern *Athyrium filix-femina*

Form 'shuttlecock'. **Frond** L to 120 cm, **delicate drooping tip**; PINNULES often deeply lobed. **Hab** damp woods, hedgebanks, mountain rocks and screes. **SS Male-ferns** (p. 306) [PINNULES less deeply toothed; SORI kidney-shaped].

may look 3-pinnate

SORI **crescent- or 'J'-shaped**

INDUSIA edge ragged

PINNULES lobed

STALK can be red

SORI **rounded; in lines along pinnule margins;** INDUSIA **soon wither to reveal black sporangia**

106 Thelypteridaceae | **Marsh Fern** family

1 sp. | 3 spp. B&I

Sori along pinnule margin

Fronds SPORANGIA black; situated along pinnule margins; otherwise shape and growth form resembles members of the Buckler-fern family (Dryopteridaceae – see below).

ⓟ Lemon-scented Fern
Oreopteris limbosperma

SPORANGIA black

J F M A M J J A S O N D

Form 'shuttlecock' tussocks; FRONDS L to 120 cm; OUTLINE broadly oval, widest at mid-frond; pinnae get progressively shorter towards the frond base. **Hab** damp woodland, hedgebanks, rocks and screes. **SS** Lady-, Male- and Shield-ferns [BASAL PINNAE not gradually shorter; SORI very differently shaped or positioned].

pinnae taper to base

minute glands on underside may give off a lemony aroma if crushed

SORI **rounded or kidney-shaped; in lines along pinnule midrib**

107 Dryopteridaceae | **Buckler-fern** family

7 spp. | 17 spp. B&I

● **Form** comprising three main groups with a ± 'shuttlecock' growth. Although the groups are fairly easy to distinguish from one another, differentiating the species within each group can be very difficult. Plant form is variable, and if in any doubt, refer to the detailed features overleaf to be sure of being in the correct group.

● **SHIELD-FERNS**
FORM tufted to loose 'shuttlecock'; OUTLINE narrowly oval; FROND **2-pinnate**; SORI **rounded.**

SORI

● **MALE-FERNS**
FORM obvious 'shuttlecock'; OUTLINE narrowly to broadly oval; FROND **2-pinnate;** SORI **kidney-shaped.**

SORI

● **BUCKLER-FERNS**
FORM tufted to loose 'shuttlecock'; OUTLINE more 'triangular' than Shield- or Male-ferns; FROND **3-pinnate*** SORI **rounded.**

* (two very localized species are 2-pinnate)

FERNS | FRONDS **twice-divided (2-pinnate)**

SORI kidney-shaped | MALE-FERNS   SORI

MALE-FERN   SCALY MALE

Male-fern and scaly male-ferns are very similar 'shuttlecock' ferns, pinnules not 'mitten-shaped'. The scaly male-ferns require an assessment of subjective, variable (*e.g.* in light or shade) and overlapping characters for identification. **Hab** woods, moors and mountains (**Borrer's** Scaly Male-fern has a preference for woodland).

J F M A M J J A S O N D   J F M A M J J A S O N D

Separating Male-fern from scaly male-ferns | The two types can usually be distinguished by the amount of scaling at the stalk base. However, a more reliable distinction is the colour of the upperside and underside of the pinna base where it meets the frond central stem.

ⓟ Male-fern *Dryopteris filix-mas*

Frond L to 120 cm.
Key ID details annotated.

UPP always green   UND typically green

PINNULES typically with sharp teeth and a rounded apex

Male-fern: pinna-bases and midrib **green on the upperside** and typically also green on the underside BEWARE: rarely, a faint dark smudge can be present.

STALK-SCALES pale brown

INDUSIA thin, shrivelled

Both **Male-fern** and **scaly male-ferns** have a 'shuttlecock' frond arrangement.

ⓟ SCALY MALE-FERNS

A complex of very similar species readily distinguishable from Male-fern by the colour of the pinna-bases, but difficult to identify from one another. The two most widespread species are shown here.

Frond L to 150 cm.
Key ID details annotated.

GOLDEN SCALY MALE-FERN

UPP dark   UND dark

STALK-SCALES usually far more conspicuous than on Male-fern

Scaly male-ferns: pinna-bases and midrib **dark on both the upperside and underside.**

ⓟ Golden Scaly Male-fern
Dryopteris affinis

Frond yellow-green to mid-green.

ⓟ Borrer's Scaly Male-fern
Dryopteris borreri

Frond yellow-green to grey-green.

Golden Scaly M-f		Borrer's Scaly M-f	
STALK-SCALES **golden-red to golden-brown**; usually dense			STALK-SCALES pale to blackish brown; usually less dense than Golden SMF
PINNULES **'crowded'**; apex **rounded** or blunt, barely toothed			PINNULES **'spaced'**; apex **squarish** with sharp teeth
INDUSIA thick, **splits radially**			INDUSIA medium, **shrivels** into 'funnel'-shape

2/2

SORI **circular** | SHIELD-FERNS

SORI

P Soft Shield-fern
Polystichum setiferum

Frond L to 150 cm; often upright; **soft**; base **broad**; PINNULES *key ID details below*. **Hab** as box (*right*).

P Hard Shield-fern
Polystichum aculeatum

Frond L to 80 cm; often ± horizontal; **leathery**; base **narrow**; PINNULES *key ID details below*.

Shield-ferns are similar, with 'mitten'-shaped pinnules. **Hab** woods, hedges, stream banks; Hard Shield-fern also in rock crevices.

PINNULES **similar size** along pinna; smaller only at apex

'thumb' prominent; ± **parallel to or overlapping** pinna midrib

MIDRIB

PINNULE

BASE broad

STALK **rounded**

margins and tips with **soft** points

PINNULES **reduce in size** along pinna

'thumb' semi-fused; **curves away from** pinna midrib

MIDRIB

PINNULE

BASE narrow

STALK **absent or flat**

margins and tips with **prickly** points

SOFT SHIELD-   HARD SHIELD-

J F M A M J J A S O N D   J F M A M J J A S O N D

SORI **kidney-shaped** (fewer per pinnule than male-ferns) | BUCKLER-FERNS

SORI

SS Other buckler-ferns; the two widespread species often occur together and also hybridize. Buckler-ferns are most easily separated by their stalk-scales.

P Broad Buckler-fern
Dryopteris dilatata

J F M A M J J A S O N D

Form 'shuttlecock' tuft. **Frond** L to 150 cm; OUTLINE ± triangular; PINNULES slightly convex. **Stalk** SCALES **golden-brown with dark centre. Hab** woodlands and other shady habitats, usually on acid soils.

PINNULES slightly convex

fronds in 'shuttlecock'

STALK-SCALES **dark centre**

FROND lowest pinnule nearest midrib L ± adjacent (usually)

Broad Buckler-fern has longer, broader and darker green fronds than Narrow Buckler-fern

P Narrow Buckler-fern
Dryopteris carthusiana

J F M A M J J A S O N D

Form creeping, fronds in irregular groups. **Frond** L to 80 cm; OUTLINE ± triangular; PINNULES almost flat. **Stalk** SCALES pale brown; no dark centre. **Hab** wet fen woodland, marshes.

PINNULES flat

fronds in irregular group

STALK-SCALES **plain**

FROND lowest pinnule nearest midrib; L > adjacent (usually)

108 Dennstaedtiaceae | **Bracken** family

| 1 sp. | 2* spp. B&I |

● **Form** the familiar patch-forming fern of heathland and woodland.

Ⓟ **Bracken** *Pteridium aquilinum*

J F M A M J J A S O N D

Form single fronds arise from **extensive, branched underground runners**. **Frond** L to 2 m, shorter (up to 50 cm) where shaded; 3-pinnate; OUTLINE broadly triangular, comprising 3, 5 or more distinct triangles; PINNULETS L to 15 mm with **downrolled margins** in which the sporangia are situated. **Hab** woods, heaths and other habitats on acid soils.

*NOTE: some authorities regard a northern form as a separate species **Pinewood Bracken** *P. pinetorum*

A stand of Bracken bordering wet heath and oak-birch woodland.

FROND broadly triangular outline, made up of smaller triangular pinnae

109 Salviniaceae | **Water Fern** family

| 1 sp. B&I |

● **Form** floating aquatic. **Fronds** branched with roots that hang in the water. Water Fern is an invasive species.

Ⓟ **Water Fern** *Azolla filiculoides* **AQ**

J F M A M J J A S O N D

Frond L to 100 mm; branched; with roots that hang in the water. **Lvs** consist of 2 oval lobes L×W to 2·5 × 1·4 mm; UPPER LOBE floating; LOWER LOBE submerged and bearing sori. Blue-green colour is due to the presence of a nitrogen-fixing blue-green alga, turns **red-brown in autumn**. **Hab** slow-moving or still lowland freshwater bodies.

Unchecked, Water Fern will form extensive continuous patches, up to 30 cm deep, outcompeting other surface plants and blocking light from reaching underwater species.

Index to species accounts

Species with brief descriptions but not illustrated are shown in light text

A

Acer campestre 128
— platanoides 128
— pseudoplatanus 128
Achillea millefolium 213
— ptarmica 213
Aconite, Winter 72
Adder's-tongue 300
Adoxa moschatellina 193
Aegopodium podagraria 226
Aesculus hippocastanum 129
Aethusa cynapium 223
Agrimonia eupatoria 99
— procera 99
Agrimony 99
—, Fragrant 99
Agrostis canina 278
— capillaris 279
— gigantea 279
— stolonifera 279
— vinealis 278
Aira caryophyllea 280
— praecox 280
Ajuga pyramidalis 183
— reptans 183
Alchemilla filicaulis 101
— glabra 101
— mollis 101
— xanthochlora 101
Alder 109
—, Grey 109
—, Italian 109
Alexanders 227
Alisma plantago-aquatica 232
Alkanet, Green 171
Alliaria petiolata 138
Allium triquetrum 242
— ursinum 242
— vineale 242
Allseed 154
Alnus cordata 109
— glutinosa 109
— incana 109
Alopecurus geniculatus 271
— myosuroides 271
— pratensis 271
Ammophila arenaria 271
Anacamptis morio 238
— pyramidalis 240
Anemone nemorosa 72
—, Wood 72
Angelica sylvestris 224
—, Wild 224
Anisantha diandra 290

Anisantha sterilis 290
Antennaria dioica 202
Anthemis cotula 214
Anthoxanthum odoratum 277
Anthriscus sylvestris 222
Antirrhinum majus 165
Anthyllis vulneraria 84
Aphanes arvensis 102
— australis 102
Apple 90
—, Crab 90
Aquilegia vulgaris 72
Arabidopsis thaliana 138
Arabis hirsuta 138
Archangel, Yellow 186
Arctium lappa 199
— minus 199
— nemorosum 199
Arctostaphylos uva-ursi 158
Arenaria leptoclados 150
— serpyllifolia 150
Armeria maritima 140
Armoracia rusticana 136
Arrhenatherum elatius 288
Arrowgrass, Marsh 233
—, Sea 233
Artemisia absinthium 201
— maritimum 201
— vulgaris 201
Arum italicum 230
— maculatum 230
Ash 179
Aspen 111
Asphodel, Bog 235
Asplenium adiantum-nigrum 304
— ceterach 303
— ruta-muraria 304
— scolopendrium 301
— trichomanes 303
— viride 303
Aster, Sea 209
Athyrium filix-femina 304
Atriplex littoralis 153
— patula 152
— portulacoides 154
— prostrata 152
Atropa belladonna 176
Aubretia 135
Aubrieta deltoidea 135
Avena fatua 289
— sativa 289
Avenella flexuosa 280
Avens, Water 97
—, Wood 99

Avenula pubescens 288
Azolla filiculoides 308

B

Ballota nigra 184
Balm 188
Balsam, Himalayan 144
Barbarea vulgaris 134
Barberry 71
Barley, Meadow 270
—, Two-rowed 270
—, Wall 270
Bartsia, Red 190
Basil, Wild 188
Bearberry 158
Bedstraw, Common Marsh- 161
—, Fen 161
—, Heath 161
—, Hedge 161
—, Lady's 160
—, Limestone 161
—, Northern 161
Beech, European 106
Beet 153
Bellflower, Adria 229
—, Trailing 229
Bellis perennis 214
Bent, Black 279
—, Brown 278
—, Common 279
—, Creeping 279
—, Velvet 278
Berberis vulgaris 71
Berula erecta 225
Beta vulgaris 153
Betonica officinalis 185
Betony 185
Betula pendula 110
— pubescens 110
Bidens cernua 199
— tripartita 199
Bilberry 158
—, Bog 158
Bindweed, Black- 144
—, Field 163
—, Hedge 162
—, Large 162
Birch, Downy 110
—, Silver 110
Bird's-foot-trefoil, Common 84
—, Greater 84
—, Narrow-leaved 84
Bird's-nest, Yellow 192
Bistort, Amphibious 143
Bitter-cress, Hairy 139

Bitter-cress, Large | 139
—, Wavy | 139
Bitter-vetch | 85
Bittersweet | 177
Black-bindweed | 144
Black-grass | 271
Black-poplar | 111
—, Hybrid | 111
Blackberry, Elm-leaved | 96
Blackstonia perfoliata | 189
Blackthorn | 92
Bladder-fern, Brittle | 304
Bladderworts | 193
Blechnum spicant | 302
Blinks | 154
Blood-drop-emlets | 190
Bluebell | 243
—, Hybrid | 243
—, Spanish | 243
Bog-myrtle | 108
Bog-rush, Black | 253
Bogbean | 230
Bolboschoenus maritimus | 253
Borage | 171
Borago officinalis | 171
Box, European | 75
Brachypodium pinnatum | 290
— *rupestre* | 290
— *sylvaticum* | 290
Bracken | 308
Bramble | 96
Brassica oleracea | 133
— *napus* | 133
— *nigra* | 133
— *rapa* | 133
Briza maxima | 276
— *media* | 276
— *minor* | 276
Brome, Barren | 290
—, False | 290
—, Great | 290
—, Lesser Wood | 290
—, Wood | 290
Bromopsis ramosa | 290
Bromus benekenii | 290
— *hordeaceus* | 290
Brooklime | 166
Broom | 81
—, Spanish | 81
Broomrape, Common | 192
—, Ivy | 192
Bryonia dioica | 114
Bryony, Black | 236
—, White | 114
Buckler-fern, Broad | 307
—, Narrow | 307
Buckthorn | 103
—, Alder | 103

Buckthorn, Sea- | 102
Buddleja davidii | 179
Bugle | 183
—, Pyramidal | 183
Bugloss | 171
—, Viper's- | 170
Bur-marigold, Nodding | 199
—, Trifid | 199
Bur-reed, Branched | 244
—, Unbranched | 244
Burdock, Greater | 199
—, Lesser | 199
—, Wood | 199
Burnet, Great | 100
—, Salad | 100
Burnet-saxifrage | 227
Butterbur | 202
Buttercup, Bulbous | 74
—, Celery-leaved | 75
—, Creeping | 74
—, Goldilocks | 74
—, Hairy | 74
—, Meadow | 74
Butterfly-bush | 179
Butterfly-orchid, Greater | 237
—, Lesser | 237
Butterwort, Common | 193
Buxus sempervirens | 75

C
Cabbage | 133
Calamagrostis spp. | 287
Calendula officinalis | 210
Callitriche brutia | 228
— *stagnalis* | 228
Calluna vulgaris | 159
Caltha palustris | 72
Calystegia sepium | 162
— *silvatica* | 162
Campanula portenschlagiana | 229
— *poscharskyana* | 229
— *rotundifolia* | 229
Campion, Bladder | 146
—, Red | 147
—, Sea | 146
—, White | 146
Canary-grass, Reed | 287
Capsella bursa-pastoris | 136
Cardamine amara | 139
— *flexuosa* | 139
— *hirsuta* | 139
— *pratensis* | 139
Carduus crispus | 197
— *nutans* | 197
Carex acuta | 264
— *acutiformis* | 265
— *arenaria* | 257
— *binervis* | 262
— *canescens* | 255

Carex caryophyllea | 258
— *demissa* | 260
— *dioica* | 254
— *disticha* | 257
— *divulsa* | 255
— *echinata* | 257
— *elata* | 264
— *extensa* | 260
— *flacca* | 263
— *hirta* | 258
— *hostiana* | 261
— *laevigata* | 262
— *lepidocarpa* | 260
— *leporina* | 257
— *muricata* | 256
— *nigra* | 264
— *otrubae* | 256
— *pallescens* | 261
— *panicea* | 263
— *paniculata* | 256
— *pendula* | 259
— *pilulifera* | 258
— *pseudocyperus* | 264
— *pulicaris* | 254
— *remota* | 255
— *riparia* | 265
— *rostrata* | 264
— *spicata* | 256
— *strigosa* | 259
— *sylvatica* | 259
— *vesicaria* | 264
Carlina vulgaris | 196
Carpinus betulus | 110
Carrot, Wild | 223
Castanea sativa | 106
Cat's-ear | 203
Cat's-tail, Smaller | 272
Catapodium rigidum | 276
Celandine, Greater | 68
—, Lesser | 72
Centaurea cyanus | 198
— *debeauxii* | 198
— *montana* | 198
— *nigra* | 198
— *scabiosa* | 198
Centaurium erythraea | 189
— *pulchellum* | 189
Centaury, Common | 189
—, Lesser | 189
Centranthus ruber | 219
Cerastium arvense | 149
— *diffusum* | 149
— *fontanum* | 149
— *glomeratum* | 149
— *semidecandrum* | 149
— *tomentosum* | 149
Ceratocapnos claviculata | 70
Ceratophyllum demersum | 67

Ceratophyllum submersum — 67
Chaenorhinum minus — 165
Chaerophyllum temulum — 222
Chamaenerion angustifolium — 128
Chamomile, Stinking — 214
Charlock — 132
Chelidonium majus — 68
Chenopodium album — 152
— *ficifolium* — 153
Cherry, Bird — 93
—, Wild — 93
Chervil, Rough — 222
Chestnut, Horse- — 129
—, Sweet — 106
Chickweed, Common — 148
Chicory — 203
Chrysosplenium alternifolium — 77
— *oppositifolium* — 77
Cicely, Sweet — 223
Cichorium intybus — 203
Cinquefoil, Creeping — 98
—, Marsh — 97
Circaea lutetiana — 125
Cirsium arvense — 196
— *dissectum* — 197
— *heterophyllum* — 197
— *palustre* — 196
— *vulgare* — 196
Claytonia sibirica — 154
Cleavers — 161
Clematis vitalba — 73
Clinopodium vulgare — 188
Clover, Alsike — 86
—, Hare's-foot — 86
—, Red — 86
—, White — 86
—, Zigzag — 86
Club-rush, Bristle — 253
—, Common — 253
—, Grey — 253
—, Sea — 253
—, Slender — 253
Clubmoss, Fir — 300
—, Krauss's — 300
—, Lesser — 300
—, Marsh — 300
Cochlearia anglica — 137
— *danica* — 137
— *officinalis* — 137
Cock's-foot — 281
Colt's-foot — 212
Columbine — 72
Comarum palustre — 97
Comfrey, Common — 170
—, Russian — 170
Conium maculatum — 224
Conopodium majus — 225
Convolvulus arvensis — 163

Cord-grass, Common — 289
—, Small — 289
Cornflower — 198
—, Perennial — 198
Cornsalad, Common — 219
—, Keeled-fruited — 219
Cornus alba — 155
— *sanguinea* — 155
— *sericea* — 155
Corydalis, Climbing — 70
—, Yellow — 70
Corylus avellana — 109
Cotoneaster, Entire-leaved — 90
—, Himalayan — 90
— *horizontalis* — 90
— *integrifolius* — 90
— *simonsii* — 90
—, Wall — 90
Cottongrass, Broad-leaved — 251
—, Common — 251
—, Hare's-tail — 251
—, Slender — 251
Couch, Bearded — 269
—, Common — 269
—, Sea — 269
Cow-wheat, Common — 190
Cowberry — 158
Cowslip — 157
Crack-willow — 112
Crane's-bill, Bloody — 122
—, Cut-leaved — 122
—, Dove's-foot — 123
—, Hedgerow — 123
—, Meadow — 122
—, Shining — 123
—, Small-flowered — 123
—, Wood — 122
Crassula helmsii — 78
— *tillaea* — 78
Crataegus laevigata — 94
— *monogyna* — 94
Creeping-Jenny — 156
Crepis capillaris — 205
— *paludosa* — 205
— *vesicaria* — 205
Cress, Field Penny- — 136
—, Hairy Bitter- — 139
—, Hoary — 137
—, Large Bitter- — 139
—, Lesser Swine- — 137
—, Swine- — 137
—, Thale — 138
—, Water- — 138
—, Wavy Bitter- — 139
—, Winter- — 134
Crocosmia × *crocosmiiflora* — 241
Crosswort — 160
Crowberry — 159

Crowfoot, Ivy-leaved — 73
—, Round-leaved — 73
Cruciata laevipes — 160
Cuckooflower — 139
Cudweed, Common — 200
—, Marsh — 200
Cupressus lawsoniana — 296
— × *leylandii* — 296
Currant, Black — 76
—, Downy — 76
—, Flowering — 76
—, Mountain — 76
—, Red — 76
Cuscuta epithymum — 163
— *europaea* — 163
Cymbalaria muralis — 165
Cynosurus cristatus — 277
Cypress, Lawson's — 296
—, Leyland — 296
Cystopteris fragilis — 304
Cytisus scoparius — 81

D
Dactylis glomerata — 281
Dactylorhiza fuchsii — 239
— *incarnata* — 239
— *maculata* — 239
— *praetermissa* — 239
— *purpurella* — 239
Daffodil — 241
Daisy — 214
—, Oxeye — 215
—, Shasta — 215
Dame's-violet — 135
Dandelions — 209
Danthonia decumbens — 277
Datura stramonium — 176
Daucus carota — 223
Dead-nettle, Cut-leaved — 187
—, Henbit — 187
—, Northern — 187
—, Red — 187
—, Spotted — 186
—, White — 186
Deergrass, Common — 252
—, Northern — 252
Deschampsia cespitosa — 280
Dewberry — 96
Digitalis purpurea — 169
Dipsacus fullonum — 220
Dittander — 136
Dock, Broad-leaved — 141
—, Clustered — 141
—, Curled — 141
—, Wood — 141
Dodder — 163
—, Greater — 163
Dog-rose — 95
Dog-violet, Common — 118

Dog-violet, Early 118
—, Heath 118
Dog's-tail, Crested 277
Dogwood 155
—, Red-osier 155
—, White 155
Dropwort 100
Drosera anglica 145
— *intermedia* 145
— *rotundifolia* 145
Dryopteris affinis 306
— *borreri* 306
— *carthusiana* 307
— *dilatata* 307
— *filix-mas* 306
Duckweed, Common 231
—, Fat 231
—, Greater 231
—, Ivy-leaved 231
—, Least 231

E
Echium vulgare 170
Elder 217
—, Ground- 226
Eleocharis palustris 252
— *quinqueflora* 252
Elm, 'Field' 104
—, Wych 104
Elodea canadensis 232
— *nuttallii* 232
Elymus athericus 269
— *caninus* 269
— *repens* 269
Empetrum nigrum 159
Enchanter's-nightshade 125
Epilobium brunnescens 126
— *ciliatum* 127
— *hirsutum* 127
— *montanum* 127
— *obscurum* 126
— *palustre* 127
— *parviflorum* 127
— *tetragonum* 126
Epipactis helleborine 237
Equisetum arvense 299
— *fluviatile* 298
— *palustre* 298
— *sylvaticum* 299
— *telmateia* 299
Eranthis hyemalis 72
Erica cinerea 159
— *tetralix* 159
Erigeron acris 209
— *canadensis* 216
— *floribundus* 216
— *karvinskianus* 214
— *sumatrensis* 216
Eriophorum angustifolium 251

Eriophorum gracilis 251
— *latifolium* 251
— *vaginatum* 251
Erodium cicutarium 124
— *moschatum* 124
Erophila verna 136
Ervilia hirsuta 83
Ervum tetraspermum 83
Erysimum cheiri 134
Erythranthe guttata 190
— *luteus* 190
Euonymus europaeus 115
Eupatorium cannabinum 201
Euphorbia amygdaloides 117
— *exigua* 116
— *helioscopia* 116
— *lathyris* 117
— *peplus* 116
Euphrasia arctica 191
— *nemorosa* 191
Evening-primrose,
Large-flowered 125
Eyebright, Arctic 191
—, Common 191

F
Fagus sylvatica 106
Fallopia baldschuanica 144
— *convolvulus* 144
False-brome, Heath 290
Fat-hen 152
Fennel 227
Fern, Borrer's Scaly Male- 306
—, Broad Buckler- 307
—, Golden Scaly Male- 306
—, Hard Shield- 307
—, Hard- 302
—, Lady- 304
—, Lemon-scented 305
—, Male- 306
—, Narrow Buckler- 307
—, Royal 304
—, Soft Shield- 307
—, Water 308
Fern-grass 276
Fescue, Giant 284
—, Meadow 284
—, Rat's-tail 285
—, Red 285
—, Squirreltail 285
—, Tall 284
Festuca ovina 285
— *rubra* 285
Feverfew 215
Ficaria verna 72
Field-rose 95
Field-speedwell, Common 168
—, Green 168
—, Grey 168

Figwort, Common 180
—, Water 180
Filago germanica 200
Filipendula ulmaria 100
— *vulgaris* 100
Fir, Douglas 293
Flax 115
—, Fairy 115
Fleabane, Bilbao 216
—, Blue 209
—, Canadian 216
—, Common 212
—, Guernsey 216
—, Mexican 214
Fluellen, Round-leaved 164
—, Sharp-leaved 164
Foeniculum vulgare 227
Forget-me-not, Changing 172
—, Creeping 173
—, Early 172
—, Field 172
—, Tufted 173
—, Water 173
—, Wood 172
Fox-and-cubs 203
Fox-sedge, False 256
Foxglove 169
Foxtail, Marsh 271
—, Meadow 271
Fragaria vesca 97
Fragrant-orchid, Chalk 240
—, Heath 240
—, Marsh 240
Frangula alnus 103
Fraxinus excelsior 179
Fuchsia 125
Fuchsia magellanica 125
Fumaria muralis 70
— *officinalis* 70
Fumitory, Common 70
—, Common Ramping- 70

G
Galanthus nivalis 242
Galega officinalis 85
Galeopsis bifida 184
— *tetrahit* 184
Galinsoga parviflora 215
— *quadriradiata* 215
Galium album 161
— *aparine* 161
— *boreale* 161
— *odoratum* 162
— *palustre* 161
— *saxatile* 161
— *sterneri* 161
— *uliginosum* 161
— *verum* 160
Gallant-soldier 215

Garlic, Three-cornered 242
Gentian, Autumn 189
Gentianella amarella 189
Geranium dissectum 122
— lucidum 123
— molle 123
— pratense 122
— pusillum 123
— pyrenaicum 123
— robertianum 124
— sanguineum 122
— sylvaticum 122
Geum rivale 97
— urbanum 99
Glaucium flavum 68
Glassworts 151
Glebionis segetum 212
Glechoma hederacea 188
Globeflower 72
Glyceria declinata 286
— fluitans 286
— maxima 287
— notata 286
Gnaphalium uliginosum 200
Goat's-beard 203
Goat's-rue 85
Golden-saxifrage,
    Alternate-leaved 77
—, Opposite-leaved 77
Goldenrod 213
—, Canadian 213
Gooseberry 76
Goosefoot, Fig-leaved 153
—, Many-seeded 153
—, Red 152
Gorse 81
—, Dwarf 81
—, Western 81
Grape-hyacinth, Garden 243
Ground-elder 226
Ground-ivy 188
Groundsel, Common 200
—, Heath 211
—, Sticky 211
Guelder-rose 217
—, Asian 217
Gymnadenia borealis 240
— conopsea 240
— densiflora 240
Gypsywort 182

H

Hair-grass, Crested 277
—, Early 280
—, Silver 280
—, Tufted 280
—, Wavy 280
Hard-fern 302
Harebell 229

Hart's-tongue 301
Hawk's-beard, Beaked 205
—, Marsh 205
—, Smooth 205
Hawkbit, Autumn 208
—, Lesser 208
—, Rough 208
Hawkweed, Mouse-eared 208
Hawkweeds 206
Hawthorn 94
—, Midland 94
Hazel 109
Heath, Cross-leaved 159
Heath-grass 277
Heather 159
—, Bell 159
Hedera helix 221
— hibernica 221
Hedge-parsley, Upright 222
Helianthemum
    nummularium 130
Helictochloa pratensis 288
Heliotrope, Winter 202
Helleborine, Broad-leaved 237
Helminthotheca echioides 205
Helosciadium nodiflorum 225
Hemlock 224
Hemlock-spruce, Western 293
Hemp-agrimony 201
Hemp-nettle, Bifid 184
—, Common 184
Henbane 176
Heracleum mantegazzianum 226
— sphondylium 226
Herb-Robert 124
Hesperis matronalis 135
Hieracium spp. 206
Hippophaë rhamnoides 102
Hippuris vulgaris 180
Hirschfeldia incana 133
Hogweed 226
—, Giant 226
Holcus lanatus 281
— mollis 281
Holly 169
Honesty 135
Honeysuckle 218
—, Box-leaved 75
—, Garden 218
—, Himalayan 218
—, Wilson's 218
Hop 103
Hordeum distichon 270
— murinum 270
— secalinum 270
Horehound, Black 184
Hornbeam 110
Horned-poppy, Yellow 68

Hornwort, Rigid 67
—, Soft 67
Horse-chestnut 129
Horse-radish 136
Horsetail, Field 299
—, Great 299
—, Marsh 298
—, Water 298
—, Wood 299
Hottonia palustris 80
House-leek 78
Humulus lupulus 103
Huperzia selago 300
Hyacinthoides × massartiana 243
— hispanica 243
— non-scripta 243
Hydrocotyle ranunculoides 221
— vulgaris 221
Hyoscyamus niger 176
Hypericum androsaemum 121
— hirsutum 121
— humifusum 120
— maculatum 121
— perforatum 120
— pulchrum 120
— tetrapterum 121
Hypochaeris radicata 203
Hypopitys monotropa 192

I

Ilex aquifolium 169
Impatiens glandulifera 144
Inula conyzae 169
Iris foetidissima 241
— pseudacorus 241
—, Stinking 241
—, Yellow 241
Isoëtes spp. 174
Isolepis cernua 253
— setacea 253
Ivy 221
—, Atlantic 221

J

Jacobaea aquatica 210
— erucifolia 210
— vulgaris 210
Jasione montana 229
Juglans regia 108
Juncus acutiflorus 247
— articulatus 247
— bufonius 246
— bulbosus 247
— conglomeratus 248
— effusus 248
— foliosus 246
— inflexus 248
— maritimus 248
— ranarius 246
— squarrosus 246

Juncus tenuis 246
Juniper, Common 295
Juniperus communis 295

K

Kickxia elatine 164
— *spuria* 164
Knapweed, Chalk 198
—, Common 198
—, Greater 198
Knautia arvensis 220
Knotgrass 142
—, Equal-leaved 142
Knotweed, Japanese 144
Koeleria macrantha 277

L

Laburnum anagyroides 82
Laburnum 82
Lactuca serriola 206
— *virosa* 206
Lady-fern 304
Lady's-mantle, Pale 101
—, Slender 101
—, Smooth 101
—, Soft 101
Lagarosiphon major 232
Lamiastrum galeobdolon 186
Lamium album 186
— *amplexicaule* 187
— *confertum* 187
— *hybridum* 187
— *maculatum* 186
— *purpureum* 187
Lapsana communis 204
Larch, European 294
—, Hybrid 294
—, Japanese 294
Larix × marschlinsii 294
— *decidua* 294
— *kaempferi* 294
Lathraea clandestina 192
— *squamaria* 192
Lathyrus linifolius 85
— *pratensis* 84
Laurel, Cherry 93
—, Portugal 93
Lemna gibba 231
— *minor* 231
— *minuta* 231
— *trisulca* 231
Leontodon hispidus 208
— *saxatilis* 208
Lepidium coronopus 137
— *didymum* 137
— *draba* 137
— *latifolium* 136
Lettuce, Great 206
—, Prickly 206
—, Wall 204

Leucanthemum × superbum 215
— *vulgare* 215
Leycesteria formosa 218
Leymus arenarius 269
Ligustrum ovalifolium 178
— *vulgare* 178
Lilac 178
Lime 129
—, Large-leaved 129
—, Small-leaved 129
Limonium spp. 140
Linaria purpurea 165
— *vulgaris* 165
Linum catharticum 115
— *usitatissimum* 115
Lipandra polysperma 153
Littorella uniflora 174
Lobelia dortmanna 174
—, Water 174
Lolium multiflorum 268
— *perenne* 268
Lonicera nitida 218
— *periclymenum* 218
— *pileata* 75
— × *italica* 218
Loosestrife, Dotted 157
—, Purple- 124
—, Tufted 157
—, Yellow 157
Lords-and-Ladies 230
—, Italian 230
Lotus corniculatus 84
— *pedunculatus* 84
— *tenuis* 84
Lousewort 191
—, Marsh 191
Lunaria annua 135
Lungwort 171
Luzula campestris 249
— *multiflora* 249
— *pilosa* 249
— *sylvatica* 249
Lycium barbarum 176
— *chinense* 176
Lycopodiella inundata 300
Lycopsis arvensis 171
Lycopus europaeus 182
Lyme-grass 269
Lysimachia arvensis 156
— *foemina* 156
— *maritima* 156
— *nemorum* 156
— *nummularia* 156
— *punctata* 157
— *tenella* 156
— *thyrsiflora* 157
— *vulgaris* 157
Lythrum portula 124

Lythrum salicaria 124

M

Madder, Field 160
Mahonia aquifolium 71
Male-fern 306
—, Borrer's Scaly 306
—, Golden Scaly 306
Mallow, Common 130
—, Dwarf 130
Malus domestica 90
— *sylvestris* 90
Malva moschata 130
— *multiflora* 130
— *neglecta* 130
— *sylvestris* 130
Maple, Field 128
—, Norway 128
Mare's-tail 180
Marigold, Corn 212
—, Pot 210
Marjoram, Wild 188
Marram 271
Marsh-bedstraw, Common 161
Marsh-marigold 72
Marsh-orchid, Early 239
—, Northern 239
—, Southern 239
Mat-grass 268
Matricaria chamomilla 214
— *discoidea* 200
Mayweed, Scented 214
—, Scentless 214
—, Sea 214
Meadow-grass, Annual 282
—, Early 282
—, Rough 283
—, Smooth 283
—, Spreading 283
—, Wood 283
Meadowsweet 100
Medicago arabica 87
— *lupulina* 87
Medick, Black 87
—, Spotted 87
Melampyrum pratense 190
Melica uniflora 276
Melick, Wood 276
Melilot, Ribbed 83
—, Tall 83
—, White 83
Melilotus albus 83
— *altissimus* 83
— *officinalis* 83
Melissa officinalis 188
Mentha aquatica 182
— *arvensis* 182
— *requienii* 105
Mentha spicata 182

Menyanthes trifoliata 230
Mercurialis annua 117
— *perennis* 117
Mercury, Annual 117
—, Dog's 117
Michaelmas-daisy, Common 209
Mignonette, Wild 131
Milium effusum 276
Milkwort, Common 88
—, Heath 88
—, Sea 156
Millet, Wood 276
Mind-your-own-business 105
Mint, Corn 182
—, Corsican 105
—, Spear 182
—, Water 182
Mistletoe 131
Moehringia trinervia 150
Molinia caerulea 278
Monkeyflower 190
Montbretia 241
Montia fontana 154
Moor-grass, Blue 277
—, Purple 278
Moschatel 193
Mountain-everlasting 202
Mouse-ear, Common 149
—, Field 149
—, Little 149
—, Sea 149
—, Sticky 149
Mugwort 201
Mullein, Great 179
Muscari armeniacum 243
Musk-mallow 130
Mustard, Black 133
—, Garlic 138
—, Hedge 134
—, Hoary 133
—, White 132
Mycelis muralis 204
Myosotis arvensis 172
— *discolor* 172
— *laxa* 173
— *ramosissima* 172
— *scorpioides* 173
— *secunda* 173
— *sylvatica* 172
Myrica gale 108
Myriophyllum alterniflorum 80
— *spicatum* 80
Myrrhis odorata 223

N

Narcissus, Poet's 241
— *poeticus* 241
Narcissus pseudonarcissus 241
Nardus stricta 268

Narthecium ossifragum 235
Nasturtium microphyllum 138
— *officinale* 138
Navelwort 78
Neottia cordata 238
— *nidus-avis* 192
— *ovata* 238
Nettle, Common 105
—, Small 105
Nightshade, Black 177
—, Deadly 176
—, Enchanter's- 125
Nipplewort 204
Nuphar lutea 67
— *pumila* 67
Nymphaea alba 67
Nymphoides peltata 67

O

Oak, Evergreen 106
—, Pedunculate 107
—, Red 107
—, Scarlet 107
—, Sessile 107
—, Turkey 107
Oat 289
Oat, Wild- 289
Oat-grass, Downy 288
—, False 288
—, Meadow 288
—, Yellow 288
Odontites vernus 190
Oenanthe crocata 224
Oenothera glazioviana 125
Onion, Wild 242
Onobrychis viciifolia 85
Ononis repens 82
— *spinosa* 82
Ophioglossum vulgatum 300
Ophrys apifera 237
Orache, Common 152
—, Grass-leaved 153
—, Spear-leaved 152
Orchid, Bee 237
—, Bird's-nest 192
—, Chalk Fragrant- 240
—, Early-purple 238
—, Greater Butterfly- 237
—, Green-winged 238
—, Heath Fragrant- 240
—, Lesser Butterfly- 237
—, Marsh Fragrant- 240
—, Pyramidal 240
Orchis mascula 238
Oregon-grape 71
Oreopteris limbosperma 305
Origanum vulgare 188
Orobanche hederae 192
— *minor* 192

Osier 112
Osmunda regalis 304
Oxalis acetosella 114
— *corniculata* 114
Oxlip 157
Oxtongue, Bristly 205
—, Hawkweed 205
Oxybasis rubra 152

P

Pansy, Dwarf 119
—, Field 119
—, Wild 119
Papaver cambricum 68
— *dubium* 69
— *lecoqii* 69
— *rhoeas* 69
— *setiferum* 69
— *somniferum* 69
Parietaria judaica 105
Parsley, Cow 222
—, Fool's 223
—, Upright Hedge- 222
Parsley-piert 102
—, Slender 102
Pear, Wild 90
Pearlwort, Annual 151
—, Knotted 150
—, Procumbent 150
—, Sea 151
—, Slender 151
Pedicularis palustris 191
— *sylvatica* 191
Pellitory-of-the-wall 105
Penny-cress, Field 136
Pennywort, Floating 221
—, Marsh 221
Pentaglottis sempervirens 171
Periwinkle, Greater 163
—, Lesser 163
Persicaria amphibia 143
— *hydropiper* 143
— *lapathifolia* 143
— *maculosa* 143
— *mitis* 143
—, Pale 143
Petasites hybridus 202
— *pyrenaicus* 202
Petrosedum forsterianum 79
— *rupestre* 79
Phalaris arundinacea 287
Phleum bertolonii 272
— *pratense* 272
Phlox 147
— *paniculata* 147
Phragmites australis 287
Picea abies 293
— *sitchensis* 293
Picris hieracioides 205

Pigmyweed, New Zealand — 78
Pignut — 225
Pilosella aurantiaca — 203
— *officinarum* — 208
Pimpernel, Blue — 156
—, Bog — 156
—, Scarlet — 156
—, Yellow — 156
Pimpinella saxifraga — 227
Pine, Black — 292
—, Lodgepole — 292
—, Scots — 292
Pineappleweed — 200
Pinguicula vulgaris — 193
Pinus contorta — 292
— *nigra* — 292
— *sylvestris* — 292
Plantago coronopus — 174
— *lanceolata* — 175
— *major* — 175
— *maritima* — 174
— *media* — 175
Plantain, Buck's-horn — 174
—, Greater — 175
—, Hoary — 175
—, Ribwort — 175
—, Sea — 174
Platanthera bifolia — 237
— *chlorantha* — 237
Ploughman's-spikenard — 169
Plum, Cherry — 92
—, Wild — 92
Poa annua — 282
— *humilis* — 283
— *infirma* — 282
— *nemoralis* — 283
— *pratensis* — 283
— *trivialis* — 283
Polygala serpyllifolia — 88
— *vulgaris* — 88
Polygonum arenastrum — 142
— *aviculare* — 142
Polypodium cambricum — 302
— *interjectum* — 302
— *vulgare* — 302
Polypody — 302
—, Intermediate — 302
—, Southern — 302
Polystichum aculeatum — 307
— *setiferum* — 307
Pond-sedge, Greater — 265
—, Lesser — 265
Pondweed, Bog — 235
—, Broad-leaved — 235
—, Curled — 234
—, Fennel — 234
—, Small — 234
Poplar, Black- — 111

Poplar, Grey — 111
—, Hybrid Black- — 111
—, White — 111
Poppy, Common — 69
—, Long-headed — 69
—, Opium — 69
—, Oriental — 69
—, Welsh — 68
—, Yellow-juiced — 69
Populus alba — 111
— *deltoides × nigra* — 111
— *nigra* — 111
— *tremula* — 111
— *× canadensis* — 111
— *× canescens* — 111
Potamogeton berchtoldii — 234
— *crispus* — 234
— *natans* — 235
— *polygonifolius* — 235
Potato — 177
Potentilla anglica — 98
— *anserina* — 99
— *erecta* — 98
— *reptans* — 98
— *sterilis* — 97
Poterium sanguisorba — 100
Primrose — 157
Primula elatior — 157
— *veris* — 157
— *vulgaris* — 157
Privet, Garden — 178
—, Wild — 178
Prunella vulgaris — 183
Prunus avium — 93
— *cerasifera* — 92
— *domestica* — 92
— *laurocerasus* — 93
— *lusitanica* — 93
— *padus* — 93
— *spinosa* — 92
Pseudofumaria lutea — 70
Pseudotsuga menziesii — 293
Pteridium aquilinum — 308
Puccinellia distans — 282
— *maritima* — 282
Pulicaria dysenterica — 212
Pulmonaria officinalis — 171
Purple-loosestrife — 124
Purslane, Pink — 154
Pyrus pyraster — 90

Q
Quaking-grass — 276
—, Greater — 276
—, Lesser — 276
Quercus cerris — 107
— *coccinea* — 107
— *ilex* — 106
— *petraea* — 107

Quercus robur — 107
— *rubra* — 107
Quillworts — 174

R
Radiola linoides — 154
Radish, Sea — 135
—, Wild — 135
Ragged-Robin — 147
Ragwort, Common — 210
—, Hoary — 210
—, Marsh — 210
—, Narrow-leaved — 211
—, Oxford — 211
Ramping-fumitory, Common — 70
Ramsons — 242
Ranunculus acris — 74
— *aquatilis* — 73
Ranunculus auricomus — 74
— *bulbosus* — 74
— *flammula* — 75
— *hederaceus* — 73
— *lingua* — 75
— *omiophyllus* — 73
— *peltatus* — 73
— *penicillatus* — 73
— *repens* — 74
— *sardous* — 74
— *sceleratus* — 75
Rape — 133
Raphanus raphanistrum — 135
Raspberry — 96
Red-cedar, Western — 296
Redshank — 143
Reed, Common — 287
Reedmace, Greater — 244
—, Lesser — 244
Reseda lutea — 131
— *luteola* — 131
Restharrow, Common — 82
—, Spiny — 82
Reynoutria japonica — 144
Rhamnus cathartica — 103
Rhinanthus minor — 190
Rhododendron — 159
— *ponticum* — 159
Ribes alpinum — 76
— *nigrum* — 76
— *rubrum* — 76
— *sanguineum* — 76
— *spicatum* — 76
— *uva-crispa* — 76
Rock-cress, Hairy — 138
Rock-rose, Common — 130
Roemeria spp. — 69
Rorippa palustris — 134
— *sylvestris* — 134
Rosa arvensis — 95
— *canina* — 95

Rosa rubiginosa 95
— *rugosa* 95
— *sherardii* 95
— *spinosissima* 94
Rose, Burnet 94
—, Dog- 95
—, Field- 95
—, Guelder- 217
—, Japanese 95
—, Sherard's Downy- 95
Rowan 91
Rubus caesius 96
— *fruticosus* agg. 96
— *idaeus* 96
— *ulmifolius* 96
Rumex acetosa 142
— *acetosella* 142
— *conglomeratus* 141
— *crispus* 141
— *obtusifolius* 141
— *sanguineus* 141
Rush, Bulbous 247
—, Compact 248
—, Frog 246
—, Hard 248
—, Heath 246
—, Jointed 247
—, Leafy 246
—, Sea 248
—, Sharp-flowered 247
—, Slender 246
—, Soft- 248
—, Toad 246
Russian-vine 144
Rustyback 303
Rye-grass, Italian 268
—, Perennial 268

S
Sage, Wood 183
Sagina apetala 151
— *filicaulis* 151
— *maritima* 151
— *nodosa* 150
— *procumbens* 150
Sainfoin 85
Salicornia spp. 151
Salix × *fragilis* 112
— *alba* 112
— *aurita* 113
— *caprea* 113
— *cinerea* 113
— *eleagnos* 112
— *purpurea* 112
— *repens* 113
— *viminalis* 112
Salsify 203
Saltmarsh-grass, Common 282
—, Reflexed 282

Sambucus nigra 217
Sandwort, Slender 150
—, Three-nerved 150
—, Thyme-leaved 150
Sanguisorba officinalis 100
Sanicle 228
Sanicula europaea 228
Saponaria officinalis 147
Sarcocornia spp. 151
Saxifraga granulata 77
— *tridactylites* 77
Saxifrage, Meadow 77
—, Rue-leaved 77
Scabiosa columbaria 220
Scabious, Devil's-bit 220
—, Field 220
—, Small 220
Schedonorus arundinaceus 284
— *giganteus* 284
— *pratensis* 284
Schoenoplectus lacustris 253
— *tabernaemontani* 253
Schoenus nigricans 253
Scorzoneroides autumnalis 208
Scrophularia auriculata 180
— *nodosa* 180
Scurvygrass, Common 137
—, Danish 137
—, English 137
Scutellaria galericulata 183
— *minor* 183
Sea-blite, Annual 154
Sea-buckthorn 102
Sea-lavender 140
Sea-purslane 154
Sea-spurrey, Greater 146
—, Lesser 146
Sedge, Bladder 264
—, Bottle 264
—, Brown 257
—, Carnation 263
—, Common 264
—, Common Yellow- 260
—, Cyperus 264
—, Dioecious 254
—, Flea 254
—, Glaucous 263
—, Greater Pond- 265
—, Green-ribbed 262
—, Grey 255
—, Hairy 258
—, Lesser Pond- 265
—, Long-bracted 260
—, Long-stalked Yellow- 260
—, Oval 257
—, Pale 261
—, Pendulous 259
—, Pill 258

Sedge, Prickly 256
—, Remote 255
—, Sand 257
——, Smooth-stalked 262
—, Spiked 256
—, Spring- 258
—, Star 257
—, Tawny 261
—, White 255
—, Wood- 259
Sedum acre 79
— *album* 79
— *anglicum* 79
— *dasyphyllum* 79
Selaginella kraussiana 300
— *selaginoides* 300
Selfheal 183
Sempervivum tectorum 78
Senecio inaequidens 211
— *squalidus* 211
— *sylvaticus* 211
— *viscosus* 211
— *vulgaris* 200
Sesleria caerulea 277
Shaggy-soldier 215
Sheep's-bit 229
Sheep's-fescue 285
Shepherd's-purse 136
Sherardia arvensis 160
Shield-fern, Hard 307
—, Soft 307
Shoreweed 174
Silene dioica 147
— *flos-cuculi* 147
— *latifolia* 146
— *uniflora* 146
— *vulgaris* 146
Silverweed 99
Sinapis alba 132
— *arvensis* 132
Sisymbrium officinale 134
Skullcap 183
—, Lesser 183
Small-reeds 287
Smyrnium olusatrum 227
Snapdragon 165
Sneezewort 213
Snow-in-summer 149
Snowberry 218
Snowdrop 242
Soapwort 147
Soft-brome 290
Soft-grass, Creeping 281
Soft-rush 248
Solanum dulcamara 177
— *lycopersicum* 177
— *nigrum* 177
— *tuberosum* 177

Soldier, Gallant- 215
—, Shaggy- 215
Soleirolia soleirolii 105
Solidago canadensis 213
— *virgaurea* 213
Sonchus arvensis 207
— *asper* 207
— *oleraceus* 207
Sorbus aria 91
— *aucuparia* 91
— *intermedia* 91
Sorrel, Common 142
—, Procumbent Yellow- 114
—, Sheep's 142
—, Wood- 114
Sow-thistle, Perennial 207
—, Prickly 207
—, Smooth 207
Sparganium emersum 244
— *erectum* 244
Spartina anglica 289
— *maritima* 289
Spartium junceum 81
Spearwort, Greater 75
—, Lesser 75
Speedwell, Blue Water- 166
—, Common Field- 168
—, Germander 167
—, Green Field- 168
—, Grey Field- 168
—, Heath 167
—, Ivy-leaved 168
—, Marsh 166
—, Pink Water- 166
—, Slender 168
—, Thyme-leaved 167
—, Wall 167
—, Wood 167
Spergula arvensis 147
Spergularia marina 146
— *media* 146
— *rubra* 147
Spike-rush, Common 252
—, Few-flowered 252
Spindle 115
Spirodela polyrhiza 231
Spleenwort, Black 304
—, Green 303
—, Maidenhair 303
Spotted-orchid, Common 239
—, Heath 239
Spring-sedge 258
Spruce, Norway 293
—, Sitka 293
Spurge, Caper 117
—, Dwarf 116
—, Petty 116
—, Sun 116

Spurge, Wood 117
Spurrey, Corn 147
—, Sand 147
St John's-wort, Hairy 121
—, Imperforate 121
—, Perforate 120
—, Slender 120
—, Square-stalked 121
—, Trailing 120
Stachys arvensis 185
— *palustris* 185
— *sylvatica* 185
Stellaria alsine 148
— *graminea* 148
— *holostea* 148
— *media* 148
— *nemorum* 148
— *palustris* 148
Stitchwort, Bog 148
—, Greater 148
—, Lesser 148
—, Marsh 148
—, Wood 148
Stonecrop, Biting 79
—, English 79
—, Mossy 78
—, Reflexed 79
—, Rock 79
—, Thick-leaved 79
—, White 79
Stork's-bill, Common 124
—, Musk 124
Strawberry, Barren 97
—, Wild 97
Stuckenia pectinata 234
Suaeda maritima 154
Succisa pratensis 220
Sundew, Great 145
—, Oblong-leaved 145
—, Round-leaved 145
Sweet-briar 95
Sweet-grass, Floating 286
—, Plicate 286
—, Reed 287
—, Small 286
Swine-cress 137
—, Lesser 137
Sycamore 128
Symphoricarpos albus 218
Symphyotrichum × salignum 209
Symphytum asperum
   × officinale 170
— *officinale* 170
Syringa vulgaris 178

T
Tamarisk 296
Tamarix gallica 296
Tamus communis 236

Tanacetum parthenium 215
— *vulgare* 198
Tansy 198
Taraxacum spp. 209
Tare, Hairy 83
—, Smooth 83
Taxus baccata 295
Teaplant, Chinese 176
—, Duke of Argyll's 176
Teasel, Wild 220
Teucrium scorodonia 183
Thistle, Carline 196
—, Creeping 196
—, Marsh 196
—, Meadow 197
—, Melancholy 197
—, Musk 197
—, Spear 196
—, Welted 197
Thlaspi arvense 136
Thorn-apple 176
Thrift 140
Thuja plicata 296
Thyme, Wild 189
Thymus drucei 189
Tilia cordata 129
— *cordata × platyphyllos* 129
— *platyphyllos* 129
—, × *europaea* 129
Timothy 272
Toadflax, Common 165
—, Ivy-leaved 165
—, Purple 165
—, Small 165
Tomato 177
Toothwort 192
—, Purple 192
Tor-grass 290
Torilis japonica 222
Tormentil 98
—, Trailing 98
Tragopogon porrifolius 203
— *pratensis* 203
Traveller's-joy 73
Tree-mallow, Smaller 130
Trefoil, Hop 87
—, Lesser 87
—, Slender 87
Trichophorum cespitosum 252
— *germanicum* 252
Trifolium arvense 86
— *campestre* 87
— *dubium* 87
— *hybridum* 86
— *medium* 86
— *micranthum* 87
— *pratense* 86
— *repens* 86

Triglochin maritima — 233
— *palustris* — 233
Tripleurospermum inodorum — 214
— *maritimum* — 214
Tripolium pannonicum — 209
Trisetum flavescens — 288
Triticum aestivum — 270
Trollius europaeus — 72
Tsuga heterophylla — 293
Tufted-sedge — 264
—, Slender — 264
Turnip — 133
Tussilago farfara — 212
Tussock-sedge, Greater — 256
Tutsan — 121
Twayblade, Common — 238
—, Lesser — 238
Typha angustifolia — 244
— *latifolia* — 244

U

Ulex europaeus — 81
— *gallii* — 81
— *minor* — 81
Ulmus glabra — 104
— *minor* agg. — 104
Umbilicus rupestris — 78
Urtica dioica — 105
— *urens* — 105
Utricularia spp. — 193

V

Vaccinium myrtillus — 158
— *uliginosum* — 158
— *vitis-idaea* — 158
Valerian, Common — 219
—, Marsh — 219
—, Red — 219
Valeriana dioica — 219
— *officinalis* — 219
Valerianella carinata — 219
— *locusta* — 219
Verbascum thapsus — 179
Vernal-grass, Sweet — 277
Veronica agrestis — 168
— *anagallis-aquatica* — 166
— *arvensis* — 167
— *beccabunga* — 166
— *catenata* — 166
— *chamaedrys* — 167
— *filiformis* — 168
— *hederifolia* — 168
— *montana* — 167
— *officinalis* — 167
— *persica* — 168
— *polita* — 168
— *scutellata* — 166
— *serpyllifolia* — 167
Vetch, Bush — 85

Vetch, Common — 82
—, Fodder — 85
—, Kidney — 84
—, Spring — 82
—, Tufted — 85
Vetchling, Meadow — 84
Viburnum lantana — 217
— *opulus* — 217
— *sargentii* — 217
— *veitchii* — 217
Vicia cracca — 85
— *lathyroides* — 82
— *sativa* — 82
— *sepium* — 85
— *villosa* — 85
Vinca major — 163
— *minor* — 163
Viola arvensis — 119
— *canina* — 118
— *hirta* — 119
— *kitaibeliana* — 119
— *odorata* — 119
— *palustris* — 119
— *reichenbachiana* — 118
— *riviniana* — 118
— *tricolor* — 119
Violet, Common Dog- — 118
—, Early Dog- — 118
—, Hairy — 119
—, Heath Dog- — 118
—, Marsh — 119
—, Sweet — 119
Viper's-bugloss — 170
Viscum album — 131
Vulpia bromoides — 285
— *myuros* — 285

W

Wall-rue — 304
Wallflower — 134
Walnut — 108
Water-cress — 138
—, Fool's — 225
—, Narrow-fruited — 138
Water-crowfoot, Common — 73
—, Pond — 73
—, Stream — 73
Water-dropwort, Hemlock — 224
Water-lily, Fringed — 67
—, Least — 67
—, White — 67
—, Yellow — 67
Water-milfoil, Alternate — 80
—, Spiked — 80
Water-parsnip, Lesser — 225
Water-pepper — 143
—, Tasteless — 143
Water-plantain — 232
Water-purslane — 124

Water-speedwell, Blue — 166
—, Pink — 166
Water-starwort, Common — 228
—, Pedunculate — 228
Water-violet — 80
Waterweed, Canadian — 232
—, Curly — 232
—, Nuttall's — 232
Wayfaring-tree — 217
—, Chinese — 217
Weld — 131
Wheat, Bread — 270
Whitebeam, Common — 91
—, Swedish — 91
Whitlowgrass, Common — 136
Wild-oat — 289
Willow, Creeping — 113
—, Eared — 113
—, Goat — 113
—, Grey — 113
—, Olive — 112
—, Purple — 112
—, White — 112
Willowherb, American — 127
—, Broad-leaved — 127
—, Great — 127
—, Hoary — 127
—, Marsh — 127
—, New Zealand — 126
—, Rosebay — 128
—, Short-fruited — 126
—, Square-stalked — 126
Winter-cress — 134
Wood-rush, Field — 249
—, Great — 249
—, Hairy — 249
—, Heath — 249
Wood-sedge — 259
—, Thin-spiked — 259
Wood-sorrel — 114
Woodruff — 162
Wormwood — 201
—, Sea — 201
Woundwort, Field — 185
—, Hedge — 185
—, Marsh — 185

Y

Yarrow — 213
Yellow-cress, Creeping — 134
—, Marsh — 134
Yellow-rattle — 190
Yellow-sedge, Common — 260
—, Long-stalked — 260
Yellow-sorrel, Procumbent — 114
Yellow-wort — 189
Yew — 295
Yorkshire-fog — 281

Acknowledgements and photographic credits

The authors would like to thanks all those that have helped behind the scenes with the many and varied tasks that have made this book possible. Thanks especially to all at BSBI for providing map and phenology data and for Rachel and Anya Still for producing the maps and phenology diagrams from these. Thanks also to Jens Christian Schou for the last-minute images.

Rachel Hamilton | Plants are my inspiration, and my earliest memories are of my botanist mother explaining and encouraging. My career was set, and my good fortune in my first job with the Field Studies Council at Flatford Mill sowed seeds for what followed. I am indebted to people with whom I shared time in the field, for my awareness and confidence as a teacher, and their generosity in sharing their knowledge: Jim Bingley, Charles Hubbard, Francis Rose, Clive Jermy, all no longer here; others who have become lifelong friends: David Streeter, Kit Grey-Wilson, Crinan Alexander, Richard Mabey. My students – at Flatford and on my degree and postgraduate programmes – have provided the most direct inspiration. What we lacked was a book that provided the springboard for beginners to become competent botanists. I am grateful for this opportunity to contribute to meeting that need, and to my husband Robin for his immeasurable support throughout.

Chris Gibson | Chris would like to thank Jude Gibson for her unfailing support and, towards the end of the project, many hours at her laptop: she has truly become 'part of the team'. Also, grand-daughter Eleanor for giving him the need to keep fighting for a better world: it is her future.

Robert Still | Rob would like to thank the many people that have assisted in the fieldwork necessary to obtain those images used in this book. In particular huge thanks to the inspirational Margaret Bradshaw for the in-the-field training me in *Alchemilla* identification and her company and time spent driving me efficiently around Teesdale. A large debt is owed to all the local BSBI recorders who were all incredibly helpful and supportive, especially Tony Mundell and Martin Rand of Hampshire. Thank also to John and Julie Moon for a morning on Salisbury Plain helping me get the last few species as well as everyone else, too many to mention, that have assisted in letting me know when target species were in flower or fruit to ensure that travel was kept to a minimum. Finally, a thanks for Sally-Ann Wilson for hosting the inaugural Plant Book Meeting way back when.

Photo Credits

All images by the authors except for the following. Special thanks to Jens Christian Schou for the supply of 4 images. 34 images are reproduced under the terms of various Creative Commons licenses (; Creative Commons Attribution 2.0 Generic; Creative Commons Attribution-Share Alike 3.0 Unported license; Creative Commons Attribution 4.0 International) and for these the photographers name is given in full followed by "(CC + relevant licence)".

p. 55 Columbine: flower **Len Worthington** [CC 2.0]; *p. 75* Box: fruit **Maurice Reille** [CC 3.0]; *p. 80* Alternate Water-milfoil: flower spike **Jens Christian Schou**; *p. 84* Greater Bird's-foot-trefoil: sepals **Daniel Cahen** [CC 4.0]; *p. 85* Goat's-rue: pod **Roger Culos** [CC 3.0], Tufted Vetch: pod **Rasbak** [CC 3.0], Bitter-vetch: pod **Krzysztof Golik** [CC 4.0]; *p. 93* Bird Cherry: fruit **Dellex** [CC 4.0]; *p. 100* Dropwort: fruit **Javier Martin** [CC 3.0]; *p. 133* Turnip: lower leaf **Gilles Ayotte** [CC 4.0]; *p. 135* Honesty: pod **Jamain** [CC 3.0]; *p. 139* Large Bittercress: leaf **Fornax** [CC 3.0]; *p. 174* Shoreweed: ♀ flower **Jens Christian Schou**; *p. 177* Potato: fruit **H. Zell** [CC 3.0]; *p. 205* Marsh Hawk's-beard: fruit **Meneerke bloem** [CC 3.0]; *p. 218* Wilson's Honeysuckle: fruit **Sten** [CC 3.0]; *p. 226* Ground-elder: fruit **Gilles Ayotte** [CC 4.0]; Giant Hogweed: leaf **Kevin Kozic / korina. info** [CC 4.0]; *p. 228* Water-starwort: fruit x2 **Jens Christian Schou**; *p. 230* Italian Lords-and-Ladies: fruit **Dominicus Johannes Bergsma** [CC 4.0]; *p. 231* Fat Duckweed: both images; Common Duckweed: upperside **Stefan.lefnaer** [CC 4.0]; *p. 232* Canadian Waterweed: family image; flower **Christian Fischer** [CC 3.0]; *p. 233* Sea Arrowgrass: ligule **Florent Beck** [CC 2.0]; Broad-leaved Pondweed: spike, flower **Frank Vincentz** [CC 3.0], key image (*right*) **Atriplexmedia** [CC 3.0]; *p. 234* Curled Pondweed: main image **Krzysztof Ziarnek, Kenraiz** [CC 4.0], fruit **Stefan.lefnaer** [CC 4.0]; *p. 236* Early Marsh-orchid: flower **Joachim Lutz** [CC 4.0]; *p. 237* Bee Orchid: main image **Björn S…**, Lesser Butterfly-orchid: flower **Armand Turpel** [CC 2.0]; *p. 240* Chalk Fragrant-orchid: main image **Flocci Nivis** [CC 4.0]; *p. 243* Common Bluebell: inflorescence **MichaelMaggs** [CC 3.0]; Spanish Bluebell: inflorescence **Jonathan Billinger** [CC 2.0], flower close-up **Andrea Moro** [CC 4.0].